STRUCTURAL AND FUNCTIONAL ASPECTS OF PHYTOCHEMISTRY

Recent Advances in Phytochemistry
Volume 5

CONTRIBUTORS

Winslow R. Briggs
Christina Chang
Ray F. Dawson
D. S. Frear
Arthur W. Galston
Gary Gardner
M. J. Jaffe
Leonard Jurd
Linda Kimler
Tom J. Mabry

Thomas S. Osdene
Carl S. Pike
Harbert V. Rice
Gary W. Sanderson
Ruth L. Satter
L. M. Schoonhoven
H. R. Swanson
F. S. Tanaka
Shu-I Tu
D. S. Van Fleet

Jui H. Wang

STRUCTURAL AND FUNCTIONAL ASPECTS OF PHYTOCHEMISTRY

Recent Advances in Phytochemistry
Volume 5

Edited by

V. C. RUNECKLES   and   **T. C. TSO**
Division of Plant Sciences   *Plant Science Research Division*
University of British Columbia   *Agricultural Research Service*
Vancouver, British Columbia, Canada *United States Department of Agriculture*
                                       Beltsville, Maryland

ACADEMIC PRESS   New York and London   1972

Copyright © 1972, by Academic Press, Inc.
ALL RIGHTS RESERVED
NO PART OF THIS BOOK MAY BE REPRODUCED IN ANY FORM,
BY PHOTOSTAT, MICROFILM, RETRIEVAL SYSTEM, OR ANY
OTHER MEANS, WITHOUT WRITTEN PERMISSION FROM
THE PUBLISHERS.

ACADEMIC PRESS, INC.
111 Fifth Avenue, New York, New York 10003

United Kingdom Edition published by
ACADEMIC PRESS, INC. (LONDON) LTD.
24/28 Oval Road, London NW1 7DD

LIBRARY OF CONGRESS CATALOG CARD NUMBER: 77-182656

PRINTED IN THE UNITED STATES OF AMERICA

CONTENTS

List of Contributors.. ix
Preface.. xi

Primary Energy Conversion Reactions in Photosynthesis
Jui H. Wang and Shu-I Tu

Saga of Previous Research.. 1
Some Unsolved Problems.. 11
Light-Driven Electron Transport... 12
Phosphorylation Coupled to Electron Transport........................... 18
References.. 27

The Nature of Purified Phytochrome
Winslow R. Briggs, Harbert V. Rice, Gary Gardner, and Carl S. Pike

Introduction.. 35
Results and Discussion.. 37
Conclusions... 47
References.. 49

A Study of the Mechanism of Phytochrome Action
Arthur W. Galston and Ruth L. Satter

Introduction.. 51
Experimental.. 53
Discussion.. 73
References.. 77

Acetylcholine as a Native Metabolic Regulator of Phytochrome-Mediated Processes in Bean Roots

M. J. Jaffe

Introduction	81
Methods	85
Identification and Measurement of Acetylcholine in Bean Roots	88
The Role of Calcium	94
Interactions with Respiration	94
Synthesis	101
References	102

The Betalains: Structure, Function, and Biogenesis, and The Plant Order Centrospermae

Tom J. Mabry, Linda Kimler, and Christina Chang

Introduction	106
Summary of Reports Pertaining to the Betalains (Tables 1–7)	107
DNA-DNA and DNA-RNA Hybridization Studies in the Centrospermae	107
Present and Proposed Research on the Betalains and the Plant Order Centrospermae	115
References	132

Recent Progress in the Chemistry of Flavylium Salts

Leonard Jurd

Introduction	135
Hydrolytic Reactions of Flavylium Salts	138
Peroxide Oxidation of Flavylium Salts	148
Flavylium Salt–Polyphenol Condensation Reactions	151
Synthesis and Some Reactions of Flavenes	155
Formation of Xanthlium Salts from Proanthrocyanidins	160
References	162

Histochemistry of Plants in Health and Disease

D. S. Van Fleet

Introduction	165
The Development of a Chemical Barrier	166
Enzyme Localization in Health and Disease	182
Analysis of the Localization of Compounds in Health and Disease	187
Conclusions	189
References	190

Secondary Plant Substances and Insects

L. M. Schoonhoven

Introduction	197
Feeding Stimulants	198
Nutrients and Host Plant Recognition	199
Chemoperception of Plant Constituents	209
Molecular Structure and Stimulating Effectiveness	210
Toxicity of Secondary Plant Substances	213
Conclusion	217
References	218

Herbicide Metabolism in Plants

D. S. Frear, H. R. Swanson, and F. S. Tanaka

Introduction	225
N-Demethylation of N-Methylphenylurea Herbicides	226
Glutathione Conjugation of 2-Chloro-s-triazine Herbicides	239
Conclusions	244
Appendix	244
References	245

The Chemistry of Tea and Tea Manufacturing

Gary W. Sanderson

Introduction	247
Tea Manufacturing Process	249
Biochemistry of the Tea Flush	251
Chemistry of Tea Manufacturing	265
Tea Cream	303
Conclusions	304
References	306

A Speculative View of Tobacco Alkaloid Biosynthesis

Ray F. Dawson and Thomas S. Osdene

Introduction	317
Correlative Information	318
Speculative Model Building	328
References	335
SUBJECT INDEX	339

LIST OF CONTRIBUTORS

Numbers in parentheses indicate the pages on which the authors' contributions begin.

WINSLOW R. BRIGGS (35), Biological Laboratories, Harvard University, Cambridge, Massachusetts

CHRISTINA CHANG (105), The Cell Research Institute and Department of Botany, The University of Texas at Austin, Austin, Texas

RAY F. DAWSON* (317), Philip Morris Research Center, Richmond, Virginia

D. S. FREAR (225), Plant Science Research Division, Agricultural Research Service, United States Department of Agriculture, Metabolism and Radiation Research Laboratory, Fargo, North Dakota

ARTHUR W. GALSTON (51), Department of Biology, Yale University, New Haven, Connecticut

GARY GARDNER (35), Biological Laboratories, Harvard University, Cambridge, Massachusetts

M. J. JAFFE (81), Department of Botany, Ohio University, Athens, Ohio

LEONARD JURD (135), Western Regional Research Laboratory, United States Department of Agriculture, Agricultural Research Service, Albany, California

LINDA KIMLER (105), The Cell Research Institute and Department of Botany, The University of Texas at Austin, Austin, Texas

TOM J. MABRY (105), The Cell Research Institute and Department of Botany, The University of Texas at Austin, Austin, Texas

THOMAS S. OSDENE (317), Philip Morris Research Center, Richmond, Virginia

* Present address: 152 Wescott Road, Princeton, New Jersey.

CARL S. PIKE* (35), Biological Laboratories, Harvard University, Cambridge, Massachusetts

HARBERT V. RICE† (35), Biological Laboratories, Harvard University, Cambridge, Massachusetts

GARY W. SANDERSON (247), Thomas J. Lipton, Inc., Englewood Cliffs, New Jersey

RUTH L. SATTER (51), Department of Biology, Yale University, New Haven, Connecticut

L. M. SCHOONHOVEN (197), Department of Entomology, Agricultural University, Wageningen, The Netherlands

H. R. SWANSON (225), Plant Science Research Division, Agricultural Research Service, United States Department of Agriculture, Metabolism and Radiation Research Laboratory, Fargo, North Dakota

F. S. TANAKA (225), Plant Science Research Division, Agricultural Research Service, United States Department of Agriculture, Metabolism and Radiation Research Laboratory, Fargo, North Dakota

SHU-I TU (1), Kline Chemistry Laboratory, Yale University, New Haven, Connecticut

D. S. VAN FLEET (165), University of Georgia, Athens, Georgia

JUI H. WANG (1), Kline Chemistry Laboratory, Yale University, New Haven, Connecticut

* Present address: Department of Biology, Franklin and Marshall College, Lancaster, Pennsylvania 17604.
† Present address: Research Department, New England Aquarium, Central Wharf, Boston, Massachusetts 02110.

PREFACE

This volume contains eleven invited papers presented at the Tenth Symposium of the Phytochemical Society of North America. This Symposium, held on October 6–9, 1970, at Beltsville-College Park, Maryland, was organized under the general title "Structural and Functional Aspects of Phytochemistry" with equal emphasis on basic and applied phases of phytochemistry.

A living plant is a dynamic system. As a living system, it depends upon free energy from its surroundings, either in the form of light or in the form of chemical free energy converted from light, to maintain and extend its life. In the first chapter of this volume, J. H. Wang and S.-I Tu discuss such an energy conversion in both photooxidation and photophosphorylation reactions of photosynthesis. As a dynamic system, the metabolic activities of plants are controlled by, so to speak, a built-in biological clock, a light-sensitive mechanism. The detection and characterization of phytochrome is one of the most outstanding works in phytochemistry in recent years. The next three chapters in this volume are devoted to recent advances in phytochrome studies, including the physiological and biochemical characterization of the pigment (W. R. Briggs, H. V. Rice, G. Gardner, and C. S. Pike), the mechanism of action (A. W. Galston and R. L. Satter), and the regulation of phytochrome-mediated processes (M. L. Jaffe).

The two following chapters are directly or indirectly related to plant pigments: the chapter by T. J. Mabry, L. Kimler, and C. Chang deals with the structure, function, and photocontrolled synthesis of betalains; the chapter by L. Jurd discusses the chemistry of flavylium salts. Flavylium salts, upon reduction, yield flavenes which are now believed to play an important role in biosynthesis of flavonoids.

The last part of this volume emphasizes applications. Several chapters are presented dealing with the chemical response of plants to disease and insect infection, or vice versa; with herbicide metabolism; and with the chemistry of two important economic crops—tea and tobacco. In the chapter on the histochemistry of plants in health and disease, by D. S. Van Fleet, there are discussions on enzyme reaction to wound injury and parasites, and the evolution of the mechanisms of disease resistance. In the following chapter, by L. M. Schoonhoven, there is a discussion of the ecological and evolutionary significance of secondary plant substances in insect–plant relationships. *In vitro* studies of herbicide metabolism in higher plants, especially those involved in several novel enzymic detoxication systems, provide a better understanding of herbicide selectivity, and are reviewed in the subsequent chapter by D. S. Frear, H. R. Swanson, and F. S. Tanaka. Reports on the chemistry of tea and tea manufacture, especially those enzyme systems involved in tea fermentation, illustrate, in the next chapter, by G. W. Sanderson, the progress made in this industry as a result of basic research. The final chapter on alkaloid biosynthesis of tobacco plants, by R. F. Dawson and T. S. Osdene, is very timely, as it presents a speculative interpretation regarding the significance of alkaloid formation in higher plant metabolism.

The organizers of this symposium and the editors of this book are grateful to the authors for their cooperation in this endeavor and for their contributions toward the advancement of phytochemistry. This volume will fill some gaps in our basic knowledge of plant chemistry, and will also contribute to the bringing together of basic information and practical applications in plant production and utilization.

V. C. RUNECKLES
T. C. Tso

PRIMARY ENERGY CONVERSION REACTIONS IN PHOTOSYNTHESIS

JUI H. WANG and SHU-I TU

Kline Chemistry Laboratory, Yale University, New Haven, Connecticut

Saga of Previous Research...	1
Early Adventures...	1
Path of Carbon..	3
Generation of Reducing Power..	4
Photophosphorylation...	4
Energy Transfer and the Photosynthetic Unit.........................	5
Reversible Bleaching of the Reaction Center.........................	7
Cooperation of Two Photoreactions...................................	7
Path of Electron Transport...	8
Some Unsolved Problems...	11
Light-Driven Electron Transport...	12
Phosphorylation Coupled to Electron Transport.........................	18
References...	27

Saga of Previous Research

EARLY ADVENTURES

Few topics in science equal photosynthesis in importance and in its widespread interest, which transcends the boundaries of the traditional disciplines. Not only does photosynthesis purify our environment and make the world beautiful, but to maintain life we all must feed on the chemical free energy converted from light by photosynthesis and stored in organic compounds in the form of food. For these reasons, the study of photosynthesis has attracted some of the best minds since the very beginning of chemistry.

For example, almost two hundred years ago Priestley (1772) discovered that green plants could convert carbon dioxide to "dephlogisticated air," which supported the respiration of mice as well as the burning of candles. The discovery inspired his young friend Lavoisier (1774) to repeat some of Priestley's experiments systematically and to conclude that the vital principle in "dephlogisticated air" was a new element which was later named oxygen. It was then that chemistry really got started. Within a short time, light (Ingen-Housz, 1779), carbon dioxide (Senebier, 1782), and water (deSaussure, 1804) were all found to be required in photosynthesis by green plants.

After this initial burst of activity, the field stayed relatively quiet for nearly half a century. Then Mayer (1845), who had enounced the law of conservation of energy three years before, treated photosynthesis as a process in which light is partially converted into chemical energy. Since in photosynthesis the volume of oxygen evolved was found equal to the volume of carbon dioxide consumed (Boussingault, 1864; Maquenne and Demoussy, 1913; Willstätter and Stoll, 1918), the net chemical change could be written as

$$n CO_2 + m H_2O + \text{light} \rightarrow C_n(H_2O)_m + n O_2 \qquad (1)$$

with $C_n(H_2O)_m$ representing the carbohydrate (Sachs, 1864) as a typical organic product. The stoichiometry of this reaction led a number of investigators to assume that the oxygen gas liberated by photosynthesis originated from the carbon dioxide.

Two important discoveries were made in the 1880's which were not fully appreciated until much later. Engelmann (1883, 1888) discovered that purple sulfur bacteria could produce organic matter by photosynthesis with the liberation of sulfur instead of oxygen. Winogradsky (1887, 1888) found that colorless sulfur bacteria could synthesize organic matter by assimilation of carbon dioxide in the dark. (The free energy required for this synthesis is derived from the oxidation of sulfide by oxygen.) Winogradsky's discovery was later interpreted as suggesting that the fixation of carbon dioxide in photosynthesis, like that in colorless bacteria, may be a secondary dark reaction (Lebedev, 1921).

The problem was greatly clarified by the work of van Niel (1931), who also pointed out that Engelmann's observation on purple sulfur bacteria and the photosynthesis by green plants can both be represented by Eq. (2)

$$CO_2 + 2 H_2A + \text{light} \rightarrow \{CH_2O\} + H_2O + 2A \qquad (2)$$

in which carbon dioxide is reduced by a general hydrogen donor H_2A to a carbohydrate represented by $\{CH_2O\}$ in Eq. (2). In the case of purple sulfur bacteria, the donor is hydrogen sulfide, which is oxidized to elementary sulfur. In the case of green plants, the donor is water, which is oxidized to oxygen gas. Recognizing that the photooxidation of a suitable donor and the dark reduction of carbon dioxide in van Niel's formulation are separate events, one may divide Eq. (2) into two separate reactions

$$2H_2A + 2X \xrightarrow{\text{light}} 2A + 2H_2X \qquad (3)$$

$$2H_2X + CO_2 \xrightarrow{\text{dark}} 2X + \{CH_2O\} + H_2O \qquad (4)$$

where X represents an endogenous acceptor.

According to Eq. (2), the oxygen liberated in photosynthesis originates from water, not from carbon dioxide. The data obtained from experiments with $H_2^{18}O$ or $CO^{18}O$ as a tracer by Ruben and co-workers (1941) and from a careful determination of the density of water synthesized from the oxygen liberated in photosynthesis by Vinogradov and Teis (1941) seemed to support this prediction. But the decisive evidence for van Niel's formulation was supplied by Hill and Scarisbrick (1940), who showed that even without carbon dioxide, isolated chloroplasts can produce oxygen when illuminated in the presence of a suitable electron acceptor, such as ferric oxalate, ferricyanide, or quinone.

PATH OF CARBON

Ruben and co-workers (1939) also made the first attempt to follow the path of carbon in photosynthesis by using $^{11}CO_2$ (half-life of ^{11}C, 22 minutes) as a tracer, and demonstrated unequivocally that the reduction of carbon dioxide in photosynthesis is a dark reaction which follows the primary process of converting light to chemical free energy. Ruben and Kamen (1940, 1941) also succeeded in making ^{14}C (half-life, 5×10^3 years), but unfortunately other duties and the subsequent death of Ruben prevented them from fully exploiting this versatile tracer.

In 1947, Calvin, Benson, Bassham, and their co-workers began to trace the path of $^{14}CO_2$ in photosynthesis with improved detection and separation techniques. By 1956, they successfully concluded the elucidation of the pathway of carbon in photosynthesis which is now known as the Calvin cycle (Calvin and Bassham, 1962). Although this pathway (in which 3-phosphoglyceric acid is the first major product) appears to describe the processes involved in carbon assimilation in most plants, more recent studies by Hatch and Slack and others (reviewed by Hatch and Slack, 1970) have shown that many tropical grasses and some members of the

Centrospermae and Geraniales are characterized by anatomical and biochemical modifications by which these plants initially incorporate CO_2 into four-carbon acids, rather than into 3-phosphoglyceric acid.

Generation of Reducing Power

During the two decades following van Niel's proposal, when the elucidation of the path of carbon in photosynthesis was in active progress and the concept of photosynthetic units was being developed (see below), the chemical identity of the endogenous acceptor X in Eqs. (3) and (4) remained a mystery. Then unexpectedly, three papers appeared independently in *Nature* within a period of 6 weeks in 1951, all reporting nicotinamide adenine dinucleotide phosphate ($NADP^+$) to be the acceptor X in the photooxidation of water (Tolmach, 1951; Vishniac and Ochoa, 1951; Arnon, 1951) as indicated by Eq. (5).

$$2H_2O + 2NADP^+ \xrightarrow{light} 2NADPH + 2H^+ + O_2 \qquad (5)$$

The NADPH generated this way is then used in the Calvin cycle to reduce carbon dioxide. Other investigators later reported that NADH plays the same role in bacterial photosynthesis as that of NADPH in green plant photosynthesis (Duysens and Sweep, 1957; Olson, 1958).

Photophosphorylation

Three years later, Arnon and co-workers (1954) made the important discovery that, when deprived of carbon dioxide and exogenous $NADP^+$, illuminated chloroplasts generate ATP from ADP and inorganic orthophosphate (P_i) but produce no other chemical change according to the simple net reaction

$$ADP + P_i \to ATP + H_2O \qquad (6)$$

Arnon and co-workers named this process cyclic photophosphorylation. Although there was no net consumption of oxygen in cyclic photophosphorylation, they observed that the process required catalytic amounts of molecular oxygen or another electron acceptor, such as FMN or vitamin K.

At the same time, Frenkel (1954) discovered independently that cyclic photophosphorylation also takes place in the photosynthetic bacteria *Rhodospirillum rubrum*. The data of Arnon and co-workers on cyclic photophosphorylation were questioned, but later experiments of improved yield (Allen *et al.*, 1958; Jagendorf and Avron, 1958) supported their earlier conclusion.

Inasmuch as the reduction of 1 mole of carbon dioxide to the oxidation level of sugar involves the oxidation of 2 moles of NADPH and the hydrolysis of 3 moles of ATP, the problem of energy conversion in photosynthesis now becomes the elucidation of the molecular mechanisms by which light energy is utilized to reduce $NADP^+$ to NADPH by water and to enable ADP and P_i to condense to form ATP and H_2O. These are the principal issues to be discussed in the present paper. But before doing so, let us complete our resume of the information obtained by previous investigators.

ENERGY TRANSFER AND THE PHOTOSYNTHETIC UNIT

Although the conversion of light to chemical free energy in photosynthesis was postulated by Mayer in 1845, a quantitative understanding of the energy conversion process had to await the advent of quantum theory (Planck, 1901; Einstein, 1905) and the development of precise gasometric (Warburg, 1919; Emerson and Lewis, 1941; Gaffron, 1942) as well as photochemical (Warburg and Schocken, 1949) techniques. During the period when van Niel's ideas were being actively tested and extended in many laboratories, quantum yield and saturation intensity measurements were also being made with the aim of unraveling the light-harvesting machinery of green plants. The wide range of quantum yields reported in the literature between 1922 and 1951, which was reviewed comprehensively by Rabinowitch (1951), corresponds to from 2.5 to 20 photons absorbed per molecule of O_2 generated or CO_2 assimilated.

By measuring the oxygen evolution from *Chlorella* in response to $\sim 10^{-5}$ second flashes of light, Emerson and Arnold (1932a, b) showed that the maximum yield per flash corresponded to one O_2 molecule generated per 2000–3000 chlorophyll molecules present. If the generation of one O_2 molecule under normal conditions requires eight photons (Kok, 1948; Franck, 1949), the data of Emerson and Arnold suggest that, in *Chlorella*, an assembly of 250–375 chlorophyll molecules serves as a photosynthetic unit (Gaffron and Wohl, 1936). Within each photosynthetic unit, the electronic excitation energy can migrate rapidly by resonant energy transfer between the pigment molecules, as suggested by Förster (1948), until it is converted to chemical free energy at the active center.

According to Förster (1948, 1949), the rate constant $k_{A^* \to A'}$ for energy transfer from an electronically excited molecule (A^*) to another molecule (A') in its ground state as represented by the chemical equation

$$A^* + A' \to A + A'^* \tag{7}$$

is given by

$$k_{A^* \to A'} = \frac{9000 (\ln 10) \kappa^2 \phi_f}{128 \pi^6 n^4 N \tau_{A^*} r^6} \int_0^\infty f_{A^*}(\nu) \epsilon_{A'}(\nu) \frac{d\nu}{\nu^4} \tag{8}$$

where $f_{A^*}(\nu)$ is the fluorescence distribution function of A*; $\epsilon_{A'}(\nu)$ is the extinction coefficient of A' as a function of frequency ν; ϕ_f is the quantum yield; n is the refractive index; N is the Avogadro's number; κ^2 is a factor that depends on the mutual orientation of the emitting and absorbing dipoles: $\kappa = 1$ if the dipoles are parallel and side by side (↑ ↑), $\kappa = 4$ if the dipoles are parallel and end to end (→→), $\kappa = 0$ if they are mutually perpendicular, $\kappa = \frac{2}{3}$ if their mutual orientation during the lifetime of A* can be considered as random; r is the distance of separation between the emitting and absorbing dipoles. Förster's theory gave a satisfactory account of the observed fluorescence depolarization of pigment molecules in solution. Recently Stryer and Haugland (1967) obtained direct supporting evidence for Eq. (8).

The number of chlorophyll molecules in each photosynthetic unit estimated photochemically agrees in its order of magnitude with that computed from the chemical composition and electron microscopical observations (Steinmann, 1952; Frey-Wyssling and Steinmann, 1953; Park and Pon, 1963; Park and Biggins, 1964) or freeze-etching data (Park and Branton, 1967). As an example, Park (1966) estimated the molecular weight of the photosynthetic unit (quantasome) to be about 2×10^6. Knowing its chemical composition, he concluded that each photosynthetic unit contains 230 chlorophyll molecules and 48 carotenoids. The elucidation of the structure of chlorophyll a as shown below was essentially completed by Fischer and Henderoth (1939), except for the

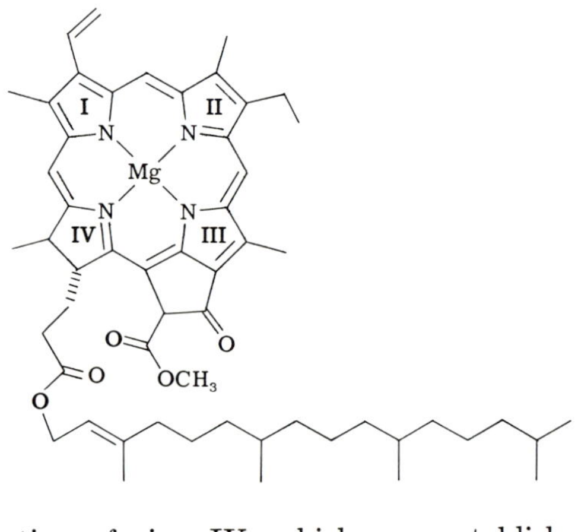

trans configuration of ring IV, which was established by Ficken and co-workers (1956) much later. The total synthesis of chlorophyll a, which was achieved relatively recently by Woodward and co-workers (1960), represents a major accomplishment in organic chemistry.

Reversible Bleaching of the Reaction Center

Duysens (1952) first observed that actinic light caused a reversible change in the absorption spectrum of purple bacteria. For example, illumination of *Rhodospirillum rubrum* decreased its absorbance at 890 nm. Subsequent studies (Duysens and co-workers, 1956) suggested that this decrease was due to an oxidative bleaching of a small fraction of the bacteriochlorophyll molecules with their normal absorption maximum at 890 nm (P 890), because the bleaching occurred more rapidly in the presence of ferricyanide (Duysens, 1953; see also Duysens, 1964). A bleaching similar to that caused by actinic light could be brought about by adding a mixture of potassium ferri- and ferrocyanide. The midpoint reduction potential of the oxidatively bleached P 890 was estimated to be 0.51 V (Thomas and Goedheer, 1953; Goedheer, 1958) or 0.47 V (Duysens, 1958).

In green plants, an analogous reversible bleaching by light occurs at about 700 nm. This bleaching can also be brought about by ferricyanide, and the midpoint reduction potential of the oxidized form of the corresponding active center (P 700) was estimated to be 0.43 V (Kok, 1961) or 0.45 V (Witt et al., 1961b; Müller et al., 1963). Kok (1961) succeeded in extracting 85 percent of the chlorophyll from chloroplasts without destroying the reversible bleaching of P 700. This last observation suggests that the observed decrease in absorbance at 700 nm was caused by the bleaching of one pigment, which may be a particular chlorophyll *a* molecule in a special environment, not by a small shift in the absorption spectrum of the bulk of the chlorophyll *a* molecules. Assuming that P 700 has a molar extinction coefficient at 700 nm of the same order of magnitude as that of ordinary chlorophyll *a* at 670 nm ($10^5\ M^{-1}\ cm^{-1}$), Kok estimated the quantum yield of its photobleaching to be about unity. The half-life of the dark recovery of the photobleached P 700 was estimated to be $\sim 10^{-2}$ second, which could be drastically altered by various oxidation–reduction reagents (Kok and Hoch, 1961).

Likewise, Clayton (1962) succeeded in removing most of the bacteriochlorophyll from the chromatophores of a blue-green mutant of *Rhodopseudomonas spheroides* by prolonged illumination followed by extraction. The 870 nm absorption band of the resulting system could then be photobleached completely and reversibly. This photobleaching of P 870 probably also corresponds to the oxidation of the responsible pigment, since a reducing environment accelerates the recovery of the bleached P 870 in the dark.

Cooperation of Two Photoreactions

In general, the fluorescence maximum occurs at a lower frequency than the corresponding absorption maximum of the same molecular species. For this reason, the transfer of electronic excitation energy from a molecule which absorbs at a higher frequency to a molecule which absorbs at a lower frequency occurs readily through dipole–dipole interaction according to Eq. 8, but not in the reverse direction, and consequently we expect the quantum yield of photosynthesis to remain fairly constant as the wavelength of the exciting light is decreased continually from 670 nm. Emerson and Lewis (1943) discovered that there is a sharp drop in the quantum efficiency of photosynthesis by green plants as the wavelength of the actinic light was increased to slightly above 685 nm, although below 685 nm the quantum yield was not very sensitive to wavelengths. This limiting wavelength is actually not far from the absorption maximum of the bulk of chlorophyll a. A careful comparison of the absorption and action spectra of a red alga by Haxo and Blinks (1950) also showed that as the wavelength of the actinic light increases beyond 700 nm, the quantum yield of photosynthesis drops rapidly to zero before the total absorbance does.

In addition, Emerson and co-workers (1956, 1957; Emerson and Lewis, 1943) observed that the rate of photosynthesis (R_c) by green algae when illuminated simultaneously by two beams of actinic light of suitably chosen wavelengths, e.g., red (650 nm) and far-red (700 nm) light, was higher than the sum of the rates of photosynthesis at each wavelength separately (R_1 and R_2), i.e.,

$$\frac{R_c}{R_1 + R_2} > 1 \tag{9}$$

These results suggest that photosynthesis may involve the cooperation of two photoreactions; one requires light of wavelength less than 685 nm, and the other may utilize light of wavelength greater than 685 nm.

Independently Blinks (1957) measured the rate of oxygen production by a red alga illuminated alternately by light beams of 580 and 675 nm. After the intensities at these two wavelengths were adjusted to give about equal steady-state rates, a sudden switching from one wavelength to the other caused rapid transients corresponding to 10–15 percent increase in the rate of photosynthesis. These chromatic transients had also been interpreted in terms of the cooperation of two photoreactions.

But what photoreactions are these and how do they cooperate?

PATH OF ELECTRON TRANSPORT

The discovery by Hill and Scarisbrick (1940) that illuminated chloroplasts could generate oxygen in the absence of carbon dioxide provided that a suitable electron acceptor was present and the discovery by Gaffron (1944) that an adapted alga under anaerobic conditions could photoreduce

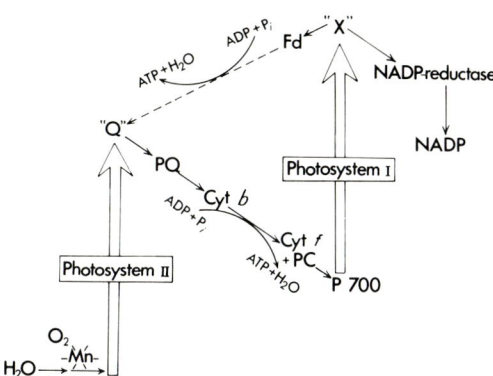

Fig. 1. The Hill-Bendall scheme of light-driven electron transport in photosynthesis by green plants.

carbon dioxide with hydrogen gas, together with the experimental results on quantum yield and enhancement effect obtained by Emerson and co-workers, led Hill and Bendall (1960) to postulate the electron transport scheme illustrated in Fig. 1. Similar schemes were proposed independently by Duysens and co-workers (1961) and by Witt and co-workers (1961a). The sequence of the electron carriers, plastoquinone (PQ) and the cytochromes (Cyt b and Cyt f), in the electron transport chain connecting Photosystem I (PS I) and Photosystem II (PS II) was determined experimentally from spectrometric studies similar to those of the respiratory chain (Chance and Williams, 1955). The phosphorylation of ADP coupled to electron transport from the unknown acceptor "X" to the unknown acceptor "Q" and from "Q" to P 700 was assumed to take place as indicated in Fig. 1 by a mechanism similar to that of mitochondrial oxidative phosphorylation.

The scheme in Fig. 1 represents a simple mechanism by which the two photosystems can cooperate in utilizing light to drive electrons from water to NADP. It also provides a plausible explanation for the enhancement effect (Emerson et al., 1956, 1957) and the partial reactions discovered by Hill and Scarisbrick (1940) and by Gaffron (1944), respectively. In addition, it is also consistent with the definitive experimental results recently obtained on mutant algae from Levine's and Bishop's laboratories (Levine, 1960; Bishop, 1962; Weaver and Bishop, 1963; Levine and Smillie, 1963; Levine and Gorman, 1966; Givan and Levine, 1967).

For cyclic photophosphorylation, the light energy absorbed by the assembly of pigment molecules in PS I is first transferred by to a particular chlorophyll *a* molecule (P 700) which utilizes it to transfer an electron to the unknown acceptor "X". Then the electron is transferred down a free-energy hill from "X" to plastocyanine (PC) or Cyt f through the electron carriers Fd, "Q", PQ, Cyt b, as indicated by the arrows in Fig. 1, and produce ATP along the way. Since, for a given number of electrons the P 700 supplies to the unknown acceptor "Q," exactly the same number is recovered from PC or Cyt f, cyclic photophosphorylation does not produce net redox change. Because at steady state the rate of oxidation of the reduced form of every carrier in an electron-transport chain is equal to the rate of reduction of its oxidized form, every carrier has to maintain a healthy proportion of its oxidized and reduced forms for the chain to function efficiently in the kinetic sense. In the event that all carriers in a particular preparation are completely in the reduced form, electron transfer from the particular chlorophyll a to the reduced "X" cannot occur. Consequently in such cases, a catalytic amount of oxygen gas, oxidized flavin, or vitamin K is needed to oxidize a sufficient fraction of the endogenous acceptor "X" to start the light-driven electron pump.

For net photosynthesis, the reduction of 1 mole of CO_2 to the oxidation level of carbohydrate involves the oxidation of 2 moles of NADPH and the hydrolysis of 3 moles of ATP according to the Calvin cycle. Because of the withdrawal of electrons at the NADP in Fig. 1 for CO_2 reduction, fewer electrons are returned to "Q" than are taken away from it by PS I through Cyt f or PC. Consequently photosynthesis cannot continue unless the electron deficit at "Q" is compensated by an exogenous electron donor. In bacterial photosynthesis with a strong reductant such as H_2S or H_2 as the donor, a single photosystem (PS I) should suffice, since each quantum of light at 870 nm is equivalent to 1.43 V. But in the photosynthesis by green plants with H_2O as the electron donor, kinetic complications may arise because of the high activation free energies of the necessary chemical mechanisms for oxidizing H_2O to O_2. The development of a second photosystem (PS II) in green plants with a manganese complex as the depolarizer could overcome these difficulties (Kessler, 1957; Wang, 1970a).

Thus in the presence of CO_2, the withdrawal of electrons at NADP by reactions of the Calvin cycle creates a deficit of electrons at "Q." This electron deficit makes "Q" a better acceptor, thereby activating the light-driven electron transfer from the chlorophyll of PS II to "Q," and the oxidized PS II can then continue to raise the oxidation state of the manganese complex until the latter starts to produce O_2 by oxidizing its own water of hydration. Extraneous electron acceptors that are capable of removing electrons rapidly from "Q," PQ, Cyt b, etc., are also expected to activate PS II and generate O_2 (Hill and Scarisbrick, 1940).

According to the scheme in Fig. 1, there should be twice as much ATP produced in cyclic photophosphorylation as in noncyclic photophosphorylation per mole of electron transferred. Laber and Black (1969) demonstrated that this is indeed the case for spinach chloroplasts.

Some Unsolved Problems

In spite of the formidable amount of supporting evidence for the simple scheme in Fig. 1, several investigators have recently proposed alternative formulations of the electron transport paths in photosynthesis by green plants. For example, McSwain and Arnon (1968) showed quite convincingly that the enhancement ratio of Eq. (9) is greater than unity only when CO_2 is used as the terminal electron acceptor. When exogenous NADP or ferricyanide was used as the electron acceptor, no enhancement was observed. From this and other related pieces of evidence, Knaff and Arnon (1969) proposed a three-photoreaction system: a noncyclic system with PS II_a and PS II_b which cooperate through an electron-transport chain containing PQ, Cyt b, and PC, and a separate cyclic system containing PS I, Cyt f, and P 700. Recently, Arnold (1970) even proposed a four-photoreaction system. It appears that the question of electron-transport path in photosynthesis is by no means settled.

In addition, the detailed structure of the photosynthetic unit is still unknown, although hypothetical models have been proposed (Avron, 1967; Williams, 1970). In fact, even the molecular identity of "X," P 700, and "Q" in the scheme of Fig. 1 is still under debate. Nor do we know the structure of the manganese complex and the mechanism by which it catalyzes the generation of molecular oxygen.

During the last decade, considerable progress has been made in the study of photochemical energy conversion. But we still do not understand clearly the specific molecular mechanisms by which chloroplasts convert light into chemical free energy.

Finally the molecular mechanism for coupling electron transport to phosphorylation has yet to be elucidated, even though it is almost universally agreed that photosynthetic phosphorylation and respiratory oxidative phosphorylation probably involve very similar molecular mechanisms.

In the remainder of this review, we shall try to summarize the main results obtained recently in our laboratory concerning the mechanism of light-driven electron transport from water to NADP and the mechanism for coupling phosphorylation to electron transport. It seems generally agreed now that these are the two primary energy conversion reactions in photosynthesis.

Light-Driven Electron Transport

In general, molecules in their excited electronic states have quite different oxidation-reduction properties than in their ground states. An electronically excited pigment molecule can be a stronger reductant than the same molecule in its ground state, because the removal of an electron from its antibonding orbitals requires less energy. It can also be a stronger oxidant, since the positive hole left in the bonding or nonbonding orbital could be filled by extracting an electron from a donor molecule. For this reason, certain pigment molecules can utilize the absorbed light energy to drive electrons up a free-energy hill (Tollin and Green, 1962, 1963; Tollin et al., 1965; Banerjee and Tollin, 1966; Chibisov et al., 1966; Fuhrhop and Mauzerall, 1968; McElroy et al., 1969; Nicolson and Clayton, 1969; Ke, 1969). Experiments with various model systems demonstrate that it is even possible to convert light to chemical free energy through rather simple photoredox reactions (Krasnovsky, 1948a, b; Ainsworth and Rabinowitch, 1960; Vernon, 1961; Vernon et al., 1965; Kassner and Kamen, 1967; Metzner, 1968; Komissarov and Shumov, 1968).

In one of the photochemical energy conversion systems developed in our laboratory (Eisenstein and Wang, 1969), *clostridial* ferredoxin (Fd) was photoreduced by glutathione (GSH) in the presence of hematoporphyrin (HP) as a sensitizer according to Eq. (10).

$$2GSH + Fd_{ox} \underset{dark}{\overset{light, HP}{\rightleftharpoons}} GSSG + Fd_{RE} + 2H^+ \qquad (10)$$

In the absence of air, illumination by white light for 20 seconds at 23°C caused an 18% reduction of the ferredoxin by glutathione. While standing in the dark in the absence of air for several hours at 23°C, the reduced ferredoxin (Fd_{RE}) was generally reoxidized by the oxidized glutathione (GSSG) back to its original form (Fd_{ox}). Since the observed spontaneous reversal of the photoreduction of Fd_{ox} in the dark must be driven by a decrease in the total free energy of the system, the preceding photoreduction reaction in the opposite direction must have resulted in an increase of total free energy, i.e., part of the absorbed light must have been converted to chemical free energy by the model system. Knowing the midpoint reduction potentials of *clostridial* ferredoxin (-0.367 and -0.398 V for the first and second reducing equivalents, respectively) and glutathione (-0.25 V) and the concentrations of all components at the end of the 20-second illumination period, we calculated the amount of light converted to chemical free energy to be 3.3 kcal per faraday of electron transferred from GSH to Fd_{ox}.

The following two simple mechanisms are consistent with our data and the above considerations.

Mechanism I

$$\begin{align} HP &\underset{}{\overset{h\nu}{\rightleftarrows}} HP^* \\ HP^* + GSH &\rightarrow HP^{\cdot}_{RE} + GS^{\cdot} + H^+ \\ HP^{\cdot}_{RE} + Fd_{ox} &\rightarrow HP + Fd_{RE} \\ 2GS^{\cdot} &\rightarrow GSSG \end{align} \quad (11)$$

where $HP^{\cdot}_{RE}$ and $GS^{\cdot}$ represent the reduced hematoporphyrin radical and the glutathione radical, respectively.

Mechanism II

$$\begin{align} HP &\underset{}{\overset{h\nu}{\rightleftarrows}} HP^* \\ HP^* + Fd_{ox} &\rightarrow HP^{\cdot}_{ox} + Fd_{RE} \\ HP^{\cdot}_{ox} + GSH &\rightarrow HP + GS^{\cdot} + H^+ \\ 2GS^{\cdot} &\rightarrow GSSG \end{align} \quad (12)$$

The difference between these two mechanisms lies in the role of the porphyrin. Mechanism I assumes that the excited hematoporphyrin molecule (HP*) is first reduced by glutathione to the free radical $HP^{\cdot}_{RE}$, whereas Mechanism II assumes that it is first oxidized by ferredoxin to the free radical $HP^{\cdot}_{ox}$. To decide between these two mechanisms it was necessary to find out which of these two hematoporphyrin radicals was formed in this model system, and, in the event that both were, which reaction path was the kinetically favored one.

By using diphenylpicrylhydrazyl as a radical trap, it was shown that Mechanism I was the principal path of this model reaction. In frozen samples, the photochemically produced $HP^{\cdot}_{RE}$ was found to exhibit a simple electron spin resonance (ESR) signal with the same g value and line width as those of the hematoporphyrin radical produced chemically in the dark.

In a second model system for photochemical energy conversion (Tu and Wang, 1969), electrons were driven by red light from cytochrome c via aggregated chlorophyllin a, $(Chl)_n$, and the flavin group of NADP-reductase to NADP according to the chemical equation

$$2 \text{ Cyt } c^{II} + NADP + H^+ \xrightarrow[\text{NADP-reductase}]{\text{red light } (Chl)_n} 2 \text{ Cyt } c^{III} + NADPH \quad (13)$$

The molar ratio of cytochrome c oxidized to NADPH produced is within experimental uncertainty equal to 2 as required of the stoichiometry of Eq. 13. Knowing the midpoint reduction potentials of cytochrome c (0.26 V) and NADP (-0.32 V) as well as the concentrations of all components at the end of each experiment, we calculated the amount of light converted to chemical free energy to be

$$\Delta G/n = -\left\{\left(-0.32F - \frac{RT}{2}\ln\frac{[\text{NADPH}]}{[\text{NADP}]}\right) - \left(-0.26F - RT\ln\frac{[\text{Cyt }c^{\text{II}}]}{[\text{Cyt }c^{\text{III}}]}\right)\right\}$$
$$= 8.4 \text{ kcal/faraday}$$

which corresponds to a photoelectromotive force of 0.36 V.

In the absence of either NADP-reductase or $(\text{Chl})_n$, very little or no NADPH was formed. The sum of the above experimental results shows that the path of electron transfer in this model system can be represented by the scheme in Fig. 2, in which either the oxidized chlorophyllin radical (Chl^+) or the reduced chlorophyllin radical (Chl^-) or both participate in the electron transfer.

Presumably energy transfer can take place between different monomer units, Chl and Chl*, of the same chlorophyllin aggregate as well as between the different chlorophyllin species in the solution. Parallel experiments showed that monomeric chlorophyllin functions better as a photoreductant and that the aggregated form of chlorophyllin functions better as a photooxidant. In view of the fact that not all the molecular species of chlorophyllin a in the equilibrium mixture represented by $(\text{Chl})_n$ are equally effective as sensitizers and that the overall photooxidation-reduction reaction discussed above involves at least three consecutive electron-transfer steps, the observed quantum yield of 0.053 is remarkably high indeed. But compared to the natural system of chloroplasts in which two photosystems in relay cooperate to drive electrons from H_2O to NADP against an electrochemical potential of at least 1.14 V, this model system still has a long way to go.

A model system has been constructed in which two photoreactions cooperate in converting light to chemical free energy (Wang, 1969). To simplify the experimental procedure, the chlorophylls and auxiliary sensitizers in the photosynthetic apparatus of green plants were replaced by multimolecular layers of Zn(II)-tetraphenylporphyrin (ZnTPP), the primary electron acceptors in both photosystems (PS I and PS II) by clean metal surfaces in close contact with their respective pigment layers, and the chain of biological electron carriers between the two light-harvesting units

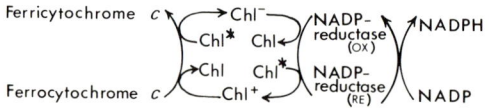

Fig. 2. A model system for the light-driven electron transfer from cytochrome c to NADP.

by platinum wires dipped in aqueous mixtures of potassium ferri- and ferrocyanide as indicated below. Of the various metal surfaces tried, it was found that clean aluminum foil adsorbed ZnTPP most satisfactorily and gave the most uniform multimolecular layers prepared by the method of Blodgett (1935).

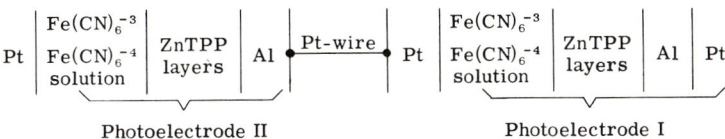

Upon illumination by orange light, charge transfer took place across more than 70 molecular layers of the pigment with a total electromotive force of more than 2 V and a total short-circuit photocurrent of 10^{-6} to 10^{-5} A. Similar results were obtained when the bare Pt electrode at the positive terminal (cathode in the electrolytic solution) was replaced by a bare Al electrode.

The electromotive force of the above two-stage photocell (2.1 to 2.4 V) is more than sufficient for oxidizing water to oxygen gas with a weak oxidant such as NADP or another water molecule which will be reduced to NADPH or hydrogen gas, respectively. To demonstrate this possibility, a microelectrolytic cell which was kept in the dark was used with its positive and negative platinum electrodes connected to the positive platinum electrode and negative photoelectrode I of the above photocell, respectively. Since NADP is not directly reducible by a platinum electrode, a small amount of NADP-reductase was added as a mediator to the electrolytic solution containing NADP in phosphate buffer at pH 7.38. The observed generation of oxygen gas by such a model system upon illumination by orange light is shown in Fig. 3.

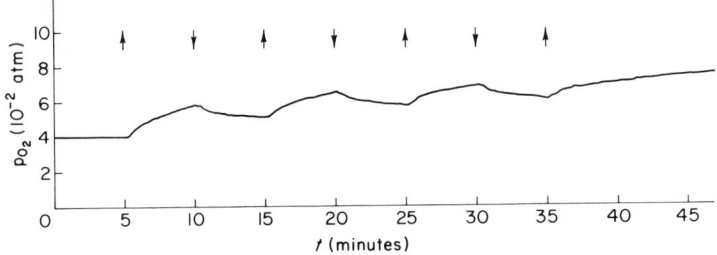

Fig. 3. Photooxidation of water by NADP in a model energy conversion system. (Reprinted from J. H. Wang, 1969. *Proc. Nat. Acad. Sci. U.S.*, **62**, 655, by permission of the copyright owner.)

But what are the primary electron acceptors "Q" and "X" in the natural system which replace the aluminum surfaces in the above model system? An obvious way to find the answer is to isolate P 700 and determine its chemical composition. Several research groups have recently made encouraging progress toward the identification of the chemical components which constitute the primary reaction centers of photosynthesis (McElroy et al., 1969; Loach and Walsh, 1969; Fuller and Nugent, 1969; Vernon et al., 1969; Clayton and Straley, 1970). While experimental elucidation of the structures of the natural photosynthetic reaction centers continues its admirable progress, we started to search for related model systems with absorption spectra and photobleaching properties resembling those of the primary reaction centers of photosynthesis. It was hoped that the supplemental studies of relevant model systems might help us to reduce the number of possible candidates for the unknown electron acceptor of P 700 and to gain a clearer understanding of the biological problem.

Recently it was discovered that chlorophyllin a and lumiflavin form a soluble complex in 0.01 percent poly-N-vinylpyrrolidone (PVP) solution in 0.01 phosphate buffer at pH 7.0 (Tu et al., 1970). While chlorophyllin a itself in 0.01 percent PVP solution has an absorption maximum at 660 nm, this molecular complex has an absorption maximum at 695 nm which can be bleached by red light. In the absence of air, the photobleached solution recovers completely in the dark with a half-life of 75 msec at 23°C.

As an alternative method to study the interaction between the chromophore and the acceptor molecules, the absorption spectrum of a film containing multimolecular layers of ethyl chlorophyllide a and lumiflavin in van der Waals contact with each other was taken and compared with that of a similar film containing only chlorophyllide a, as shown in Fig. 4.

The solid curve in Fig. 4 represents the absorption spectrum of such a composite film prepared by the dipping method of Blodgett (1935). The overlapping broken curve represents the absorption spectrum of a similarly prepared film of pure ethyl chlorophyllide a but normalized in such a way that it would contain as much ethyl chlorophyllide a per square centimeter as the composite film. The difference in absorbance, ΔA, between these two curves is plotted with expanded scale in the upper inset of Fig. 4. The minimum at 670 nm and the maximum at 700 nm of this difference curve correspond to the absorption maxima at 670 nm and 700 nm for the unperturbed and lumiflavin-perturbed ethyl chlorophyllide a molecules, respectively. As in the case of soluble chlorophyllin a–lumiflavin complex, the positive peak at 700 nm in the difference spectrum can be bleached reversibly by red light. The photobleached film recovers in the dark with a half-life of 46 msec at 23°C.

Rumberg and Witt (1964) measured the photobleaching of chloroplasts at 703 nm and found it to recover in the dark with a half-life of $\sim$20 msec at 20°C. Yamamoto and Vernon (1969) observed that highly purified P 700 particles prepared from spinach chloroplasts showed maximum bleaching at 698 nm and that the bleached sample recovered at room temperature in the dark with a half-life of the same order of magnitude.

The above observations suggest that the P 700 of photosynthesis may be a molecular complex of chlorophyll a and an acceptor molecule with flavin-like structure (Fl) which utilizes light energy to drive electrons from cytochrome f or plastocyanine up a free-energy hill to ferredoxin or NADP-reductase as represented by Scheme 1.

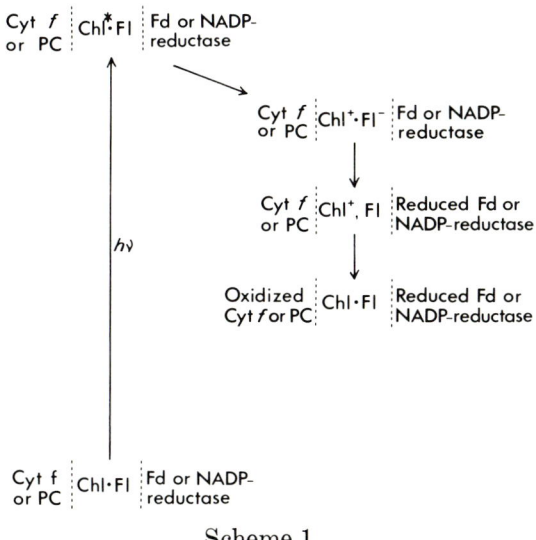

Scheme 1

(Reprinted from S. I. Tu, J. H. Wang, Y. J. Tan, 1971. *Bioinorg. Chem.* **1**, 94, by permission of American Elsevier Publishing Co., Inc., New York.)

ESR studies of the model complexes of lumiflavin and chlorophyllide a or chlorophyllin a show that the radicals Chl$^+$ and Fl$^-$ can indeed be generated readily by red light of moderate intensity (Tu et al., 1970). The ESR spectra of chloroplasts under illumination are too complex for detailed interpretation, but are compatible with the above scheme.

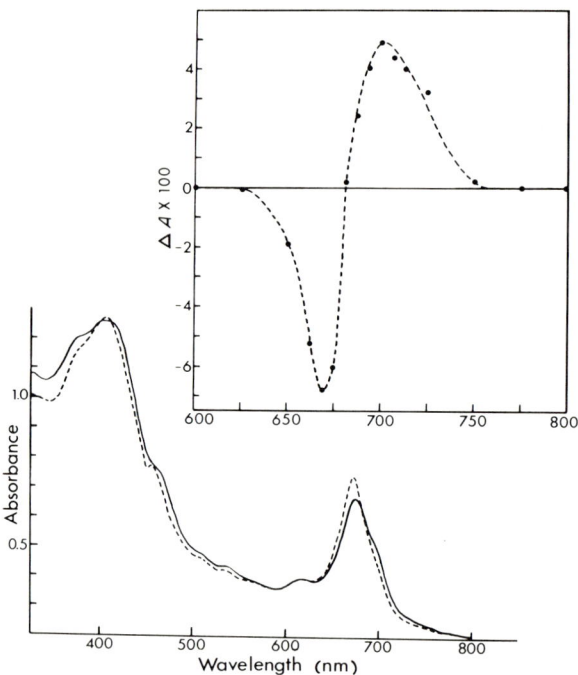

Fig. 4. Absorption spectra of multimolecular layers of ethyl chlorophyllide a and lumiflavin. (Reprinted from S. I. Tu, J. H. Wang, and Y. J. Tan, 1971. *Bioinorg. Chem.* **1**, 83, by permission of American Elsevier Publishing Co., Inc., New York.)

Phosphorylation Coupled to Electron Transport

In both photosynthesis and respiration, the endergonic synthesis of ATP is coupled to electron transport. But by what molecular mechanism is the free energy liberated in an electron transfer process utilized to drive a phosphorylation reaction in which P_i and ADP condense to form ATP and water? There is not yet a universally accepted answer.

Believing in the ultimate simplicity of nature, but frustrated by the complexity of chloroplasts and mitochondria, we started in 1961 to search for relevant model oxidative phosphorylation reactions to guide our thinking on the biological system. It was found, when ferrohemochrome was oxidized by air in N,N-dimethylacetamide solution containing AMP, P_i, and imidazole, that ADP and ATP were formed (Brinigar et al., 1967). Additional experiments showed that in this model reaction 1-phosphoimidazole was first produced which subsequently transferred its phosphoryl group to AMP or ADP to form ADP or ATP, respectively. But how was 1-phosphoimidazole formed?

Experiments with methylated imidazoles (Cooper et al., 1968) suggest that O_2 first extracts two electrons from the ferrohemochrome to produce a complex of Fe(III)-heme and the reactive imidazolyl radical ($C_3H_3N_2$), i.e.

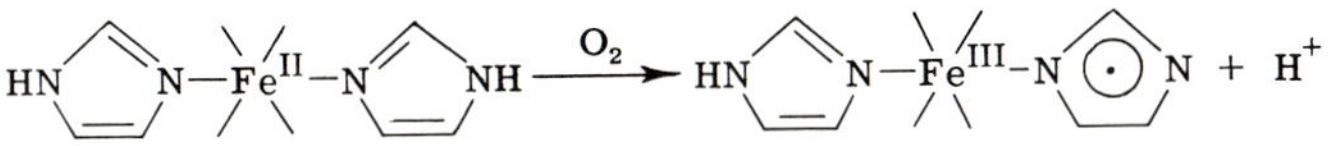

$$HN\text{—}N\text{—}Fe^{II}\text{—}N\text{—}NH \xrightarrow{O_2} HN\text{—}N\text{—}Fe^{III}\text{—}N(\cdot)N + H^+$$

This radical can then react with P_i to form an unstable phosphoimidazole radical, which is subsequently reduced by another ferrohemochrome molecule to produce 1-phosphoimidazole and water as illustrated below:

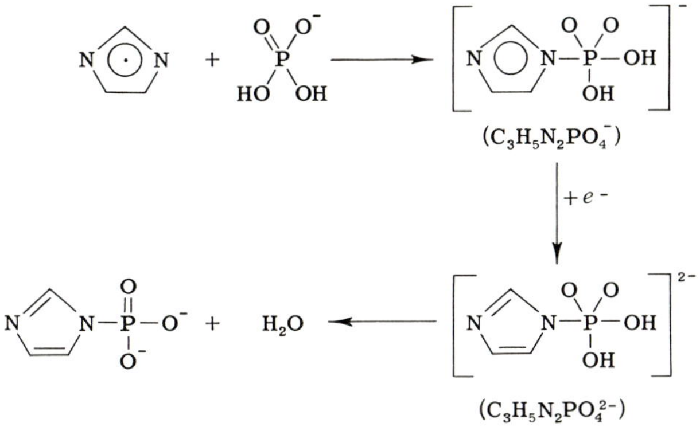

$$N(\cdot)N + \overset{O}{\underset{HO}{\overset{\|}{P}}}\overset{O^-}{\underset{OH}{}} \longrightarrow \left[N\bigcirc N\text{—}\overset{O}{\underset{OH}{\overset{\|}{P}}}\text{—}OH \right]^-$$

$$(C_3H_5N_2PO_4^-)$$

$$\downarrow +e^-$$

$$N\bigcirc N\text{—}\overset{O}{\underset{O^-}{\overset{\|}{P}}}\text{—}O^- + H_2O \longleftarrow \left[N\bigcirc N\text{—}\overset{O}{\underset{OH}{\overset{\|}{P}}}\text{—}OH \right]^{2-}$$

$$(C_3H_5N_2PO_4^{2-})$$

The formation of the trigonal-bipyramidal intermediate compound 1-orthophosphoimidazole ($C_3H_5N_2PO_4^{2-}$) by the usual nucleophilic attack at the P atom is not only very slow but thermodynamically unfavorable. But since radical reactions generally require much lower activation free energy, the trigonal-bipyramidal phosphoimidazolyl radical ($C_3H_5N_2PO_4^-$) can be formed much more rapidly (through the radical addition to the P=O double bond). In a subsequent step driven by the oxidation-reduction free energy, this phosphoimidazolyl radical is reduced to the unstable 1-orthophosphoimidazole, which then spontaneously eliminates H_2O to form -phosphoimidazole. In this way, electron transfer can be coupled to phosphorylation.

With this model oxidative phosphorylation reaction as a guide, a molecular mechanism for mitochondrial oxidative phosphorylation has been proposed (Wang, 1967, 1970b). The mechanism is summarized in Fig. 5, where the direction of electron transport under normal conditions is

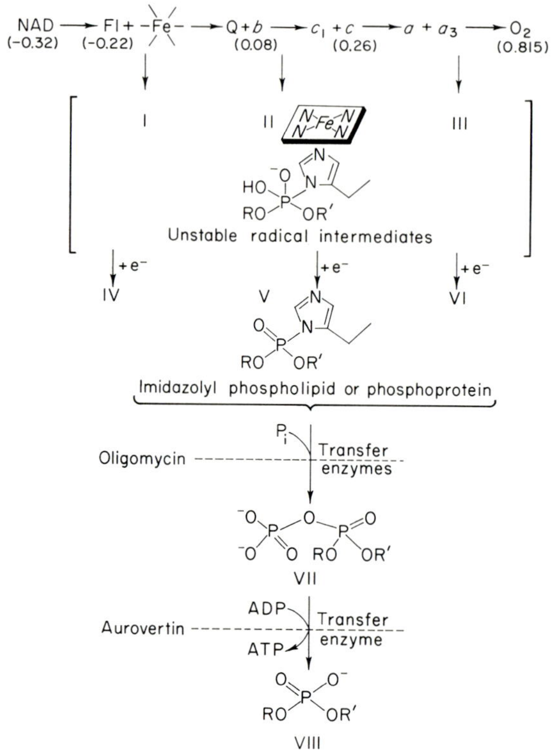

FIG. 5. A proposed molecular mechanism of mitochondrial oxidative phosphorylation.

represented by the boldface arrows between the respiratory electron carriers, i.e., NAD, flavin (Fl), quinone (Q), and the cytochromes (b, c_1, c, a, a_3).

The oxidative phosphorylation scheme in Fig. 5 utilizes a coupling mechanism similar to that of the model reactions, with the important difference that in mitochondria the imidazolyl radical cannot diffuse away and cause biological damage because it is a part of the structure of the protein, e.g., the cytochromes. If in the mitochondrial system the cytochrome is adjacent to a suitable phospholipid (or phosphoprotein) molecule, represented by (RO) (R'O) PO_2^- in Fig. 5, where R denotes an organic group and R' denotes either an organic group or a hydrogen atom, the imidazolyl radical can readily react with the latter to form a phosphoimidazolyl radical which can subsequently be reduced to the "energy-rich" intermediates IV, V, or VI. Each of these intermediates can likewise react with P_i in the presence of a specific enzyme or coupling factor to give VII, which can subsequently transfer its phosphoryl group to ADP in the presence of another enzyme or coupling factor to give ATP and regenerate the phospholipid for the next round of oxidative phosphorylation. Since the phosphoryl group in IV, V, or VI is not from inorganic phosphate but from endogenous phospholipid (or phosphoprotein), each of these "energy-rich" compounds should behave as a "nonphosphorylated intermediate" according to the conventional criterion (Lee and Ernster, 1968). A theory based on the scheme in Fig. 5 has been developed to account for the observed respiratory control and changes in ionic concentration gradients (Wang, 1970b) which seems to have gained direct experimental support from the recent work of Wilson and Dutton (1970).

Recently we observed that the same molecular mechanism can be utilized to couple light-driven electron transfer to phosphorylation in aqueous solution when molecular oxygen is carefully excluded (Tu and Wang, 1970). For example, yellow light absorbed by hematoporphyrin can be utilized to drive the condensation of imidazole with inorganic phosphate in aqueous solution to form 1-phosphoimidazole. The data of a typical experiment are illustrated in Fig. 6.

The mechanism of this photophosphorylation reaction is essentially the same as that for the model oxidative phosphorylation reaction discussed

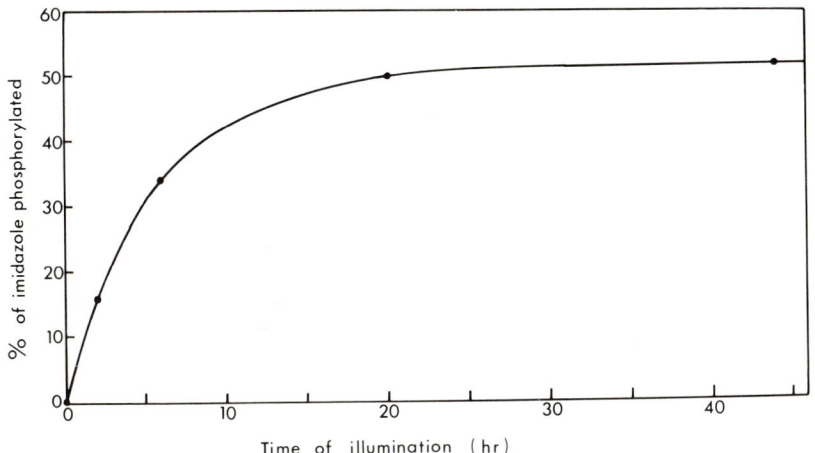

FIG. 6. Phosphorylation of imidazole coupled to the light-driven electron transfer in aqueous phosphate buffer at pH 7.0. The success of this experiment requires the complete removal of oxygen. (Reprinted from S.-I. Tu and Y. J. Wang, 1970. *Biochemistry* **9**, 4506, by permission of the American Chemical Society, Washington, D.C.)

above, except that the electronically excited hemotoporphyrin molecule, HP*, is used instead of molecular oxygen as the oxidant. The initial step is

$$\text{HN} \underset{}{\overset{}{\diagdown}}\text{N} + \text{HP}^* \longrightarrow \text{H}^+ + \text{N}\underset{}{\overset{}{\diagdown}}\text{N} + \text{HP}^-$$

The imidazolyl radical can then readily react with $H_2PO_4^-$ to form phosphoimidazolyl radical, which can subsequently be reduced by the HP$^-$ radical to form 1-phosphoimidazole and water.

The production of HP$^-$, imidazolyl, and phosphoimidazolyl radicals was demonstrated by electron spin resonance (ESR) measurements described below.

When a solution of hematoporphyrin and a suitable electron donor in aqueous phosphate buffer at pH 7 was frozen at $-150°C$ and illuminated with yellow light in the absence of imidazole, the sample did not show any detectable ESR signal during the first 30 seconds, but then an ESR signal at $g = 2.0026$ (half-width, 7.5 G) gradually appeared and grew to substantial intensity over a period of about 20 minutes. This ESR signal was identical to that of the reduced hematoporphyrin radical characterized previously (Eisenstein and Wang, 1969).

TABLE 1

Light-Induced Electron Spin Resonance Signals in Aqueous Mixtures at $-150°C$

Generating reaction[a]	g Value	Line width (G)
HP + HPhlorine $\xrightarrow[\text{slow}]{h\nu}$	2.0026	7.5
HP + EDTA $\xrightarrow[\text{slow}]{h\nu}$	2.0026	7.5
HP + imidazole $\xrightarrow[\text{fast}]{h\nu}$	2.0036	6.1
HP + imidazole-d_4 $\xrightarrow[\text{fast}]{h\nu}$	2.0036	5.5[b]

[a] *Abbreviations*: HP, hematoporphyrin; HPhlorine, hematophlorine; EDTA, ethylenediaminetetracetate.
[b] Same value in either H_2O or D_2O solution.

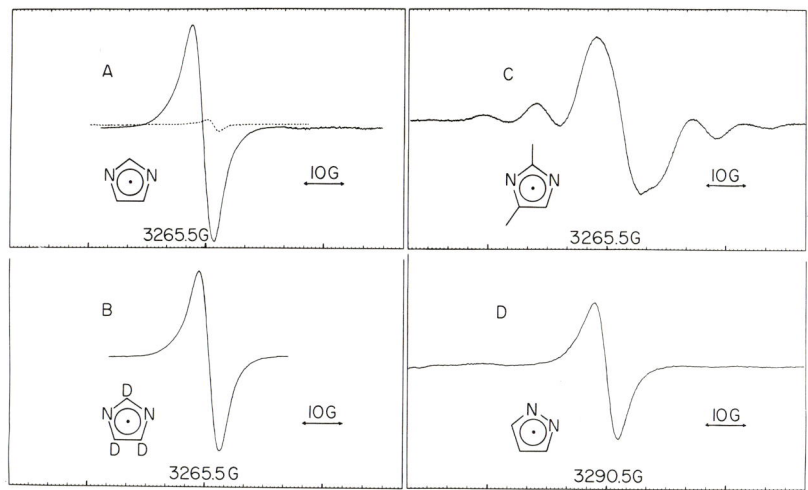

Fig. 7. Electron spin resonance (ESR) spectra of photogenerated imidazolyl and related radicals trapped in frozen mixtures at $-150°C$. All samples contain $2.0 \times 10^{-4} M$ hematoporphyrin in $0.1 M$ phosphate buffer at pH 7.0. Additional reactant in each sample: (A) imidazole ($1.5 \times 10^{-2} M$) (solid curve, frozen sample under illumination; broken curve, sample melted for a few seconds and refrozen in the dark, residual signal due to the empty ESR tube); (B) imidazole-d_3 or -d_4 ($1.5 \times 10^{-2} M$); (C) 2,4-dimethylimidazole ($1.5 \times 10:^2 M$); (D) pyrazole ($1.5 \times 10^{-2} M$). The microwave frequencies are (A) 9.418, (B) 9.168, (C) 9.160, (D) 9.223 GHz, respectively. (Reprinted from S.-I. Tu and Y. J. Wang, 1970. *Biochemistry*, **9**, 4507, by permission of the American Chemical Society, Washington, D.C.)

By contrast, when a solution of hematoporphyrin and imidazole in phosphate buffer at pH 7 was frozen at $-150°C$ and illuminated by yellow light, an ESR signal with $g = 2.0036$ (half-width = 6.1 G) was detected immediately (within 1 second). We attributed this fast signal to imidazolyl radical, since its line width decreased to 5.5 G when the normal imidazole was replaced by deuterated imidazole (imidazole-d_3 or d_4). These experimental results are summarized in Table 1.

The ESR spectra of the illuminated frozen samples containing imidazole or its analogs are summarized in Fig. 7.

As expected from our previous work (Cooper *et al.*, 1968), the illuminated frozen samples of HP + imidazole, HP + 2,4-dimethylimidazole, and HP + pyrazole yielded the imidazolyl radical-type of ESR spectra, and the illumination of a frozen sample of HP + 1-methylimidazole did not. The imidazolyl radical trapped in frozen samples is unstable in liquid solution, since its ESR spectrum disappeared completely when the sample was melted for even a few seconds and then quickly refrozen.

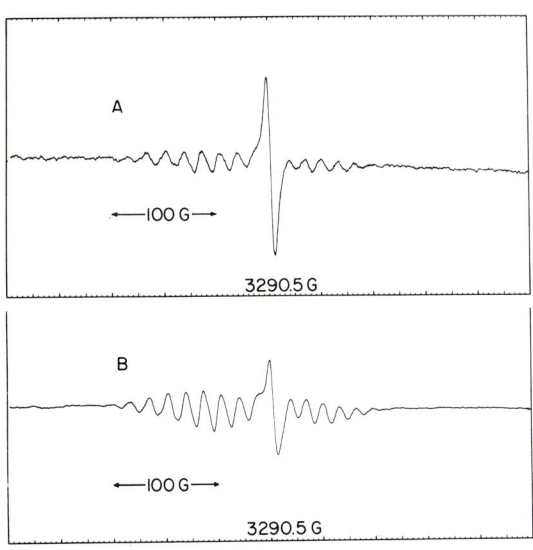

Fig. 8. Electron spin resonance (ESR) spectra of radicals derived from 1-phosphoimidazole or 1-phosphopyrazole in frozen mixtures at −196°C. All samples contain 2.0×10^{-4} M hematoporphyrin in 0.1 M phosphate buffer at pH 7.0. Special features of each experiment: (A) The mixture containing 1.5×10^{-2} M imidazole was illuminated by yellow light in an evacuated and sealed tube for 48 hours at room temperature, subsequently frozen in liquid nitrogen, and again illuminated for a few minutes, then the light was turned off and the ESR spectrum was taken in the dark at −196°C. Microwave frequency, 9.219 GHz. (B) The mixture containing 9.0×10^{-3} M 1-phosphoimidazole was sealed in an evacuated tube, frozen in liquid nitrogen, and illuminated by yellow light for a few minutes; then the ESR spectrum was taken in the dark at −196°C. Microwave frequency, 9.222 GHz. (Reprinted from S.-I. Tu and Y. J. Wang, 1970. *Biochemistry* **9**, 4508, by permission of the American Chemical Society, Washington, D.C.)

The usually narrow line width and the absence of hyperfine structure in the ESR spectra of imidazolyl and pyrazolyl radicals shown in Fig. 7 are probably due to rapid hydrogen atom exchange. It is likely that imidazole molecules form aggregates in the frozen aqueous sample. If the imidazolyl radical in an aggregate can extract a hydrogen atom from a neighboring imidazole molecule within the relaxation time of its unpaired electron, a narrowing of the ESR signal similar to that caused by spin exchange may result. It was hoped that by replacing imidazole with the less symmetrical 2,4-dimethylimidazole in the above-described experiment the aggregation of the base in the frozen sample could be minimized. The observed ESR spectrum of such a system (Fig. 7C) indeed exhibits some hyperfine structure. If it can be assumed that the hyperfine splittings due to contact interaction of the unpaired electron with the protons of 2,4-dimethylimidazole radical are obscured by local anisotropy in the frozen sample, then Fig. 7C is consistent with the ESR spectrum of isolated radicals with two approximately equivalent N atoms superposed on the narrow ESR signal due to the radicals in molecular aggregates.

After the frozen HP + imidazole in phosphate buffer sample was illuminated for many hours, side bands began to appear in the ESR spectrum. The generation of these side bands can be accelerated by preilluminating the liquid sample with yellow light above 0°C. If the preilluminated sample was frozen in the dark, no ESR spectrum was detectable. But subsequent illumination in the frozen state at −150°C readily generated a complex ESR spectrum shown in Fig. 8A.

The strong single-derivative peak in Fig. 8A is due to imidazolyl radical in the mixture, since it is identical with the ESR spectrum shown in Fig. 7A. The complex spectrum in Fig. 8A is probably due to radical species produced through the reaction of 1-phosphoimidazole, since its intensity increases with the concentration of 1-phosphoimidazole produced during the preillumination period. This conclusion is further supported by the observation that by freezing a solution of hematoporphyrin and synthetic 1-phosphoimidazole and illuminating with yellow light, the ESR spectrum in Fig. 8B, which is identical with that of Fig. 8A, appeared within a few seconds.

A mechanistic scheme for the probable reaction which generate the phosphoimidazolyl radicals is given in Fig. 9.

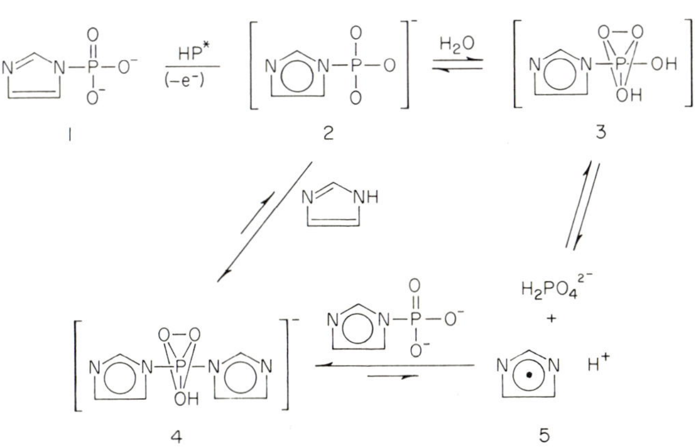

FIG. 9. A mechanistic scheme for the probable reactions that generate the phosphoimidazolyl radicals in the model system for coupling phosphorylation to light-driven electron transfer. (Reprinted from S.-I. Tu and Y. J. Wang, 1970. *Biochemistry* **9**, 4507, by permission of the American Chemical Society, Washington, D.C.)

Initially when an aqueous solution of hematoporphyrin and imidazole in phosphate buffer was illuminated with yellow light, only imidazolyl radical was formed by the light-driven electron transfer from imidazole to hematoporphyrin. But as the concentration of 1-phosphoimidazole continued to increase prolonged illumination, the radical **2** may be formed through the direct extraction of an electron from 1-phosphoimidazole (**1**) by the excited hematoporphyrin (HP*). In view of its electron deficiency, we expect **2** to be unusually susceptible to the nucleophilic attack by imidazole and H_2O (or OH^-) to form the radicals **4** and **3**, respectively. The imidazolyl radical **5** may also react directly with $H_2PO_4^-$ and 1-phosphoimidazole to form **3** and **4**, respectively.

If the ^{14}N-nuclei in **4** interact approximately similarly with the unpaired electron, the ESR spectrum of **4** should consist of two slightly overlapping nonets, each with relative intensities 1:4:6:10:16:19:16:10:6:4:1.

In addition, if we assume that the radicals **4** and **5** are of greater stability and hence at considerably higher steady-state concentrations than **2** and **3**, we would expect the frozen mixture to exhibit, upon illumination, an ESR spectrum equivalent to the superposition of the spectra of **4** and **5**. That is, the superposition of a spectrum of two slightly overlapping nonets with the spectrum shown in Fig. 7A. Although local anisotropy in the frozen samples prevent us from making reliable assignments of the ESR lines, the observed spectra in Fig. 8 are consistent with this interpretation. It is interesting to note that the hyperfine lines in the ESR spectrum of Fig. 8B are even more clearly defined than those in Fig. 8A. This last observation is quite consistent with the mechanistic scheme given in Fig. 9, since the sample in Fig. 8B contains a higher concentration of 1-phosphoimidazole.

These results show that the imidazolyl radical generated by electron transfer can rapidly form a labile P—N bond with inorganic orthophosphate and be subsequently reduced to the "energy-rich" 1-phosphoimidazole in aqueous solution at pH 7. In this model system of light-driven electron transfer, the only appreciable overall net reaction is the condensation of inorganic orthophosphate with imidazole to form 1-phosphoimidazole and water. The remarkable cleaness and efficiency of this coupling mechanism suggest its possible application to the coupling of phosphorylation to electron transport in chloroplasts.

The most obvious weakness of any which involves chemical intermediates is that, in spite of the intensive efforts during the last 25 years, no "energy-rich" intermediate has ever been isolated from either chloroplasts or mitochondria. The failure to isolate such an intermediate could either be due to the fact that the intermediate does not exist or that it cannot be isolated by conventional methods. Suggestions were made from studies of mitochondria-catalyzed ^{18}O-exchange between P_i and H_2O (Boyer and Harrison, 1954; Boyer, 1967) and measurements of ionic concentration gradients generated by electron transport (Mitchell, 1961, 1966) that ATP is produced directly from ADP and P_i by a special ATPase without the formation of phosphorylated intermediates.

However, by using $^{32}P_i$ as a tracer and trapping the hypothetical compound VII in Fig. 5 with the antibiotics oligomycin and aurovertin, we have recently detected the phosphorylated intermediate in rat liver mitochondria (Cross et al., 1970). Therefore it would appear that the proposed direct formation of ATP from P_i and ADP in chloroplasts may also be unnecessary.

ACKNOWLEDGMENT

This work has been supported by a research grant (GB 7459X) from the National Science Foundation.

REFERENCES

Ainsworth, S., and E. I. Rabinowitch. 1960. Electron transfer and absorption spectra of complexes. *Science* **131**:303.
Allen, M. B., F. R. Whatley, and D. I. Arnon. 1958. Photosynthesis by isolated chloroplasts. *Biochim. Biophys. Acta* **27**:16–23.
Arnold, W. 1970. Mechanism of delayed light production by photosynthetic organisms. *International Conference on the Photosynthetic Unit, Gatlinburgh, Tennessee.*
Arnon, D. I. 1951. Extracellular photosynthetic reactions. *Nature (London)* **167**:1008–1010.
Arnon, D. I., M. B. Allen, and F. R. Whatley. 1954. Photosynthesis by isolated chloroplasts. *Nature (London)* **174**:394–396.
Avron, M. 1967. Mechanism of photoinduced electron transport in isolated chloroplasts. *Curr. Top. Bioenerg.* **2**:17–22.
Banerjee, A. K., and G. Tollin. 1966. Reversible light-induced single-electron transfer reactions between chlorophyll and hydroquinone in solution. *Photochem. Photobiol.* **5**:315–322.
Bishop, N. I. 1962. Separation of the oxygen evolving system of photosynthesis from photochemistry in a mutant of scenedesmus. *Nature (London)* **195**:55–57.
Blinks, L. R. 1957. Chromatic transients in photosynthesis of red algae. *In* "Research in Photosynthesis" (H. Gaffron, ed.), pp. 444–449. Wiley (Interscience), New York.
Blodgett, K. B. 1935. Films built by depositing successive monomolecular layers on a solid surface. *J. Amer. Chem. Soc.* **57**:1007–1022.
Boussingault, T. B. 1864. La végétation dans l'obscurite. *Ann. Sci. Nat. Bot. Biol. Vegetale* **1**:314–324.
Boyer, P. D. 1967. O^{18} and related exchanges in enzymic formation and utilization of nucleoside triphosphates. *Curr. Top. Bioenerg.* **2**:99–149.
Boyer, P. D., and W. H. Harrison. 1954. On the mechanism of enzymic transfer of phosphate and other groups. *In* "Mechanism of Enzyme Action" (W. D. McElroy and B. Glass, eds.), pp. 658–674. Johns Hopkins Press, Baltimore, Maryland.
Brinigar, W. S., D. B. Knaff, and Jui H. Wang. 1967. Model reactions for coupling oxidation to phosphorylation. *Biochemistry* **6**:36–42.

Calvin, M., and J. A. Bassham. 1962. "The Photosynthesis of Carbon Compounds." Benjamin, New York.

Chance, B., and G. R. Williams. 1955. Respiratory enzymes in oxidative phosphorylation 1. Kinetics of oxygen utilization. *J. Biol. Chem.* **217**:383–393.

Chibisov, A. K., A. V. Karyakin, and V. B. Yevstigneyev. 1966. Primary processes in the reaction of interaction of chlorophyll with methyl viologen. *Biofizika* **11**:983.

Clayton, R. K. 1962. Primary reactions in bacterial photosynthesis II. The quantum requirement for bacteriochlorophyll conversion in the chromatophore. *Photochem. Photobiol.* **1**:305–311.

Clayton, R. K., and S. C. Straley. 1970. An optical absorption change that could be due to reduction of the primary photochemical electron acceptor in photosynthetic reaction centers. *Biochem. Biophys. Res. Commun.* **39**:1114–1119.

Cooper, T. A., W. S. Brinigar, and Jui H. Wang. 1968. Characterization of the reaction intermediates in a model oxidative phosphorylation reaction. *J. Biol. Chem.* **243**:5854–5858.

Cross, R. L., B. A. Cross, and Jui H. Wang. 1970. Detection of a phosphorylated intermediate in mitochondrial oxidative phosphorylation. *Biochem. Biophys. Res. Commun.* **40**:1155.

deSaussure, N. T. 1804. "Recherches sur la végétation." Nyon, Paris

Duysens, L. N. M. 1952. Transfer of excitation energy in photosynthesis. Thesis, Univ. of Utrecht, Utrecht.

Duysens, L. N. M. 1953. *Carnegie Inst. Wash., Yearb.* **52**:157.

Duysens, L. N. M. 1958. The path of light energy in photosynthesis. *Brookhaven Symp. Biol.* **11**:10–25.

Duysens, L. N. M. 1964. Photosynthesis. *Progr. Biophys. Mol. Biol.* **14**:1–104.

Duysens, L. N. M., and G. Sweep. 1957. Fluorescence spectrophotometry of pyridine nucleotide in photosynthesizing cells. *Biochim. Biophys. Acta* **25**:13–16.

Duysens, L. N. M., W. J. Huiskamp, J. J. Vos, and J. M. van der Hart. 1956. Reversible changes in bacteriochlorophyll in purple bacteria upon illumination. *Biochim. Biophys. Acta* **19**:188–190.

Duysens, L. N. M., J. Amesz, and B. M. Kamp. 1961. Two photochemical systems in photosynthesis. *Nature (London)* **190**:510–511.

Einstein, A. 1905. Über einen die Erzeugung und Verwandlung des Lichtes betreffenden heuristischen Gesichtspunkt. *Ann. Phys. (Leipzig)* **17**:132–148.

Eisenstein, K. K., and Jui H. Wang. 1969. Conversion of light to chemical free energy I. porphyrin-sensitized photoreduction of ferredoxin by glutathione. *J. Biol. Chem.* **244**:1720–1728.

Emerson, R., and W. Arnold. 1932a. A separation of the reactions in photosynthesis by means of intermittent light. *J. Gen. Physiol.* **15**:391–420.

Emerson, R., and W. Arnold. 1932b. The photochemical reaction in photosynthesis. *J. Gen. Physiol.* **16**:191–205.

Emerson, R., and C. M. Lewis. 1941. Carbon dioxide exchange and the measurement of the quantum yield of photosynthesis. *Amer. J. Bot.* **28**:789–804.

Emerson, R., and C. M. Lewis. 1943. The dependence of the quantum yield of *Chlorella* photosynthesis on wavelength of light. *Amer. J. Bot.* **30**:165–178.

Emerson, R., R. V. Chalmers, C. Cederstrand, and M. Brody. 1956. Effect of temperature on the long-wave limit of photosynthesis. *Science* **123**:673.

Emerson, R., R. V. Chalmers, and C. Cederstrand. 1957. Some factors influencing the longwave limit of photosynthesis. *Proc. Nat. Acad. Sci. U.S.* **43**:133–143.

Engelmann, T. W. 1883. *Arch. Gesamte Physiol. Menschen Tiere* **30**:95–124.
Engelmann, T. W. 1888. Die Purpurbacterien und ihre Bezichungen zum Licht. *Bot. Zentralbl.* **46**:661–676.
Ficken, G. E., R. B. Johns, and R. P. Linstead. 1956. Chlorophyll and related compounds. Part IV. The position of the extra hydrogens in chlorophyll. The oxidation of pyrophaeophorbide-a. *J. Chem. Soc., London* pp. 2272–2280.
Fischer, H., and H. Henderoth. 1939. Zur Kenntis von Chlorophyll. *Justus Liebigs Ann. Chem.* **537**:170–177.
Förster, T. 1948. Zwischenmolekulare Energiewanderung und Fluoresenz. *Ann. Phys. (Leipzig)* **2**:55–76.
Förster, T. 1949. Experimentelle und theoretische Untersuchung des Zwischenmolekularen Übergangs von Elektronenanregungsenergie. *Z. Naturforsch. A* **4**:321–327.
Franck, J. 1949. An interpretation of the contradictory results in measurements of the photosynthetic quantum yields and related phenomena. *Arch. Biochem.* **23**:297–314.
Frenkel, A. W. 1954. Light-induced phosphorylation by cell-free preparations of photosynthetic bacteria. *J. Amer. Chem. Soc.* **76**:5568–5569.
Frey-Wyssling, A., and E. Steinmann. 1953. *Vierteljahresschr. Naturforsch. Ges. Züerich* **98**:20.
Fuhrkop, J. H., and D. Mauzerall. 1968. The one-electron oxidation of magnesium octaethylporphrin. *J. Amer. Chem. Soc.* **90**:3875–3876.
Fuller, R. C., and N. A. Nugent. 1969. Pteridine and the function of the photosynthetic reaction center. *Proc. Nat. Acad. Sci. U.S.* **63**:1311–1318.
Gaffron, H. 1942. The effect of specific poisons upon the photo-reduction with hydrogen in green algae. *J. Gen. Physiol.* **26**:195–217.
Gaffron, H. 1944. Photosynthesis, photoreduction and dark reduction of carbon dioxide in certain algae. *Biol. Rev. Cambridge Phil. Soc.* **19**:I.
Gaffron, H., and K. Wohl. 1936. Zur Theorie der Assimilation. *Naturwissenschaften* **24**:81–96, 103–107.
Givan, A. L., and R. P. Levine. 1967. The photosynthetic electron transport chain of chlamydomonas reinhardt. VII. Photosynthetic phosphorylation by a mutant strain of chlamydomonas reinhardi deficient in active P700. *Plant Physiol.* **42**:1264–1268.
Goedheer, J. C. 1958. Reversible oxidations of pigments in bacterial chromophores. *Brookhaven Symp. Biol.* **11**:325–331.
Hatch, M. D., and C. R. Slack. 1970. Photosynthetic CO_2-fixation pathways. *Annu. Rev. Plant Physiol.* **21**:141–162.
Haxo, F. T., and L. R. Blinks. 1950. Photosynthetic action spectra of marine algae. *J. Gen. Physiol.* **33**:389–422.
Hill, R., and F. Bendall. 1960. Function of the two cytochrome components in chloroplasts: A working hypothesis. *Nature (London)* **186**:136–137.
Hill, R., and R. Scarisbrick. 1940. Production of oxygen by illuminated chloroplasts. *Nature (London)* **146**:61–62.
Ingen-Housz, J. 1779. "Experiments upon Vegetables, Discovering their Great Power of Purifying the Common Air in Sunshine and Injuring It in the Shade and at Night." Elmsly & Payne, London.
Jagendorf, A. T., and M. Avron. 1958. Cofactors and rates of photosynthetic phosphorylation by spinach chloroplasts. *J. Biol. Chem.* **231**:277–290.
Kassner, R. J., and M. D. Kamen. 1967. The photoreduction of spinach ferredoxin in the presence of porphyrin and an electron donor. *Proc. Nat. Acad. Sci. U.S.* **58**:2445–2450.

Ke, B. 1969. Nature of the primary electron acceptor in bacterial photosynthesis. *Biochim. Biophys. Acta* **172**:583–585.

Kessler, E. 1957. Stoffevechselphysiologische Untersuchungen an Hydrogenase enthaltenden Grünalgen 1. Über die Rolle des Mangans bei Photoreduktion und Photosynthese. *Planta* **49**:435–454.

Knaff, D. B., and D. I. Arnon. 1969. A concept of three light reactions in photosynthesis by green plants. *Proc. Nat. Acad. Sci. U.S.* **64**:715–722.

Kok, B. 1948. A critical consideration of the quantum yield of chlorella-photosynthesis. *Enzymologia* **13**:1–56.

Kok, B. 1961. Partial purification and determination of oxidation-reduction potential of the photosynthetic complex absorbing at 700 mμ. *Biochim. Biophys. Acta* **48**:527–533.

Kok, B., and G. Hoch. 1961. *In* "Light and Life" (W. D. McElroy and B. Glass, eds.), p. 404. Academic Press, New York.

Komissarov, G. G. and Y. S. Shumov. 1968. *Dokl. Akad. Nauk SSSR* **182**:1226.

Krasnovsky, A. A. 1948a. *Dokl. Akad. Nauk SSSR* **60**:421.

Krasnovsky, A. A. 1948b. *Dokl. Akad. Nauk SSSR* **61**:91.

Laber, L. J., and C. C. Black. 1969. Site-specific uncoupling of photosynthetic phosphorylation in spinach chloroplasts. *J. Biol. Chem.* **244**:3463–3467.

Lavoisier, A. 1774. *Memoires Academie de Paris, 1770* (see Lavoisier, A. 1789. *Traite de Chimie, Paris*).

Lebedev, F. 1921. *Izvest. Donskovo Gosudarst. Univ.* **3**:25–55.

Lee, C. P., and L. Ernster. 1968. Studies of the energy transfer system of submitochondrial particles. 1. Competition between oxidation phosphorylation and the energy-linked nicotinamide-adenine dinucleotide transhydrogenase reaction. *Eur. J. Biochem.* **3**:385–390.

Levine, R. P. 1960. Genetic control of photosynthesis in chlamydomonas reinhardi. *Proc. Nat. Acad. Sci. U.S.* **46**:972–978.

Levine, R. P., and D. S. Gorman. 1966. Photosynthetic electron transport chain of chlamydomonas reinhardi. III. Light-induced absorbance changes in chloroplast fragments of the wild type and mutant strains. *Plant Physiol.* **41**:1293–1300.

Levine, R. P., and R. M. Smillie. 1963. The photosynthetic electron transport chain of Chlamydomonas reinhardi. *J. Biol. Chem.* **238**:4052–4062.

Loach, P. A., and K. Walsh. 1969. Quantum yield for the photoproduced electron paramagnetic resonance signal in chromatophores from rhodospirillum rubrum. *Biochemistry* **8**:1908–1913.

McElroy, J. D., G. Feher, and D. Mauzerall. 1969. On the nature of the free radical formed during the primary process of bacterial photosynthesis. *Biochim. Biophys. Acta* **172**:180–183.

McSwain, B. D., and D. I. Arnon. 1968. Enhancement effects and the identity of the two photochemical reactions of photosynthesis. *Proc. Nat. Acad. Sci. U.S.* **61**:989–996.

Maquenne, L., and E. Demoussy. 1913. Sur la valeur des coefficients chlorophylliens et leurs rapports avec les quotients respiratores réels. *C. R. Acad. Sci.* **156**:506–512.

Mayer, J. R. 1845. "Die Organische Bewegung in ihrem Zusammenhang mit dem Stoffwechsel." (see Rabinowitch, E. I. 1945. "Photosynthesis and Related Processes," Vol. I, pp. 24–26. Interscience, New York).

Metzner, H. 1968. Lichtinduzierte Wasserspaltung in Photosynthese-Modell. *Hoppe-Seyler's Z. Physiol. Chem.* **349**:1586–1588.

Mitchell, P. 1961. Coupling of phosphorylation to electron and hydrogen transfer by a chemi-osmotic type of mechanism. *Nature (London)* **191**:144–148.

Mitchell, P. 1966. Metabolic flow in the mitochondrial multiphase system: An appraisal of the chemi-osmotic theory of oxidative phosphorylation. *Regul. Metab. Processes Mitochondria, Proc. Symp., Bari, 1965* pp. 65–84.

Müller, A., B. Rumberg, and H. T. Witt. 1963. On the mechanism of photosynthesis. *Proc. Roy. Soc., Ser. B* **157:**313–332.

Nicolson, G. L., and R. K. Clayton. 1969. The reducing potential of the bacterial photosynthetic reaction center. *Photochem. Photobiol.* **9:**395–399.

Olson, J. M. 1958. Fluorometric identification of pyridine nucleotide changes in photosynthetic bacteria and algae. *Brookhaven Symp. Biol.* **11:**316–324.

Park, R. B. Chloroplast structure. *In* "The Chlorophylls" (L. P. Vernon and G. R. Seely, eds.), p. 307. Academic Press, New York.

Park, R. B., and J. Biggins. 1964. Quantasome: Size and composition. *Science* **144:** 1009–1011.

Park, R. B., and D. Branton. 1967. Freeze-etching of chloroplasts from glutaraldehyde-fixed leaves. *Brookhaven Symp. Biol.* **19:**341–352.

Park, R. B., and N. G. Pon. 1963. Chemical composition and the structure of lamellae isolated from spinacea oleracea chloroplasts. *J. Mol. Biol.* **6:**105–114.

Planck, M. 1901. Ueber das Gesetz der Energieverteilung im Normalspectrum. *Ann. Phys. (Leipzig)* **4:**553–563.

Priestley, J. 1772. Observations on different kinds of air. *Phil. Trans. Roy. Soc., London* **62:**147–264.

Rabinowitch, E. I. 1951. "Photosynthesis and Related Processes," Vol. II, Part 1, pp. 1083–1141. Wiley (Interscience), New York.

Ruben, S., and M. D. Kamen. 1940. Radioactive carbon of long half-life. *Phys. Rev.* **57:**549.

Ruben, S., and M. D. Kamen. 1941. Long-lived radioactive carbon: C^{14}. *Phys. Rev.* **59:**349–354.

Ruben, S., W. Z. Hassid, and M. D. Kamen. 1939. Radioactive carbon in the study of photosynthesis. *J. Amer. Chem. Soc.* **61:**661–663.

Ruben, S., M. Rendall, M. D. Kamen, and J. L. Hyde. 1941. Heavy oxygen (O^{18}) as a tracer in the study of photosynthesis. *J. Amer. Chem. Soc.* **63:**877–879.

Rumberg, Von B., and H. T. Witt. 1964. Analyse der Photosynthese mit Blitzlicht. I. Die Photooxydation von Chlorophyll-a_1-430-703. *Z. Naturforsch. B* **19:**693–707.

Sachs, J. 1864. Ueber die Auflösung und Wiederbildung de Amylums in den Chlorophyllkörnern bei wechselnder Beleuchtung. *Bot. Zentralbl.* **22:**289–296.

Senebier, J. 1782. "Mémoires physico chimiques sur l'influence de la lumière solaire pour modifier les être de trois regnes, surtout ceux du regne végétal." 3 Vols. Chirol, Geneva.

Steinmann, E. 1952. Contribution to the structure of granular chloroplasts. *Experientia* **8:**300–301.

Stryer, L., and R. P. Haugland. 1967. Energy transfer: A spectroscopic ruler. *Proc. Nat. Acad. Sci. U.S.* **58:**719–726.

Thomas, J. B., and J. C. Goedheer. 1953. Relative efficiency of light absorbed by carotenoids in photosynthesis and phototaxis of rhodospirillum rubrum. *Biochim. Biophys. Acta* **10:**385–390.

Tollin, G., and G. Green. 1962. Light-induced single electron transfer reactions between chlorophyll *a* and quinones in solution. I. Some general features of kinetics and mechanism. *Biochim. Biophys. Acta* **60:**524–538.

Tollin, G., and G. Green. 1963. Light-induced single electron-transfer reactions between chlorophyll *a* and quinones in solution. II. Some effects of non-quinonoid donors and acceptors: Riboflavin, thioetic acid and NADH. *Biochim. Biophys. Acta* **66:**308–318.

Tollin, G., K. K. Chatterjee, and G. Green. 1965. Light-induced single electron transfer reactions between chlorophyll a and quinone in solution. III. Absolute quantum yields as a function of temperature. *Photochem. Photobiol.* **4**:593–601.

Tolmach, L. J. 1951. Effects of triphosphopyridine nucleotide upon oxygen evolution and carbon dioxide fixation by illuminated chloroplasts. *Nature (London)* **167**: 946–948.

Tu, S.-I, and J. H. Wang. 1969. Conversion of light into chemical free energy through chlorophyllin-sensitized photoreduction of oxidized nicotinamide-adenine dinucleotide phosphate by cytochrome c. *Biochemistry* **8**:2970–2974.

Tu, S.-I, and J. H. Wang. 1970. Phosphorylation coupled to electron transport in aqueous solutions. *Biochemistry* **9**:4505–4509.

Tu, S.-I, J. H. Wang, and Y. J. Tan. 1971. Synthetic model complexes for studying light-driven electron transfer in photosynthesis. *Bioinorg. Chem.* **1**:79–94.

van Niel, C. B. 1931. *Arch. Mikrobiol.* **3**:1–112 (see van Niel, C. B. 1941. The bacterial photosyntheses and their importance for the general problem of photosynthesis. *Adv. Enzymol.* **1**:263–328).

Vernon, L. P. 1961. Photoreduction of pyridine nucleotides by ascorbate: Catalyses by chlorophylls and porphyrins in aqueous media. *Acta Chem. Scand.* **15**:1651–1659.

Vernon, L. P., A. San Pietro, and D. Limbach. 1965. Photoreduction of NAD and NADP by chlorophyllin a in the presence of ascorbate: Requirement for NADP reductase. *Arch. Biochem. Biophys.* **109**:92–97.

Vernon, L. P., H. Y. Yamamoto and T. Ogawa. 1969. Partially purified photosynthetic reaction centers from plant tissue. *Proc. Nat. Acad. Sci. U.S.* **63**:911–917.

Vinogradov, A. P., and R. V. Teis. 1941. *C. R. Acad. Sci. URSS* **33**:490.

Vishniac, W., and S. Ochoa. 1951. Photochemical reduction of pyridine nucleotides by spinach grana and coup ed carbon dioxide fixation. *Nature (London)* **167**:768–769.

Wang, J. H. 1967. The molecular mechanism of oxidative phosphorylation. *Proc. Nat. Acad. Sci. U.S.* **58**:37–44.

Wang, J. H. 1969. Synthesis of a model system for the primary energy conversion reactions in photosynthesis. *Proc. Nat. Acad. Sci. U.S.* **62**:653–660.

Wang, J. H. 1970a. Synthetic biochemical models. *Accounts Chem. Res.* **3**:90–97.

Wang, J. H. 1970b. Oxidative and photosynthetic phosphorylation mechanisms. *Science* **167**:25–30.

Warburg, O. 1919. Über die Geschwindigkeit der Photochemischen Kohlensaurezerstezung in lebenden Zellen. *Biochem. Z.* **100**:230–270.

Warburg, O., and V. Schocken. 1949. A manometric actinometer for the visible spectrum. *Arch. Biochem.* **21**:363–369.

Weaver, E. C., and N. I. Bishop. 1963. Photosynthetic mutants separate electron paramagnetic resonance signals of scenedesmus. *Science* **140**:1095–1097.

Williams, W. P. 1970. Spatial organization and interaction of the two photosystems in photosystems in photosynthesis. *Nature (London)* **225**:1214–1217.

Willstätter, R., and A. Stoll. 1918. "Untersuchungen über die Assimilation der Kohlensäure," pp. 315–341. Springer-Verlag, Berlin.

Wilson, D. F., and P. L. Dutton. 1970. The oxidation-reduction potentials of cytochromes a and a_3 in intact rat liver mitochondria. *Arch. Biochem. Biophys.* **136**:583–584.

Winogradsky, S. 1887. Ueber Schwefelbacterien. *Bot. Zentralbl.* **45**:489–512.

Winogradsky, S. 1888. "Zur Morphologie und Physiologie der Schwefelbacterien." Felix, Leipzig.

Witt, H. T., A. Müller, and B. Rumberg. 1961a. Experimental evidence for the mechanism of photosynthesis. *Nature (London)* **191**:194–195.

Witt, H. T., A. Müller, and B. Rumberg. 1961b. Oxidized cytochrome and chlorophyll $c_2{}^+$ in photosynthesis. *Nature (London)* **192:**967–969.

Woodward, R. B., W. A. Ayer, J. M. Beaton, F. Bickelhaupt, R. Bonnett, P. Buchschacher, G. L. Closs, H. Dutler, J. Hannah, F. P. Hauck, S. Itô, A. Langemann, E. LeGoff, W. Leimgruber, W. Lwowski, J. Sauer, Z. Valenta, and H. Volz. 1960. The total synthesis of chlorophyll. *J. Amer. Chem. Soc.* **82:**3800–3802.

Yamamoto, H. Y., and L. P. Vernon. 1969. Characterization of a partially purified photosynthetic reaction center from spinach chloroplasts. *Biochemistry* **8:**4131–4137.

THE NATURE OF PURIFIED PHYTOCHROME

WINSLOW R. BRIGGS, HARBERT V. RICE,* GARY GARDNER, and
CARL S. PIKE†

Biological Laboratories, Harvard University, Cambridge, Massachusetts

Introduction...	35
Results and Discussion..	37
Conclusions...	47
References..	49

Introduction

A large number of physiological and developmental phenomena in plants are potentiated by very brief irradiation with red light. In almost every case, this potentiation, whether it be for seed germination, leaf expansion, or chloroplast movement, can be reversed with a similar brief irradiation with far-red light, provided the far-red treatment is given sufficiently soon after the red. The vast amount of information on such responses that has accumulated in the past 35 years is covered in several reviews (Borthwick and Hendricks, 1965; Hillman, 1967; Smith, 1970). The physiological studies led to efforts to determine the nature of the photoreceptor(s) involved, and in 1959, a group working for the U.S. Department of Agriculture at Beltsville first detected the predicted photoreversible pigment. They accomplished this by some ingenious spectrophotometry with intact tissue samples and made the first aqueous extract

---

* Present address: Research Department, New England Aquarium, Central Wharf, Boston, Massachusetts 02110.

† Present address: Department of Biology, Franklin and Marshall College, Lancaster, Pennsylvania 17604.

(Butler et al., 1959). Their immediate objective was then to purify and characterize the pigment. The 1959 work had suggested that the pigment was a chromoprotein. By 1964, the Beltsville group had achieved approximately 60-fold purification of the pigment (given the name phytochrome) from dark-grown oat seedlings, using standard methods of protein purification (Siegelman and Firer, 1964). Two years later, Mumford and Jenner (1966), working at the E. I. du Pont de Nemours Experimental Station, at Wilmington, Delaware, published a purification procedure from oat seedlings which in their hands yielded 750-fold purification of phytochrome; this product was homogeneous by a number of different physical criteria. In another two years, another purification procedure appeared in the literature, this time from the Radiation Biology Laboratory of the Smithsonian Institution, as Correll et al. (1968a) extracted and purified phytochrome from dark-grown rye seedlings. Finally (after yet another two years), Hopkins and Butler (1970) published observations relating to possible protein conformational changes that occurred during phytochrome phototransformation. They used phytochrome purified from dark-grown oat seedlings and followed yet another procedure.

Although one might expect to obtain a product showing reasonable uniformity from two closely related plants, oats and rye, there were clearly some serious discrepancies, the primary one being in estimates of molecular weight. In the first purification paper, Siegelman and Firer (1964) estimated from the sedimation coefficient a molecular weight between 90,000 and 150,000, but commented on the heterogeneity of the preparation as preventing a more precise determination. Mumford and Jenner (1966), utililizing gel exclusion, bracketed their purified oat phytochrome between bovine serum albumin (68,000) and egg albumin (44,000), obtaining an estimated molecular weight of 62,000. They mentioned briefly that preliminary ultracentrifuge studies also yielded a molecular weight near 60,000, but with some evidence for aggregation into higher molecular weight species. Correll et al. (1968b) did both velocity and equilibrium ultracentrifugation experiments with their purified rye phytochrome. Preparations that were reduced and carboxymethylated for complete separation of putative subunits yielded a monomeric molecular weight of 42,000, while their native protein yielded a value of about 180,000. In addition, they reported evidence for a higher molecular weight aggregate. They proposed tetrameric and hexameric aggregates of the monomer, but conceded that the precise molecular weight and aggregate structure of phytochrome should await further analysis.

We had found in our own laboratory several years ago (Briggs et al., 1968) that if one passed crude solutions of oat phytochrome through a Sephadex G-200 column, one obtained two different molecular weight species. A rough calibration of the Sephadex column gave estimates of 80,000 and 180,000 for the two forms seen. More highly purified preparations, however, gave only the smaller species. Correll and Edwards (1970) subsequently showed that rye phytochrome exhibited the same phenomenon, although their Sephadex columns were not calibrated. They also showed that the large and small forms differed in their behavior in diethylaminoethyl cellulose (DEAE)-chromatography, and that the larger molecular weight species could be salted out by lower ammonium sulfate concentrations than the smaller.

The rather tangled nature of the evidence concerning the molecular weight of phytochrome raises several questions. First, what is the large molecular weight species? Is it an aggregate of the small, or is it in fact a different protein? Second, is the protein moiety of phytochrome in different genera quite different, since rye phytochrome seems so different from oat phytochrome? Third, what is native phytochrome? If one could show that large comes from small, or that the two are in equilibrium, then one must ask whether both are biologically active. The importance of detailed biochemical or physical studies of purified phytochrome is clearly vitiated in the absence of clear answers to these questions.

We had a number of things which we wished to do with purified phytochrome, including immunological, structural, and spectral studies. However, we felt that we should try to resolve the discrepancies noted above before such studies could be considered as giving representative data for the native phytochrome molecule. In this paper, we present and discuss some of the evidence that we have obtained which points to a reasonable resolution. The detailed procedures involved seem inappropriate for presentation here and will appear in standard journals (Gardner *et al.*, 1971; Pike and Briggs, 1972a,b).

Results and Discussion

Our initial efforts were directed toward purification of phytochrome from etiolated oat seedlings (*Avena sativa* L.). Utilizing between 12 and 15 kg of oat tissue, we developed a procedure for obtaining approximately 25 mg purified phytochrome, with a final yield between 15 and 25 percent. The procedure involved extracting the tissue in a Waring Blendor with tris (hydroxymethyl) aminomethane buffer (Tris) at pH 8.5, followed by four column-chromatographic steps. All operations were done in a cold room under dim green light. The initial column was calcium phosphate gel (brushite), to which the crude protein was bound after clarification by centrifugation. Binding occurred at 0.01 M phosphate buffer concentration; after first a 0.010 and then a 0.015 M phosphate buffer wash, the phytochrome was eluted by phosphate buffer at 0.062 M. Mercaptoethanol was always present from initial extraction through elution, although the concentration was lowered stepwise from 0.7 percent in the extraction buffer to 0.1 percent in the elution buffer. The pH was kept between 7.2 and 7.8. After ammonium sulfate fractionation, the protein was bound to DEAE-cellulose at pH 7.5, and eluted with a salt gradient, from 0.0 to 0.30 M KCl. A cation exchanger, carboxymethyl-Sephadex C-50, was used next, binding occurring at pH 6.0, and elution following a step to pH 7.8. Finally a molecular sieve step was incorporated, either Sephadex G-200 or Bio-Gel P-150, to yield the final product. Concentration between columns was achieved by ammonium sulfate precipitation. The phytochrome was contained in the pellet, which was then redissolved in the appropriate buffer and dialyzed for several hours to remove any residual ammonium sulfate before it was applied to the next column. Only the first ammonium sulfate fractionation yielded some degree of purification.

The final product was remarkably similar to that obtained by Mumford and Jenner (1966). Its absorption spectrum is shown in Fig. 1. The ratio

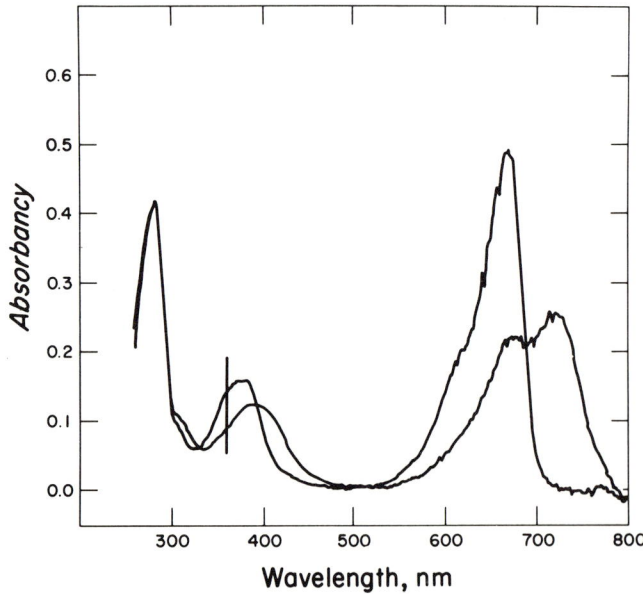

Fig. 1. Absorption spectrum of purified oat phytochrome after 5 minutes of saturating far-red light (P_r) and 5 minutes of saturating red light (P_{fr}). Spectra obtained on a Cary Model 14 scanning spectrophotometer.

of protein absorption at 280 nm to chromophore absorption at 667 nm (see spectrum for P_r) was about 0.8, with slightly less protein absorption (280 nm) per unit chromophore absorption (667 nm) than their final product. We then applied a number of tests for homogeneity to determine whether we had a single protein species or more than one. Sedimentation in the ultracentrifuge on a 5 to 20 percent sucrose gradient yielded a single band, as did velocity sedimentation on the analytical ultracentrifuge. Equilibrium ultracentrifugation suggested a molecular weight of about 60,000. Electrophoresis in Tris-glycine buffer on polyacrylamide gels yielded a single band. Electrophoresis on sodium dodecyl sulfate (SDS) acrylamide gels (Weber and Osborn, 1969) also yielded a single major band, although in this case, there was evidence for minor smaller contaminants. This technique is extremely useful for determining subunit structure, since the detergent is strongly electronegative. Thus, in forming a micelle around the protein, SDS not only separates subunits, but also swamps out minor charge differences. The technique can therefore be used for estimating subunit molecular weight: relative mobility decreases inversely as the log of the molecular weight (for a detailed treatment, see Reynolds and Tanford, 1970). Comparison with several known marker proteins yielded a molecular weight estimate of approximately 60,000 for our purified phytochrome.

Finally, phytochrome is an excellent antigen (Hopkins and Butler, 1970). Rabbits were challenged with the purified material, appropriately bled 6 weeks later to obtain serum containing antibody, and a series of analyses were done using the Ouchterlony technique. Serum containing antibody was placed in the center well of a plate of agarose gel, and various concentrations of purified phytochrome were placed in the circle of wells around the center. The precipitin bands were formed as antibody diffused centrifugally and phytochrome (antigen) diffused centripetally, and could be stained for permanent records. Over a wide range of phytochrome concentrations, only a single precipitin band was formed. An important test for the homogeneity of the original antigen is that of equivalence. Phytochrome samples were taken after each of the four columns in the purification procedure, adjusted to equal phytochrome concentration, and used in the Ouchterlony plate against the antibody. Once again, regardless of the level of purity, each of the fractions yielded a single precipitin band which was confluent from one purification step to the next. Thus, at least immunologically, the original purified phytochrome used as antigen seemed homogeneous.

On the basis of these and other results, we felt that our purification procedure was a good one, that the product was pure, and that we could perhaps proceed with confidence on other studies with the protein. However, there were a number of disquieting observations. First, we had made no progress in determining the nature of the large molecular weight phytochrome. Like other authors (Mumford and Jenner, 1966; Correll et al., 1968b; Correll and Edwards, 1970), we had treated it as an aggregation phenomenon. However, this assumption was weakened by the fact that we were unable to do anything that would make our purified material aggregate. Ultracentrifuge studies revealed that a concentration range of 1.8 to 16 mg of phytochrome per milliliter, a pH range between 6 and 8, and a range of salt concentrations between 0 and 0.5 M KCl all failed to give any evidence for reaggregation of the 60,000 molecular weight form. Similar experiments using a molecular sieving gel column, Sephadex G-200, as the analytical procedure instead of the ultracentrifuge were likewise negative. Though such negative evidence is hardly conclusive, three possibilities seemed worth considering. First, perhaps phytochrome was indeed an aggregating system, but during purification, it became slightly denatured in such a way that its aggregating properties were lost. Second, perhaps the large and small molecular weight forms observed early in purification were in fact unrelated proteins. Those workers who chose oats as their source of phytochrome selectively purified the small molecular weight form, and those who chose rye selectively purified the large. Third, perhaps the small was a breakdown product of the large—brought about by the cleavage of a covalent bond rather than by simple disaggregation. This third possibility seemed unlikely at first, since we had observed no differences in the visible absorption spectra between large and small.

Another disquieting observation was that different seed lots of oats gave rather different ratios of large to small phytochrome at similar stages of purification. The Smithsonian group had also reported substantial differences in the behavior of phytochrome from different seed lots of rye, or from the same seed lot after storage. Although their differences were in yield rather than ratio of size classes, it seemed that the two observations might be related.

The final disturbing observation came from an attempt to determine the amino-terminal amino acid from our own purified phytochrome. We used the Sanger (1945) procedure as modified by Burgess (1969), following the chromatographic procedures of Wang and Wang (1967). Instead of a single amino acid, we found four, although a run in parallel with a known protein—bovine serum albumin—yielded only a single product. The question was clear: Are there four different chains, all of equal size and charge, or has the preparation been somewhat degraded by enzymes or other agents during purification?

One of the questions raised above, concerning the relationship between large and small molecular weight phytochrome could be readily answered. A known mixture of large and small molecular weight rye phytochrome (partially purified by the oat procedure, through the first ammonium sulfate fractionation) was allowed to remain on the cold room shelf for about 96 hours. An aliquot was taken at zero time, layered on a 5 to 20 percent sucrose gradient, and sedimented in the SB-283 head of an International B-60 ultracentrifuge at 40,000 rpm for 20 hours. The gradient was then analyzed for distribution of phytochrome, and clearly revealed both large and small molecular weight forms (Fig. 2). A second aliquot of exactly the same volume was taken after 96 hours, and similarly treated. All the large molecular weight material had disappeared and was almost quantitatively replaced by small (Fig. 3). One can conclude, therefore, that the large and small forms are not in fact totally different proteins. Large does indeed give rise to small over time. The same experiment with oat phytochrome gave similar results. The question as to how the small molecular weight form was generated—whether by the breaking of a

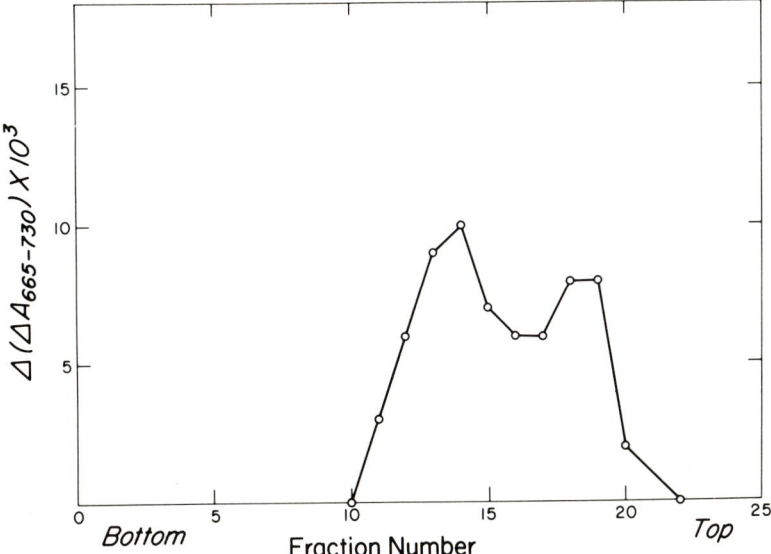

FIG. 2. Distribution of crude rye phytochrome immediately after calcium phosphate (brushite) chromatography and ammonium sulfate fractionation, in a 5 to 20 percent sucrose gradient. The sample was layered on a gradient and centrifuged for 20 hours in an International B-60 ultracentrifuge at 40,000 rpm in the SB-283 head (4°C). The gradient was analyzed by puncturing tubes, collecting fractions, and measuring phytochrome photoreversibility in a Ratiospect R2 difference spectrophotometer (Agricultural Specialties Co.).

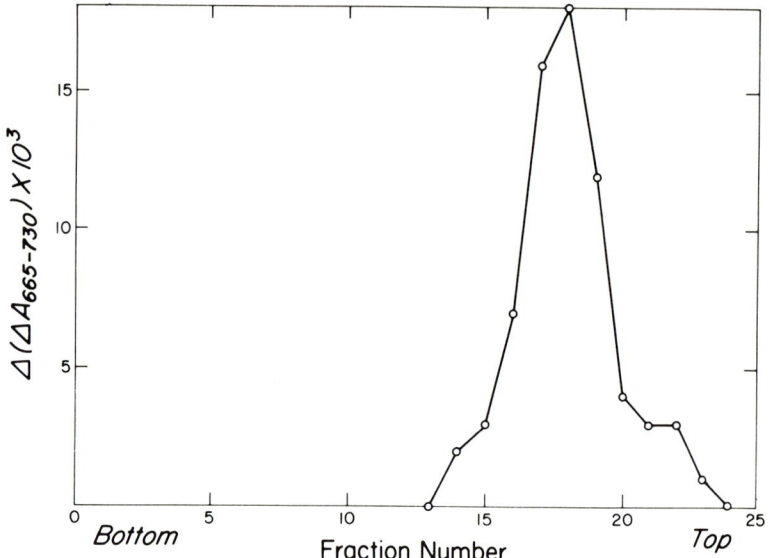

Fig. 3. Same as Fig. 2, but the sample was allowed to remain on cold-room shelf (4°C) in the dark for 96 hours prior to layering on 5 to 20 percent gradient.

covalent bond or by disaggregation—remained unanswered, although other experiments showing that the process takes hours or days made disaggregation seem unlikely, as did the absence of any evidence for equilibrium mixtures of the two forms.

While this work was being carried out, a related study was under way in another corner of the laboratory. One of us was studying a protease originally obtained from dark-grown oat seedlings under conditions somewhat similar to those for phytochrome purification in the hope that its action on phytochrome might shed some light on differences between P_r and P_{fr}. After partial purification of the protease, characterization studies were initiated. The enzyme was assayed by using Azocoll (Calbiochem), an insoluble denatured collagen to which a dye has been linked. Any endopeptidase will cleave small polypeptides (with their attached dye molecules) from the insoluble material, the colored polypeptides enter solution, and one can then simply remove the insoluble substrate and measure polypeptide release spectrophotometrically. The protease turned out to have a number of interesting properties. First, it showed a broad pH optimum, with maximal activity at about pH 6.5, but still quite active at pH 8.0. Second, it obviously required that its sulfhydryl groups be reduced for full activity, since it showed substantial activation by mercaptoethanol or dithiothreitol (Cleland's reagent). Maximal activation occurred with the mercaptoethanol concentration between about 0.1 and 1.0 percent. Third, it was inhibited by sulfhydryl reagents such as mercuric chloride ($HgCl_2$). It was also inhibited by phenylmethylsulfonyl fluoride (PMSF), an agent that does inhibit some proteases known to require sulfhydryl groups for activity (Whitaker and Perez-Villaseñor, 1968) but is thought to react with serine in its inhibition of trypsin (Fahrney and Gold, 1963). Finally, it was inhibited by high salt concentrations (e.g., 0.2 M KCl).

An immediate question is thus raised: Is a low level of this protease contaminating the phytochrome preparation during purification and in fact cleaving the large molecular weight chains into small? To answer this question, we obtained rye seed and made a crude preparation of rye phytochrome (through the brushite step and first ammonium sulfate fractionation only). Two aliquots were taken, and to one was added a small amount of the protease preparation. The other aliquot remained as a control. After incubation for 20 hours in the dark, and in the cold room at 4°C, the samples were layered on the usual sucrose gradients, centrifuged for 20 hours, and examined the next day. The results are shown in Fig. 4.

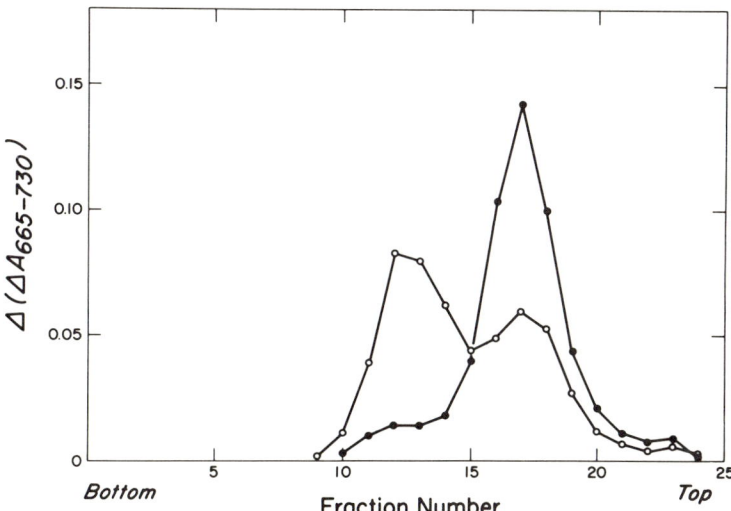

Fig. 4. Influence of oat protease on distribution of rye phytochrome in 5 to 20 percent sucrose gradients. Open circles: control. Filled circles: sample incubated for 20 hours with oat protease (partially purified). Control sample shows both large and small molecular weight phytochrome after 20 hours at 4°C in the dark, while identical sample to which protease was added at the start of the 20 hours shows only the small species. Sample preparation, centrifugation, and phytochrome assay as in Fig. 2.

Clearly, in the absence of the protease preparation, both large and small molecular weight species were present. In the presence of the protease, however, we recovered only small species, with no loss of total photoactivity.

Since the protease preparation was not highly purified, the possibility persisted that this enzyme had nothing to do with the large-to-small breakdown observed. We therefore attempted the same experiment with highly purified commercial trypsin (Worthington TRTPCK—chymotrypsin inhibited). An amount of trypsin equal to less than 0.1 percent of the total phytochrome was incubated with a preparation of relatively pure large molecular weight rye phytochrome. Once again, a 20-hour incubation in the cold room and in the dark, was sufficient to convert the large phytochrome entirely to small. The control remained entirely large.

As a final test for the role of proteases in the breakdown of large phytochrome to small, we tested the influence of PMSF on the spontaneous breakdown observed in relatively crude phytochrome obtained just after the brushite column. With both oat and rye phytochrome preparations, substantial large molecular weight phytochrome remained after incubation for 31 and 96 hours, respectively, although the untreated controls had broken down completely to the small form. The results with oat phytochrome are shown in Fig. 5. In both cases the Azocoll assay revealed protease activity present in the controls, but not in the inhibited samples. The conclusion seemed almost inescapable that the origin of small phytochrome is proteolytic cleavage of large.

Since the levels of protease were lower in rye than in oat preparations, we then attempted the purification of large molecular weight rye phytochrome. It was necessary to alter conditions somewhat to obtain reasonable yields and purification, and among other things a hydroxyapatite step replaced the carboxymethyl-Sephadex column. The final product appeared single-banded on sodium dodecyl sulfate gel electrophoresis, and calibration with known protein markers gave an estimated molecular weight of about 120,000 for the single chain. A preliminary equilibrium ultracentrifugation experiment, on the other hand, gave a value of about 245,000, suggesting that the native material was composed of at least two subunits. It is not clear at this time whether these two subunits are identical. An amino-terminal end group assay yields two amino acids—glutamate and aspartate—but it would be unwise at this point to dismiss the possibility that one of the subunits is a slightly degraded version of the other. Thus the matter is at present unresolved.

Evidently the large molecular weight phytochrome obtained from rye is not a globular protein. It shows anomalous behavior in systems for molecular weight determination in which molecular shape can be a factor.

The Nature of Purified Phytochrome

Thus on a calibrated Sephadex G-200 column, the phytochrome elutes with an apparent molecular weight of almost 400,000. On the other hand, velocity sedimentation in the analytical ultracentrifuge and in sucrose density gradients gives an s value of about 9 S, consistent with a globular protein of molecular weight 180,000. Both these values are in the right direction for a nonglobular protein, since such a molecule would be more retarded by molecular sieving gels than a globular one of the same molecular weight (and hence would appear larger), but would sediment more slowly in the ultracentrifuge (and hence would appear smaller).

The analytical SDS gels documenting the influence both of the oat protease and two commercial endopeptidases, trypsin and chymotrypsin, are shown in Fig. 6. From left to right, one first sees small molecular weight purified oat phytochrome; large molecular weight purified rye phytochrome; the oat protease alone at 1.3 times the concentration used in reaction mixtures with phytochrome; rye phytochrome incubated for 24 hours with the oat protease; the same, with trypsin; and finally, the same, with chymotrypsin. The main product of digestion with the oat protease or with trypsin is a single band whose relative mobility is iden-

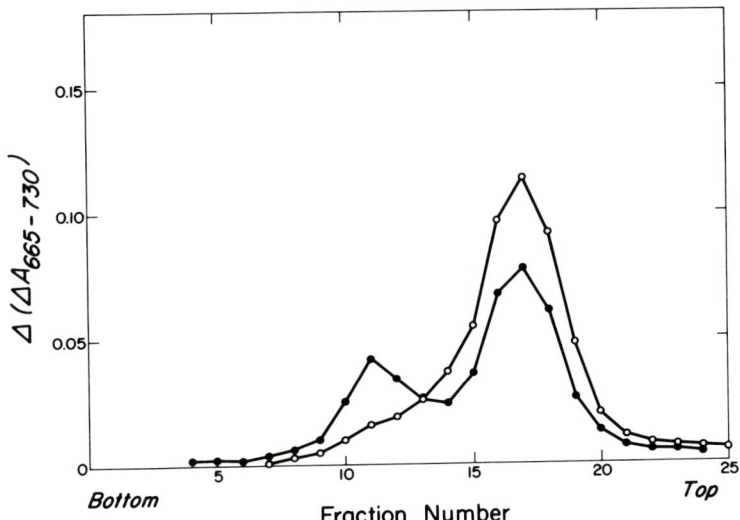

FIG. 5. Influence of phenylmethylsulfonyl fluoride (PMSF) on breakdown of oat phytochrome in absence of any added protease. Open circles: control. Filled circles: identical phytochrome sample, with 5×10^{-3} M PMSF. Both samples were incubated for 31 hours in the dark at 4°C. Sample preparation, centrifugation, and phytochrome assay as in Fig. 2. Control sample shows only small molecular weight phytochrome; PMSF-treated sample shows both large and small.

tical with that of the oat phytochrome (60,000). The main product from chymotrypsin digestion is somewhat larger (90,000). Under the conditions of incubation used, time-course studies showed rapid production of the principal bands, and then only very slow degradation of these bands, suggesting that there must be a labile region of the longer chain, highly susceptible to endopeptidase activity.

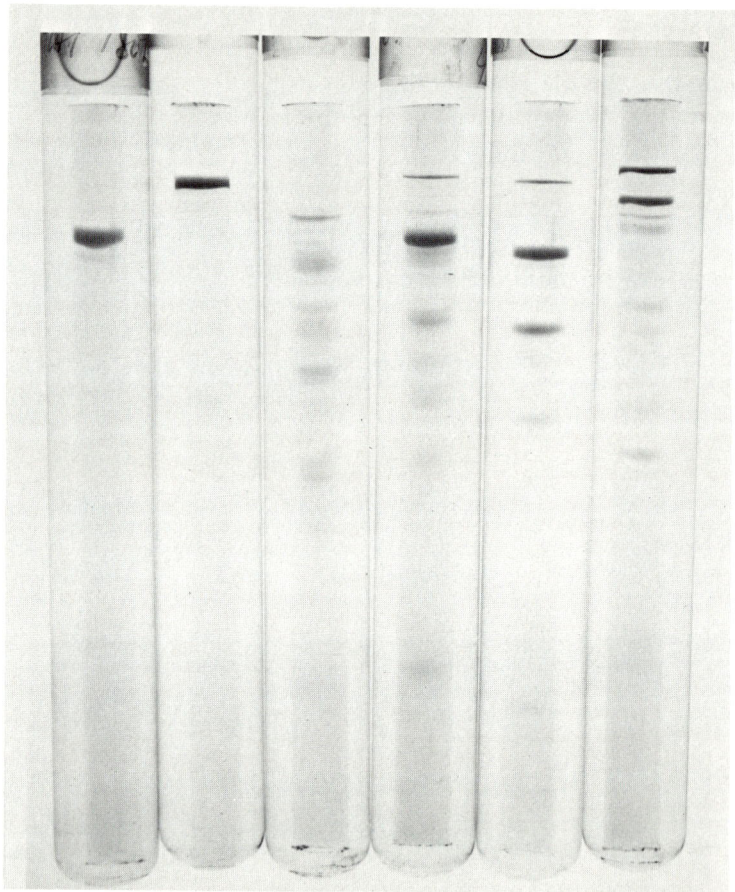

Fig. 6. Analytical SDS-gel electrophoresis of purified rye phytochrome samples after incubation with various proteases. Left to right: purified oat phytochrome; purified rye phytochrome; oat protease at 1.3 times the concentration used in the reaction mixture with rye phytochrome; rye phytochrome plus oat protease after 24-hour incubation in dark at 4°C; same, but plus trypsin instead of oat protease; same, but plus chymotrypsin instead of oat protease. Electrophoretic migration from top to bottom. Small molecular weight proteins migrate farther than large.

A final point should be made: the absorption spectra of purified oat phytochrome (small), purified rye phytochrome (large), oat protease-generated small molecular weight rye phytochrome, and trypsin-generated small rye phytochrome, show no obvious qualitative differences. All are fully photoreversible, all show the same details of the visible spectra for both forms of the pigment, and all show a bleaching of P_r by red light that is quite closely matched by the absorbancy increase seen in P_{fr}. The only difference one can detect is in the relative absorbancies of the protein peak at 280 nm versus the 667 nm peak of P_r. The ratio is about 1.25 (280:667) for purified large molecular weight material, and is about 0.80 for purified small, regardless of how it was obtained. Furthermore, the product of trypsinized rye phytochrome is indistinguishable from the purified oat phytochrome on gel filtration (Sephadex G-200).

Conclusions

It is clear from the results discussed above that the presence of a small amount of protease as a contaminant during the isolation and purification of a protein can lead to disastrously erroneous conclusions. It was the recent work of Pringle (1970a,b) that first raised the ugly possibility that what we initially purified as native phytochrome from oats might in fact be merely a large and relatively stable chromophoric polypeptide (or perhaps several such polypeptides) produced by endopeptidase activity during the purification procedure. Pringle (1970b) showed dramatically that even some of the most highly touted commercially purified enzymes contained trace amounts of protease contaminants. Under certain conditions, the protease might represent only one part in 10^{10} of the total protein present—an amount that no physical techniques known at present could detect—but would still slowly degrade the purified protein it contaminated, as detected on SDS gels. All the phytochrome purification procedures published to date involve use of a reducing agent at least in the early steps, a pH between 6.0 and 8.0 most of the time, and low salt conditions except occasionally. All of these conditions optimize the activity of the oat protease.

The unfortunate aspect of the situation with phytochrome is that large chains can be cleaved into small without significant qualitative alteration of absorption spectra. Thus in the absence of continual monitoring of molecular size during purification, one has no way of detecting proteolytic activity. Furthermore, since the principal proteolytic product is relatively stable, there is no continual breakdown to make one suspicious. Correll et al. (1968a) probably came the closest to purifying the large molecular weight material, and one can only speculate that the smaller values they obtained for the molecular weight of "monomer" and "tetramer" using equilibrium ultracentrifugation techniques (Correll et al., 1968b) reflect either continued proteolytic degradation or the presence of a major contaminant protein. In fairness, however, we must admit that our differences with the Smithsonian group regarding molecular weight for rye phytochrome remain to be resolved.

There are two general conclusions that one could draw from the work described above. The first, of fairly broad implication, is that normal spectral properties for phytochrome or normal enzymatic properties for some other protein do not necessarily indicate native state: the assay one uses may not detect chain cleavage distant from chromophore or active site. It is possible (though as yet unproved) that native phytochrome consists of a minimum of two different subunits, each close to 120,000 molecular weight, and each with two chromophores. Even if the chains were identical, it is clear that separation of the subunits and proteolytic cleavage near their centers is insufficient to alter their behavior in the assay (in this case spectral). Such cleavage could certainly generate enough different amino-terminal end groups to account for our first results with oat phytochrome. One now has four different pieces, each chromophoric, but sufficiently similar in size and charge that conventional techniques fail to separate them.

The second conclusion is that almost everything done to date in characterizing phytochrome, either as a protein or a pigment, must be repeated on the larger molecule. Thus amino acid composition (Mumford and Jenner, 1966; Correll et al., 1968b), studies bearing on a possible light-induced protein conformational change (Briggs et al., 1968; Hopkins and Butler, 1970; Pratt and Butler, 1970), intermediates in phototransformation (Linschitz et al., 1966; Linschitz and Kasche, 1967; Cross et al., 1968; Pratt and Butler, 1968; Briggs and Fork, 1969; Everett and Briggs, 1970), thermodynamics of phototransformation (Pratt and Butler, 1970), and dark reversion of P_{fr} to P_r (Mumford, 1966; Taylor, 1968; Anderson et al., 1969), all require careful reexamination. The composition and behavior of one or several large peptide fragments of a protein do not necessarily reflect the composition and behavior of the native molecule.

ACKNOWLEDGMENTS

The authors gratefully acknowledge the excellent technical assistance of Mr. Cecil Jackson-White. The work was carried out while two authors (Harbert Rice and Carl S. Pike) held NSF predoctoral fellowships, and one (Gary Gardner) held a USPHS predoctoral fellowship. It was supported by NSF grant GB 15572 and a grant from E. I. du Pont de Nemours & Co. to Winslow R. Briggs. We are grateful for this support.

References

Anderson, G. R., E. L. Jenner, and F. E. Mumford. 1969. Temperature and pH studies on phytochrome *in vitro*. *Biochemistry* **8**:1182–1187.
Borthwick, H. A., and S. B. Hendricks. 1965. *In* "Chemistry and Biochemistry of Plant Pigments" (T. W. Goodwin, ed.), pp. 405–436. Academic Press, New York.
Briggs, W. R., and D. C. Fork. 1969. Long-lived intermediates in phytochrome transformation I: *In vitro* studies. *Plant Physiol.* **44**:1081–1088.
Briggs, W. R., W. D. Zollinger, and B. B. Platz. 1968. Some properties of phytochrome isolated from dark-grown oat seedlings (*Avena sativa* L.). *Plant Physiol.* **43**:1239–1243.
Burgess, R. R. 1969. The subunit structure of RNA polymerase. Ph.D. Thesis, Harvard Univ., Cambridge, Massachusetts.
Butler, W. L., K. H. Norris, H. W. Siegelman, and S. B. Hendricks. 1959. Detection, assay, and preliminary purification of the pigment controlling photoresponsive development of plants. *Proc. Nat. Acad. Sci. U.S.* **45**:1703–1708.
Correll, D. L., and J. L. Edwards. 1970. Aggregation states of phytochrome from etiolated rye and oat seedlings. *Plant Physiol.* **45**:81–85.
Correll, D. L., J. L. Edwards, W. H. Klein, and W. Shropshire, Jr. 1968a. Phytochrome in etiolated annual rye. III. Isolation of photoreversible phytochrome. *Biochim. Biophys. Acta* **168**:36–45.
Correll, D. L., E. Steers, Jr., K. M. Towe, and W. Shropshire, Jr. 1968b. Phytochrome in etiolated annual rye. IV. Physical and chemical characterization of phytochrome. *Biochim. Biophys. Acta* **168**:46–57.
Cross, D. R., H. Linschitz, V. Kasche, and J. Tenenbaum. 1968. Low-temperature studies on phytochrome: light and dark reactions in the red to far-red transformation and new intermediate forms of phytochrome. *Proc. Nat. Acad. Sci. U.S.* **61**:1095–1101.
Everett, M. S., and W. R. Briggs. 1970. Some spectral properties of pea phytochrome *in vivo* and *in vitro*. *Plant Physiol.* **45**:679–683.
Fahrney, D. E., and A. M. Gold. 1963. Sulfonyl fluorides as inhibitors of esterases. I. Rates of reaction with acetylcholinesterase, α-chymotrypsin, and trypsin. *J. Amer. Chem. Soc.* **85**:997–1000.
Gardner, G., C. S. Pike, H. V. Rice, and W. R. Briggs. 1971. "Disaggregation" of phytochrome *in vitro*—a consequence of proteolysis. *Plant Physiol.* **48**: 686–693.
Hillman, W. S. 1967. The physiology of phytochrome. *Annu. Rev. Plant Physiol.* **18**: 301–324.
Hopkins, D. W., and W. L. Butler. 1970. Immunochemical and spectroscopic evidence for protein conformational changes in phytochrome transformations. *Plant Physiol.* **45**:567–570.
Linschitz, H., and V. Kasche. 1967. Kinetics of phytochrome conversion: multiple pathways in the Pr to Pfr pathway, as studied by double flash technique. *Proc. Nat. Acad. Sci. U.S.* **58**:1059–1064.
Linschitz, H., V. Kasche, W. L. Butler, and H. W. Siegelman. 1966. The kinetics of phytochrome conversion. *J. Biol. Chem.* **241**:3395–3403.
Mumford, F. E. 1966. Studies on the phytochrome dark reaction *in vitro*. *Biochemistry* **5**:522–524.
Mumford, F. E., and E. L. Jenner. 1966. Purification and characterization of phytochrome from oat seedlings. *Biochemistry* **5**:3657–3662.
Pike, C. S., and W. R. Briggs. 1972a. The dark reactions of rye phytochrome *in vivo* and *in vitro*. *Plant Physiol.* In press.

Pike, C. S., and W. R. Briggs. 1972b. Partial purification and characterization of a phytochrome-degrading neutral protease from etiolated oat shoots. *Plant Physiol.* In press.

Pratt, L. H., and W. L. Butler. 1968. Stabilization of phytochrome intermediates by low temperature. *Photochem. Photobiol.* **8**:477–485.

Pratt, L. H., and W. L. Butler. 1970. The temperature dependence of phytochrome transformations. *Photochem. Photobiol.* **11**:361–369.

Pringle, J. R. 1970a. The molecular weight of the *undegraded* chain of yeast hexokinase. *Biochem. Biophys. Res. Commun.* **39**:46–52.

Pringle, J. R. 1970b. Studies of malate dehydrogenase and of proteases. Ph.D. Thesis, Harvard Univ., Cambridge, Massachusetts.

Reynolds, J. A., and C. Tanford. 1970. The gross conformation of protein-sodium dodecyl sulfate complexes. *J. Biol. Chem.* **245**:5161–5165.

Sanger, F. 1945. The free amino groups of insulin. *Biochem. J.* **39**:507–515.

Siegelman, H. W., and E. M. Firer. 1964. Purification of phytochrome from oat seedlings. *Biochemistry* **3**:418–423.

Smith, H. 1970. Phytochrome and photomorphogenesis in plants. *Nature (London)* **227**:665–668.

Taylor, A. O. 1968. *In vitro* phytochrome dark reversion process. *Plant Physiol.* **43**:767–774.

Wang, K., and I. S. Y. Wang. 1967. Chromatographic identification of dinitrophenyl amino acids on polyester film supported polyamide layers. *J. Chromatogr.* **27**:318–320.

Weber, K., and M. Osborn. 1969. The reliability of molecular weight determination by dodecyl sulfate-polyacrylamide gel electrophoresis. *J. Biol. Chem.* **244**:4406–4412.

Whitaker, J. R., and J. Perez-Villaseñor. 1968. Chemical modification of papain. I. Reaction with the chloromethyl ketones of phenylalanine and lysine and with phenylmethylsulfonyl fluoride. *Arch. Biochem. Biophys.* **124**:70–78.

A STUDY OF THE MECHANISM OF PHYTOCHROME ACTION

ARTHUR W. GALSTON and RUTH L. SATTER

Department of Biology, Yale University, New Haven, Connecticut

Introduction	51
Experimental	53
The Cellular Basis for Leaflet Closure	53
Control of Closure by Rhythms and Phytochrome	53
Aerobicity and Temperature Dependence	55
Potassium Flux as a Basis for Leaflet Movement	59
Organic Acid Promotion of Leaflet Closure	67
Discussion	73
References	77

Introduction

The impressive recent advances in our understanding of the nature of the phytochrome molecule have been well reviewed in the previous paper (Briggs *et al.*, 1972). Much less is known with certainty concerning the mechanism through which the phytochrome molecule acts in controlling plant development. This ignorance results partly from the difficulty involved in collating and sorting out the significance of data from the many different kinds of physiological systems involved in phytochrome experiments. Partly, also, our lack of understanding results from the inherent difficulty of specifying the function of any substance which occurs in the plant in such minute quantities.

There are at present two main theories of phytochrome action. One, proposed and developed by Mohr (1969), considers that phytochrome exerts control on developmental processes through direct repression and derepression of genes. He bases this theory on the demonstrated appearance of new enzymatic activities in tissue whose phytochrome has been converted from the P_r to the P_{fr} form and the inhibition of these effects by substances like puromycin and actinomycin D, which prevent, respectively, translation and transcription (Lange and Mohr, 1965). From this he considers it reasonable to conclude that *de novo* protein synthesis leading to new enzyme activity is the major process ultimately controlled by phytochrome. A contrary view has been advanced by the Beltsville group (Fondéville *et al.*, 1966) on the basis of the demonstrated phytochrome control of much more rapid phenomena, such as the turgor changes leading to nyctinastic (sleep) movements of the sensitive plant, *Mimosa pudica*. The rapidity of such effects, together with the insensitivity of analogous nyctinastic movements to agents like actinomycin D (Jaffe and Galston, 1967), makes a gene derepression mechanism difficult to accept, and favors a theory that would place phytochrome in membranes, where it could control various processes by regulating the rate of passage of specific materials into and out of cellular compartments. This theory has received support from the experiments of Haupt (Haupt *et al.*, 1969) whose use of polarized red and far-red light in the control of chloroplast rotation in the alga *Mougeotia* leads him to believe that phytochrome is located in an oriented layer (presumably the plasma membrane) at the periphery of the cell. Direct microspectrophotometric evidence (Galston, 1968) also indicates the possible occurrence of phytochrome in the nuclear envelope of etiolated oat coleoptile and pea epicotyl cells. Finally, the recent demonstration that phytochrome controls virtually instantaneous changes in the ability of tissue fragments to adhere to glass (Tanada, 1967), and the suggestion that these changes result from changes in bioelectric potential (Jaffe, 1968) mediated by acetylcholine-like substances (Jaffe, 1970) has shifted attention to membrane permeability as the process most probably controlled by phytochrome conversion.

We propose to describe in some detail our experiments on the phytochrome-controlled leaflet closure in the leguminous plant, *Albizzia julibrissin*, a nonthigmonastic relative of *Mimosa pudica*. Our choice of this system was based on several considerations, including the rapidity of the response, the convenience of its measurement, and the susceptibility of the system to biochemical and biophysical analyses (Hillman and Koukkari, 1967; Jaffe and Galston, 1967; Satter *et al.*, 1970a, b). It is our hope that a satisfactory explanation of the role of phytochrome in this process will provide some basis for an understanding of its action in such diverse processes as seed germination (Borthwick et al., 1952), flower initiation (Parker et al., 1950), enzyme induction (Attridge and Smith, 1967), chloroplast movements (Haupt, 1965) and the generation of bioelectric potentials (Jaffe, 1968).

Experimental

THE CELLULAR BASIS FOR LEAFLET CLOSURE

Albizzia, like *Mimosa*, has doubly compound leaves. Each pinna is attached to the midrib (rachis) by a pulvinus. In turn each pinna is subdivided into pinnules (leaflets), oppositely arranged and attached to the rachilla by a pulvinule (Fig. 1). In the light the leaflets are usually wide open, the angle between them being approximately 180°. Shortly after transfer to darkness the leaflets close by an upward folding together, and in the process move forward toward the rachilla. These changes in orientation are due to competitive alterations of the turgor pressure of the motor cells on the ventral and dorsal sides of the pulvinule. When the ventral cells are highly turgid and the dorsal cells less than optimally turgid, the leaflets are open; when, on the other hand, the dorsal cells are maximally turgid and the ventral cells are flaccid, the leaflets close (Fig. 2, p. 54). The motor cells are anatomically unusual in that they have many vacuoles, some containing presumptive tannin bodies. They also have unusually thick cell walls, numerous spherosomes and oriented fibrillar cytoplasm (Fig. 3, p. 56).

CONTROL OF CLOSURE BY RHYTHMS AND PHYTOCHROME

Leaflet closure after transfer to darkness is measurable within 5 to 10 minutes, and generally goes to completion within 30 to 90 minutes, de-

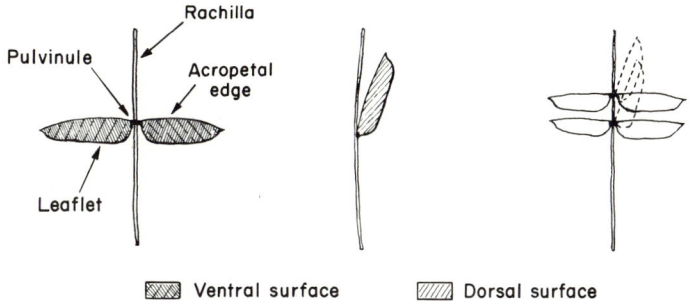

FIG. 1. Open (left) and closed (middle) pinnule pairs of *Albizzia julibrissin*. When leaflets close, their tips move toward each other and toward the rachilla (right). From Satter et al. (1970a), by courtesy of the *American Journal of Botany*.

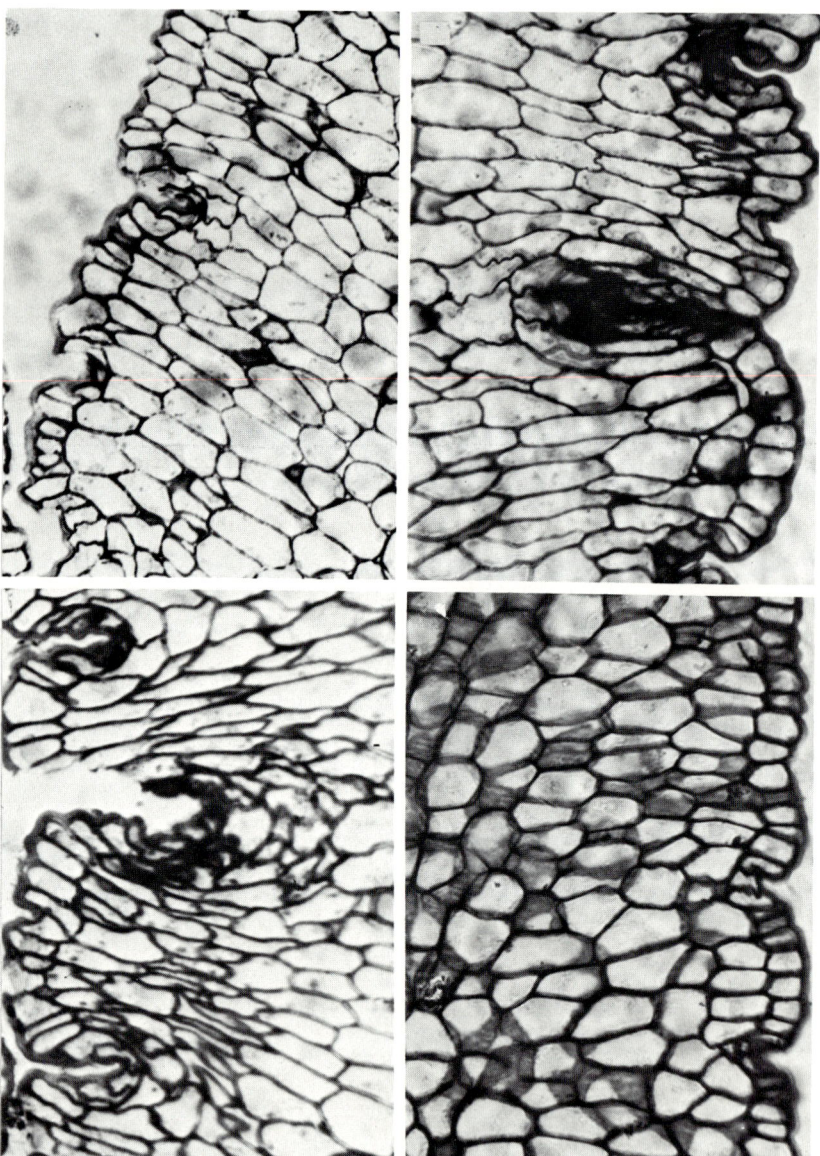

Fig. 2. *Albizzia julibrissin* motor cells. Ventral cells of open pulvinules (upper left) and closed pulvinules (lower left); dorsal cells of open pulvinules (upper right) and closed pulvinules (lower right). × 400. The motor cells are located in the 2 or 3 layers immediately below the epidermis. From Satter *et al.* (1970a), by courtesy of the *American Journal of Botany*.

pending on the leaf age and the plant. The various leaves of *Albizzia* behave somewhat differently, and each is controlled by an endogenous rhythm causing periodic fluctuations in the leaflet angle (Fig. 4, p. 57). Under some conditions these rhythms are so powerful that leaflets will close even during full illumination if the light period is extended for a sufficiently long time; similarly, leaflets will open even in darkness if the dark period is extended sufficiently.

The partial control of leaflet movement by phytochrome can be demonstrated at appropriate parts of the diurnal cycle. We have found that median pinnules of the fourth through the sixth leaves of 4-month-old plants respond well to phytochrome control if they are darkened after 1–4 hours of light. If the last light before darkness is red, the leaflets close promptly, but if it is far-red, their closure is slower and incomplete. Red and far-red irradiations are mutually antagonistic in this response, and even after several cycles of red, far-red treatment the leaflets respond only to the last irradiation, indicating phytochrome control. The photoreceptor is located in the pulvinule (Koukkari and Hillman, 1968), which is also the locus of turgor changes causing movement.

Physiological experiments may be conveniently performed by excising pinnule pairs attached to a small bit of rachilla and floating such explants in experimental solutions in petri dishes. These explants show reproducible movements for several cycles of treatment over a period of several hours after excision; we usually use them once, to study the rate of closure over a 30-120-minute period after excision.

In such excised systems the state of the phytochrome molecule controls the rate of closure throughout the closure period. Thus, a slowly closing pinnule pair exposed to far-red light prior to darkness begins to close rapidly soon after the administration of red light; conversely, a rapidly closing pinnule treated with red light prior to darkness may stop closing if given far-red at any time prior to the completion of closure. The lag period for such changes is approximately 10 minutes (Fig. 5, pp. 58, 59). *This means that the state of phytochrome controls the rate of the reaction throughout the period of response*, and that phytochrome does not act simply as a trigger.

Aerobicity and Temperature Dependence

Leaflet closure is obligately aerobic, and introduction of a nitrogen atmosphere at any time during the process diminishes the rate of closure sharply (Fig. 6, p. 60). In the same way, reintroduction of oxygen into a slowly closing anaerobic system results in a more rapid closure. This obligate aerobicity is ostensibly linked to the necessity for ATP synthesis,

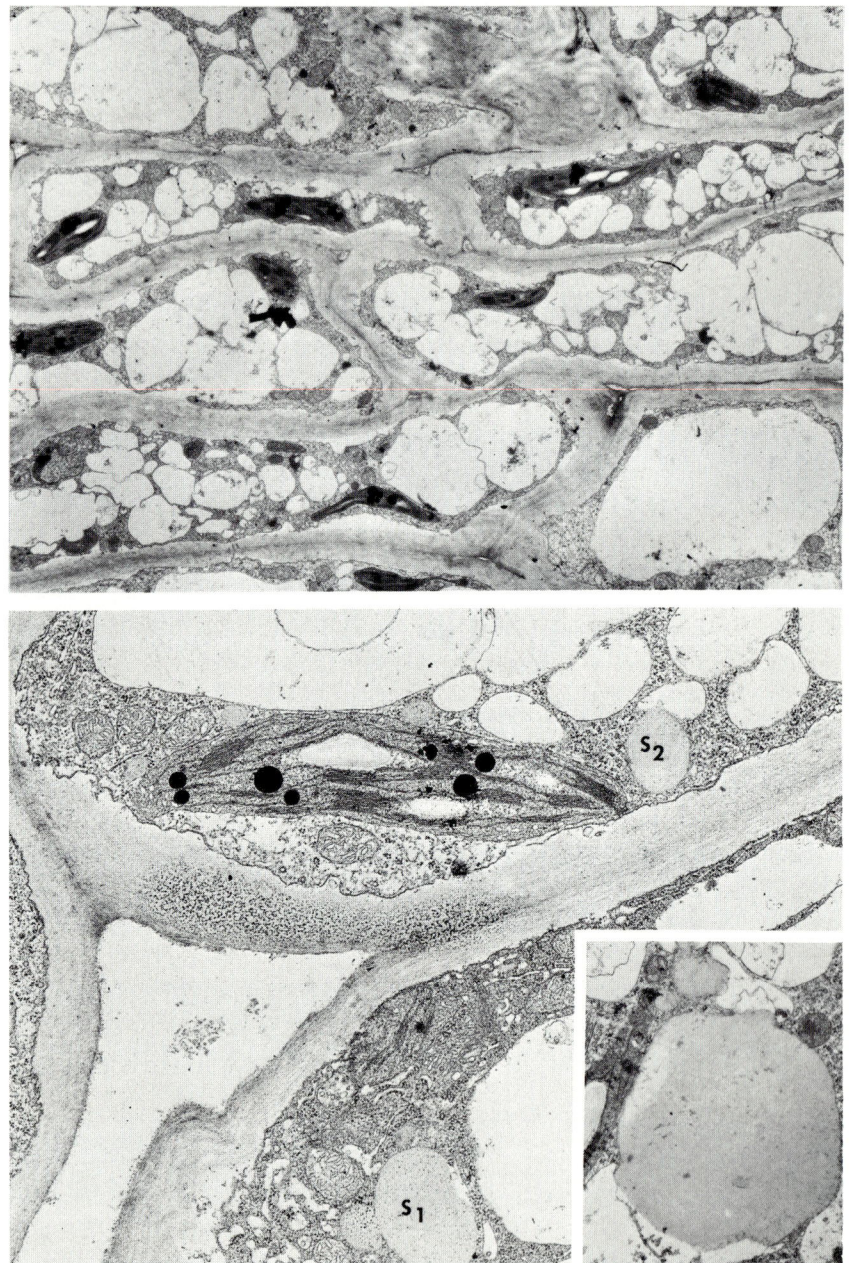

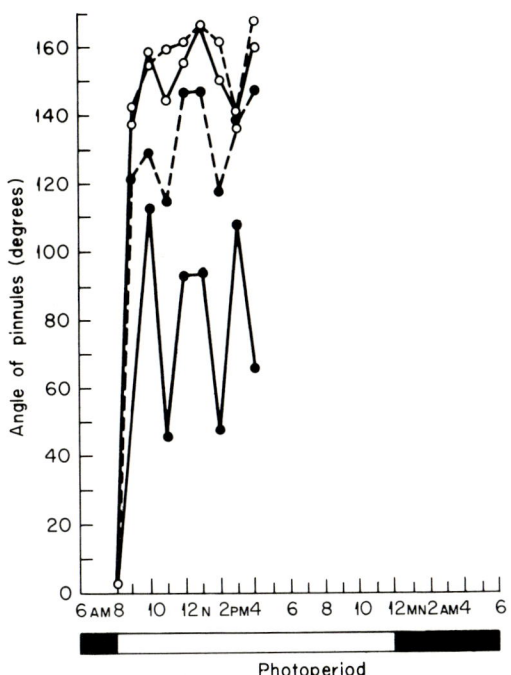

Fig. 4. Periodic fluctuations in the angle formed by leaflets on different leaves after opening at the start of the light period. The first leaf is the youngest one. ●———●, 1st leaf; ●- - -●, 2nd leaf; ○———○, 3rd leaf; ○- - -○, 4th leaf. From Jaffe and Galston (1967), by permission of Springer-Verlag, Berlin and New York.

since such agents as sodium azide and 2,4-dinitrophenol also inhibit leaflet closure in air (Fig. 7, pp. 62, 63). Control by phytochrome is shown best at median temperatures, such as 24°C, even though the absolute rate of closure is greater at lower temperatures, such as 18°C (Fig. 8, pp. 64, 65). The differences between red and far-red pretreated pinnule pairs are usually not great at 18°C and are nonexistent at 8°C. At 29°C, on the other hand, red before darkness is required for leaflet closure; leaflets open rather than

---

Fig. 3. Electron micrographs of the motor cells. Upper: A longitudinal section through the subepidermal tissue reveals the polyvacuolate condition of the motor cells. × 3600. Lower: Numerous spherosomes (S) are found within the cytoplasm. Some appear to enlarge by coalescence (S_1) and others are frequently found in close association with vacuoles (S_2). × 11,500. The inset shows a spherosome that has enlarged considerably and appears to be giving rise to a vacuole. From Satter et al. (1970a), by courtesy of the *American Journal of Botany*.

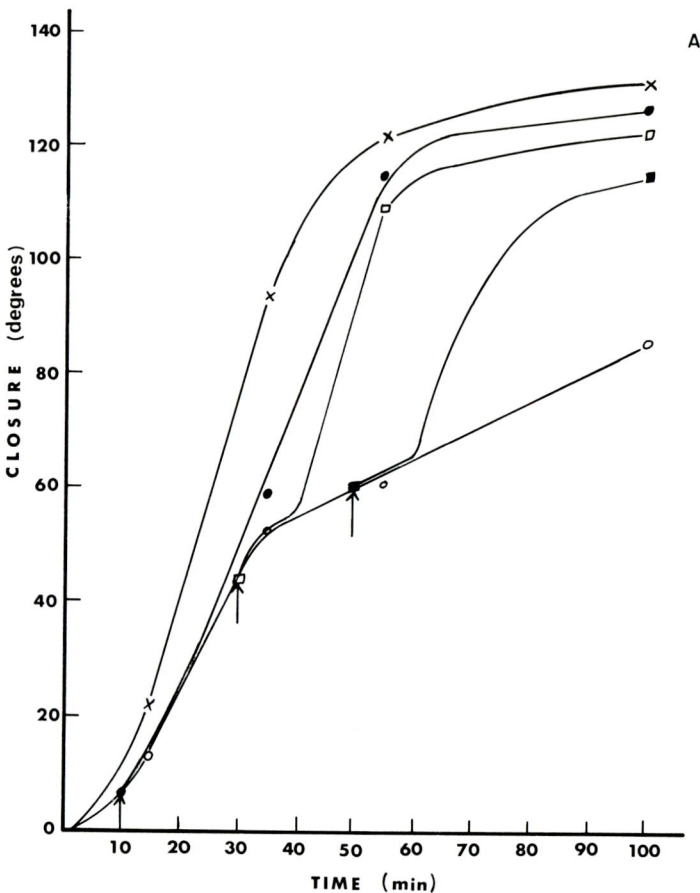

FIG. 5. (A) Red promotion following far-red inhibition of leaflet closure. Pinnule pairs were irradiated with far-red light and then darkened. The dark period was interrupted by red (arrows) after 0 minutes (×———×), 10 minutes (●———●), 30 minutes (□———□) or 50 minutes (■———■) or not at all (control, ○———○). (B) Far-red reversal of the leaflet closure potentiated by red light. Pinnule pairs were irradiated with red light and then darkened. The dark period was interrupted by far-red (arrows) after 0 minutes (○———○), 10 minutes (●———●), 20 minutes (□———□), or 40 minutes (■———■), or not at all (control, ×———×). Lag data for both (A) and (B) were obtained in a separate experiment and are incorporated here. From Satter et al. (1970b), by courtesy of the *American Journal of Botany*.

close if pretreated with far-red at this temperature. It appears that leaflet closure is controlled by at least two systems and that the phytochrome-controlled system predominates when the temperature is 24°C or higher. For these reasons, we make our observations at temperatures near 24°C.

Mechanism of Phytochrome Action

POTASSIUM FLUX AS A BASIS FOR LEAFLET MOVEMENT

Several lines of evidence indicated that flux of ions, especially of potassium, across membranes of the motor cells must be responsible for the movement of water which ultimately controls turgor pressure, and thus the closure of leaflets. In the first place, when excised pinnae are placed

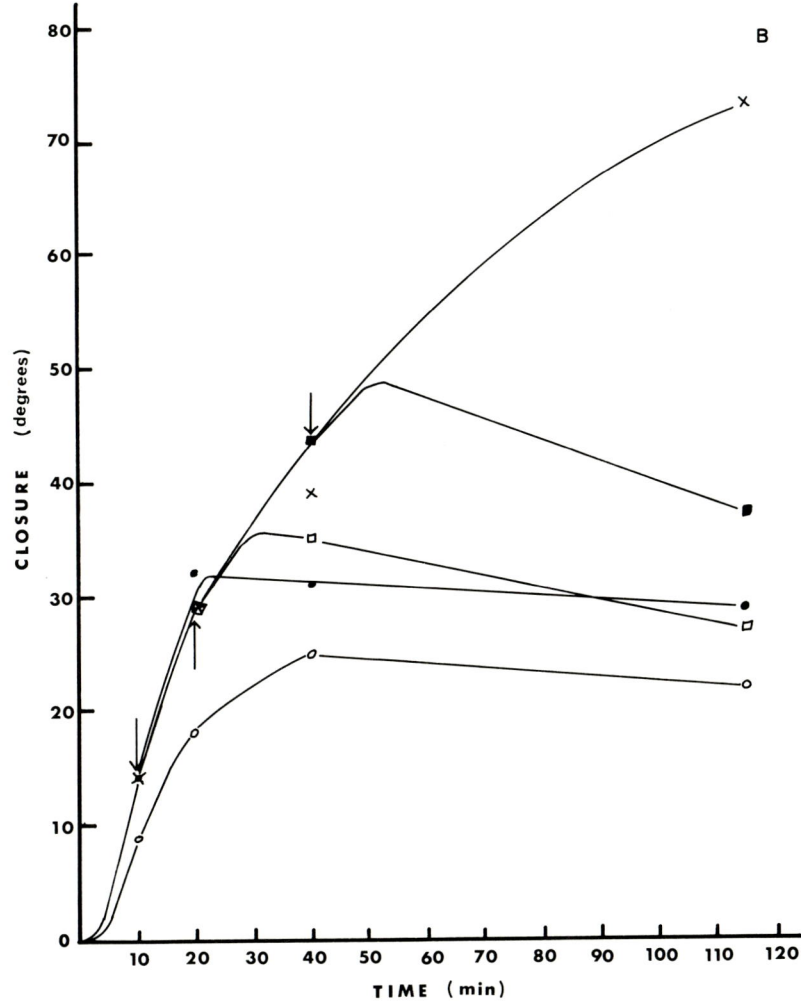

FIG. 5B. See caption under Fig. 5A.

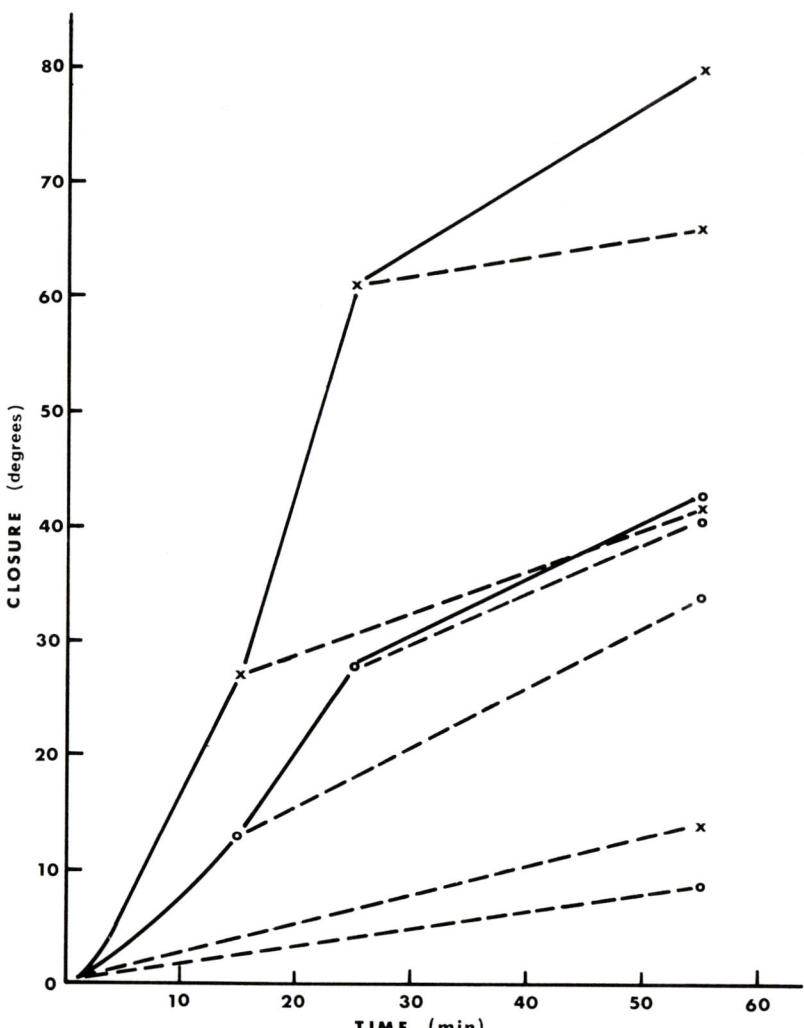

FIG. 6. Dark closure of leaflets in a nitrogen atmosphere. Open pinnule pairs were floated on water in Lucite chambers, irradiated with red or far-red, and then darkened. Nitrogen was flushed through the chambers (1) just before phytochrome irradiation, (2) from minute 10 to minute 20 of the dark period, or (3) from minute 20 to minute 30 of the dark period. The chambers were then sealed. Control leaflets were in petri dishes in air. – – –, N_2; ——, air; ×, red, dark; ○, far-red, dark. From Satter et al. (1970b), by courtesy of the American Journal of Botany.

with their bases in small volumes of water, conductometrically measurable materials are excreted in greater quantities during closure than when the pinnules remain open (Jaffe and Galston, 1967). Second, elemental analyses by flame photometer indicate an extraordinarily high potassium content in the pulvinule, almost 0.5 M, as compared with 0.24 M in the rachilla and 0.16 M in the lamina. We know of no other solute cation or molecule which approaches this concentration. Third, closure is repressed if the external bathing solution contains 0.2 to 1.0 N salts of monovalent cations, especially of potassium (Fig. 9, p. 67). Divalent cations seem to be completely ineffective. The inorganic anion accompanying potassium seems relatively unimportant, although KI and KNO_3 give appreciably greater inhibition than either KBr or KCl.

Verification of the hypothesis that the movement of potassium salts into and out of key ventral and dorsal pulvinule cells might be responsible for the leaflet movements is obviously dependent on an analysis of potassium within the cells themselves. Excision of sufficient quantities of ventral and dorsal motor cells for potassium analyses seemed impractical. We therefore set about to develop techniques to measure the changes *in situ*, finally achieving success by the use of an electron microprobe. Tissue was prepared for analysis by the following technique: Pulvinules were quickly frozen in modified Tissue-Tek embedding medium so as to kill the cells and immobilize all materials. Thin slices, approximately 24 μ thick, were then cut in a cryostat microtome, and the sections were lyophilized carefully in such a way as to preserve the morphology of the cells (Satter *et al.*, 1970b). Sections so prepared, in which no potassium movement had ostensibly occurred during or after freezing, were subjected to analysis in an Acton electron microprobe. In this instrument, a high energy beam of electrons impinges upon the tissue at the focal plane. Elements in the path of the beam are excited to emit X-rays whose frequency is characteristic of the excited element. An X-ray monochromator then selects the radiations characteristic of each element, such as potassium, and leads them to an exit slit, at which point they are converted to scintillations which are counted by the instrument. By this means it is possible to measure four elements simultaneously, for example, sodium, potassium, calcium, and phosphorus. The instrument also contains a light microscope that enables the operator to select appropriate cells for measurement (Anderson, 1967).

If one wishes to determine the changes in potassium content per pulvinule cell during closure, one is faced by the problem that more cells are in the path of the electron beam when cells are compressed than when they are expanded. This is particularly pronounced for the ventral motor cells,

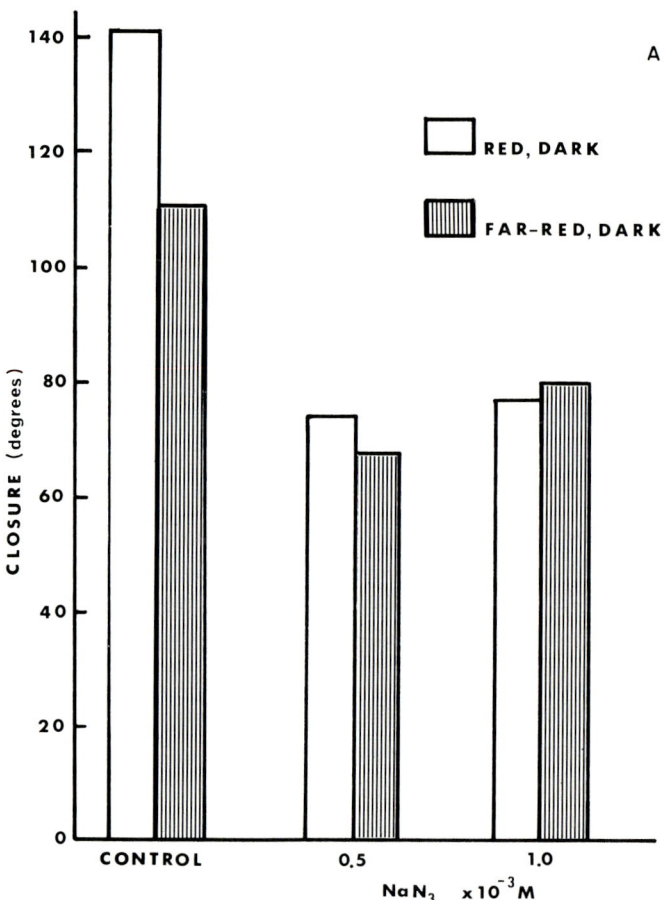

Fig. 7. (A) Dark closure (during 40 minutes) of leaflets floating on sodium azide. (B) Dark closure (during 35 minutes) of leaflets floating on 2,4-dinitrophenol. From Satter et al. (1970b), by courtesy of the *American Journal of Botany*.

whose areas, in dorsoventral longitudinal sections, shrink approximately 50 percent during leaflet closure. The problem is less severe for dorsal cells, whose areas change relatively little during closure (Fig. 2). Partial compensation is provided by the instrument itself, since the depth to which the beam penetrates is inversely proportional to tissue density. One can further correct for alterations in cell volume during leaflet movement by calculating the ratio of potassium to an element which is primarily immobile and can serve as an internal standard. Calcium was the most reliable internal standard; phosphorus was also used with some success. Further considerations of internal corrections will be published elsewhere.

Once the technique was satisfactorily developed (Fig. 10, p. 68), it became very clear that the potassium content of ventral motor cells diminished, and that of dorsal motor cells increased, during closure (Table 1, p. 66). The reverse changes occurred during opening. Similar changes in

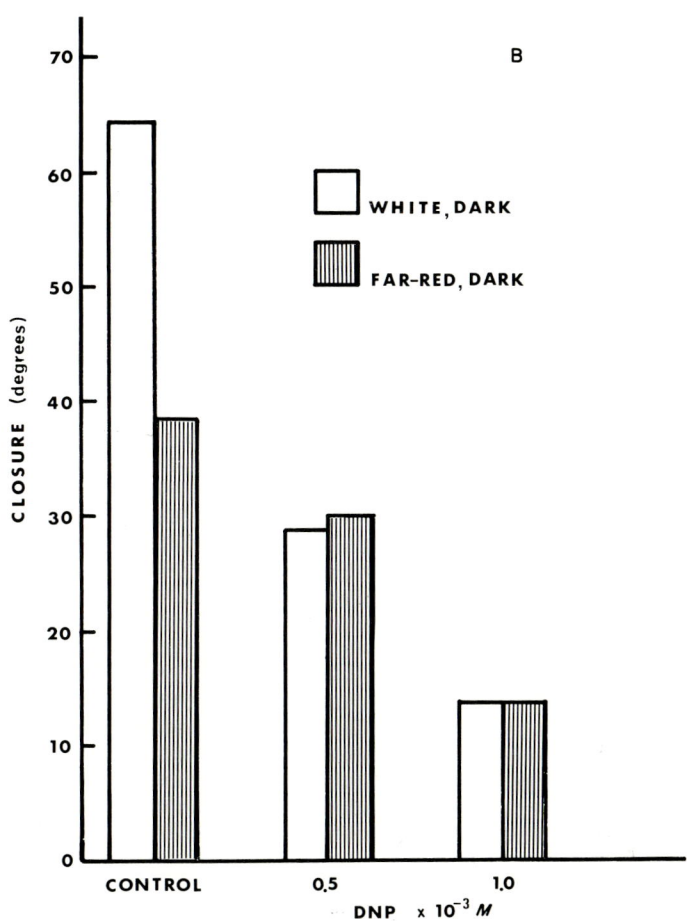

FIG. 7B. See caption under Fig. 7A.

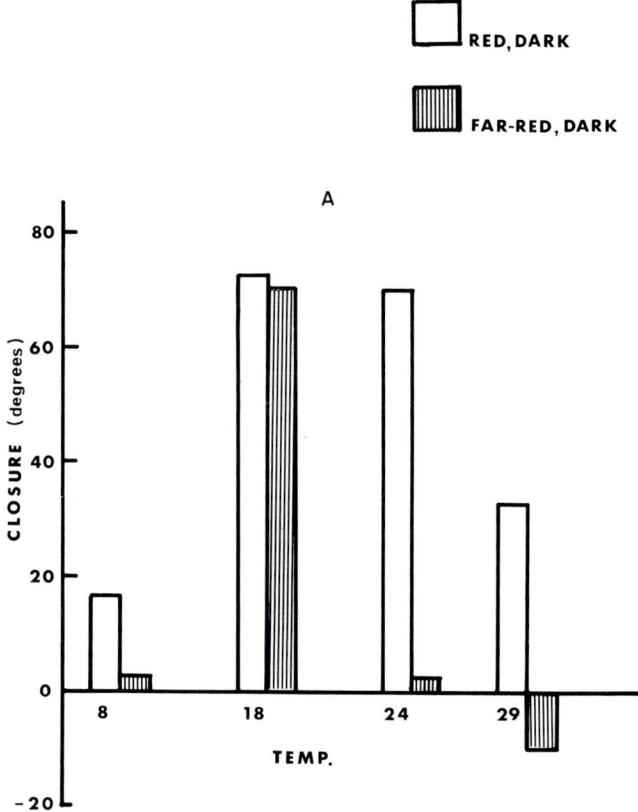

FIG. 8. Effect of temperature on dark closure. Leaflet angles were measured 30 minutes (A) and 60 minutes (B) after red- or far-red preirradiation. From Satter et al. (1970b), by courtesy of the *American Journal of Botany*.

the potassium content of motor cells occurred when closure was controlled by phytochrome (Table 2, p. 71), by endogenous rhythm (Fig. 11, p. 69) or by chemicals, such as sodium acetate (Fig. 12, p. 71), whose action we shall discuss below. Furthermore, such changes were well correlated in time with the leaf movements. From these data it appears that potassium flux in pulvinule motor cells is the crucial event controlling leaflet orientation.

Phytochrome exerts considerable control over potassium movement at certain times in the diurnal cycle. If leaflets are darkened from 1 to 4 hours after the beginning of the photoperiod, a high P_{fr} level is required for potassium efflux from ventral cells. On the other hand, potassium moves into the dorsal cells whether leaflets were preirradiated with red or far-red (Table 2). Since leaflet angle is determined by the relative turgor of cells on the two sides of the pulvinule, darkened leaflets previously exposed to far-red light close partially, while those previously exposed to red light close completely. Here too, one notes the excellent correlation between leaflet angle and potassium flux data.

A kinetic study of potassium fluxes during the dark period revealed

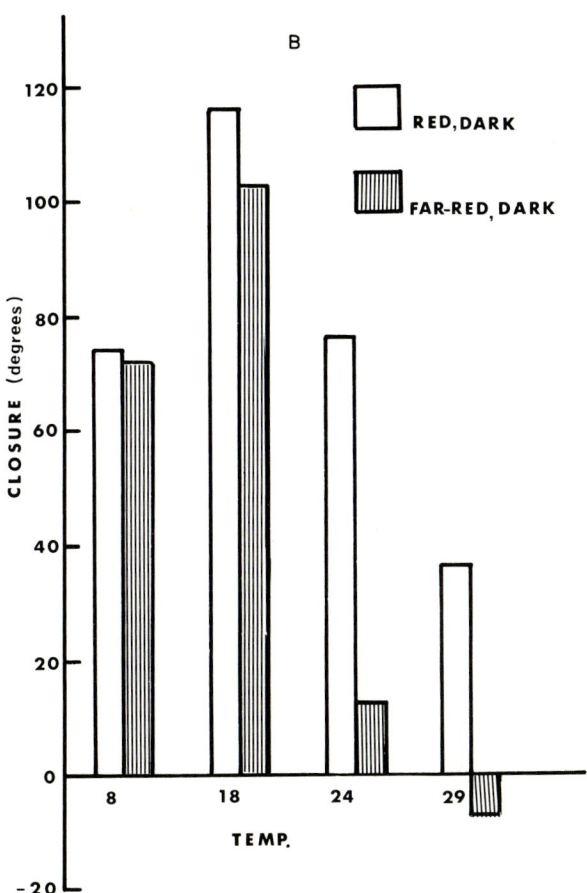

FIG. 8B. See caption under FIG. 8A.

TABLE 1

Microprobe Analysis of K, K/Ca, and K/P in Motor Cells of Open and Closed *Albizzia* Pulvinules[a]

Pulvinule	Ventral			Dorsal		
	K	K/Ca	K/P	K	K/Ca	K/P
Open (light)	293 ± 68[b]	1.82 ± 0.40	0.73 ± 0.20	155 ± 35	0.84 ± 0.21	0.33 ± 0.08
Closed (dark)	121 ± 24	0.65 ± 0.11	0.28 ± 0.06	400 ± 63	2.34 ± 0.31	0.98 ± 0.19

[a] The electron beam intercepted circular areas (20 μ diameter) in the motor cell regions of longitudinal dorsoventral sections. Data show K scintillation and the ratio of K/Ca and K/P scintillations during 25 seconds.
[b] Standard deviation.

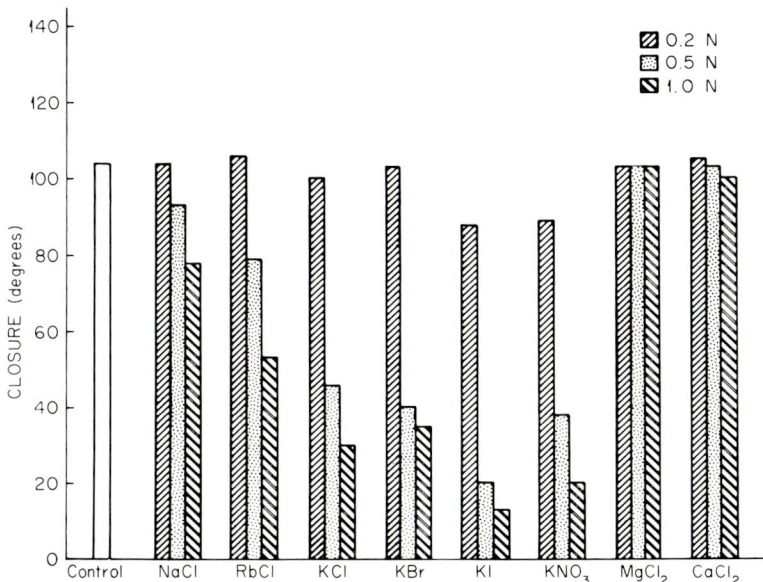

Fig. 9. Dark closure (during 50 minutes) of leaflets floating on inorganic salt solutions. From Satter et al. (1970b), by courtesy of the *American Journal of Botany*.

that far-red preirradiation not only prevents potassium efflux but actually permits potassium to enter the ventral cells (Fig. 13, p. 72). This knowledge clarified some data that had previously been puzzling; i.e., the partial reopening late in the dark period of leaflets pretreated with far-red (Fig. 5B). Leaflet opening in the dark is promoted by high temperature (Fig. 8); our current experiments suggest that this is due to increased potassium movement into the ventral cells.

Thus the state of the phytochrome molecule determines the direction of net potassium movement in the ventral motor cells. We favor the view that potassium movement *into* the ventral cells is largely independent of phytochrome, but that such movement can be masked or overcome by a phytochrome-controlled active secretion of potassium.

ORGANIC ACID PROMOTION OF LEAFLET CLOSURE

Since certain organic acids are known to control the osmotic swelling and shrinkage of some cell organelles (Packer et al., 1967), we investigated

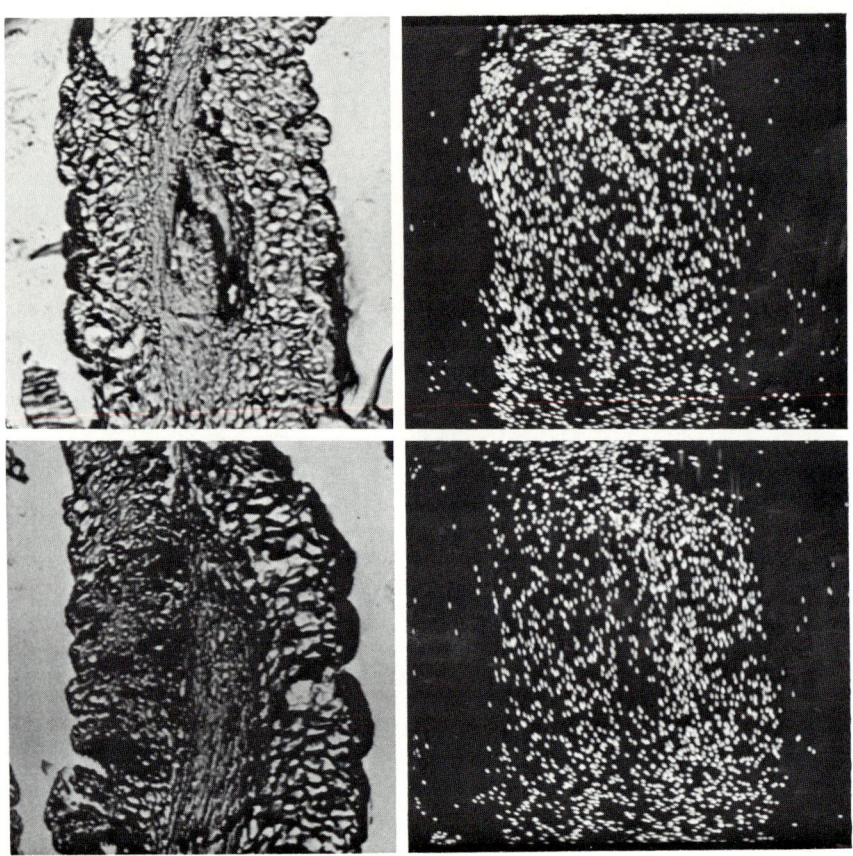

Fig. 10. Specimen morphology (left) and potassium scintillations (right) in longitudinal dorsoventral sections of *Albizzia* pulvinules. Specimen current photos, on the left, show surface morphology; scintillation photos, on the right, show the potassium distribution. Open pulvinules are shown above, and closed pulvinules below. Orientation of each specimen: left to right = ventral to dorsal.

their possible control of turgor changes in *Albizzia* pulvinule cells. Buffered sodium and potassium salts (0.05 to 0.10 N) of a variety of organic acids introduced into the bathing solution of explants were very effective in facilitating leaflet closure. This occurred both in the light (Fig. 14, p. 73) and in darkness preceded by either red or far-red (Fig. 15, p. 74). Acetate was the most effective anion tested, and its effectiveness was independent of pH between 4.5 and 8.0. An aerobic atmosphere was required for its promotive effect. Promotion by red light and sodium acetate are additive, even in the optimal dose range for each agent alone, indicating that these two agents control different biochemical processes.

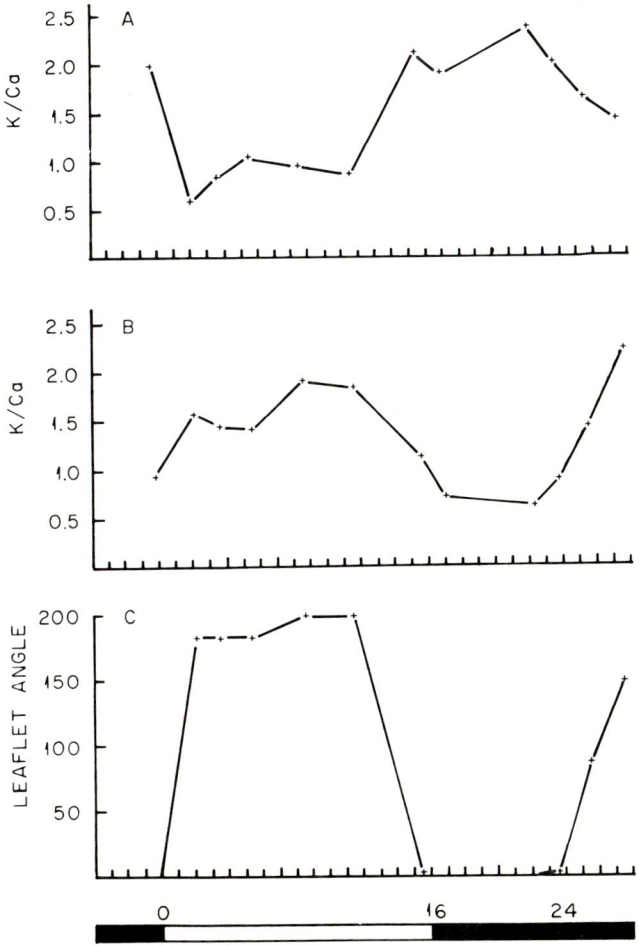

FIG. 11. Comparative kinetics of potassium flux and leaflet movement during 16 hours of white light followed by 11 hours of darkness. Pulvinules were excised from an intact plant every 1 to 4 hours and were frozen immediately in preparation for microprobe analysis. The plant was maintained on its usual photoperiod except that the final dark period was extended to permit leaflet opening. (A) Dorsal; (B) ventral; (C) leaflet movement.

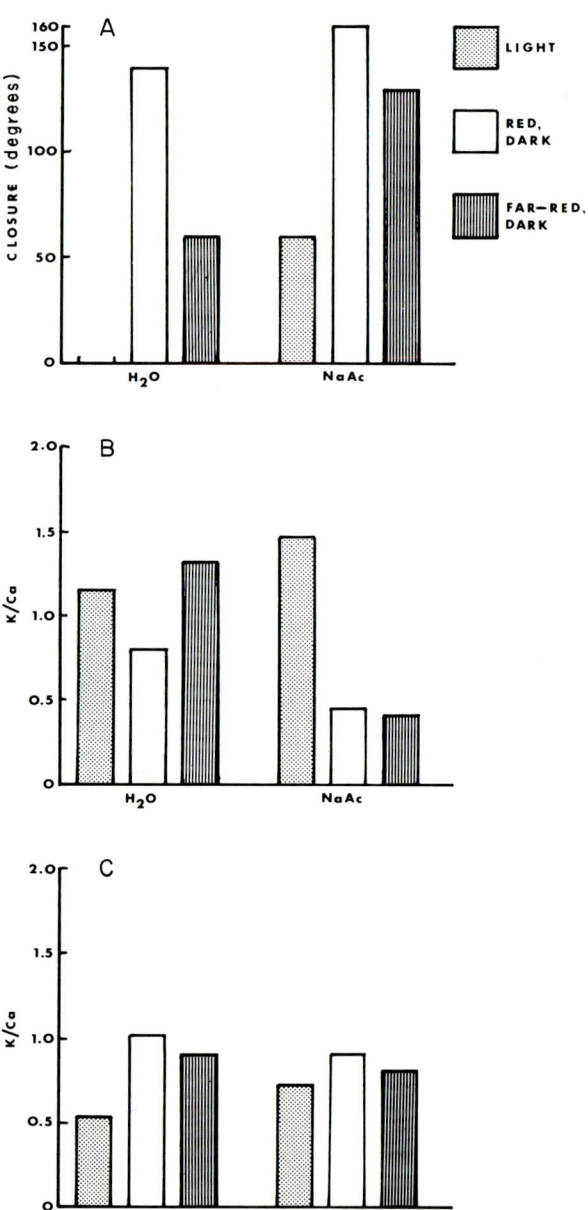

TABLE 2

Effect of Red and Far-red Preirradiation on K Flux in *Albizzia* Motor Cells during a Subsequent Dark Period[a,b]

Treatment	Angle between pulvinules		Ventral			Dorsal		
	Final	Closure	K	K/Ca	K/P	K	K/Ca	K/P
Light (zero time)	180°	—	156	1.14	0.57	135	0.81	0.40
Light (100 min)	180°	0°	140	0.91	0.39	134	0.72	0.29
R, dark (80 min)	20°	160°	104	0.62	0.35	218	1.55	0.72
FR, dark (90 min)	80°	100°	135	0.90	0.46	195	1.40	0.63

[a] From Satter et al. (1970b).
[b] Data, obtained with an electron microprobe, show K scintillations and the ratio of K/Ca and K/P scintillations during 15 seconds.

A comparison of the activity of acetate with that of the homologous alkylcarboxylic acids from formate to caproate is shown in Fig. 16 (p. 75). The series acetate-glycolate-glyoxylate showed a decreasing effect on closure with increasing oxidation state of the α-carbon atom. Our survey of the relation between structure and activity of other organic acids is still too fragmentary to yield generalizations concerning the role of the anions.

Potassium analyses of pulvinules floated on NaAc revealed that acetate facilitated the egress of potassium from ventral pulvinule cells, but did not affect the dorsal cells. In the light, however, acetate promoted closure by moving potassium into the dorsal cells. Thus, while the net result of both effects was pulvinule closure, this closure occurred for fundamentally different reasons in the two situations.

Fig. 12. Effect of 0.05 M sodium acetate on leaflet movement and potassium flux in the light and in darkness preceded by red or far-red light. Pulvinules were taken from pinnae whose cut ends were in NaAc or H_2O. The length of the experimental period varied with the time required for excision and embedding. Red, dark: NaAc, 70 minutes; H_2O, 80 minutes. Far-red, dark: NaAc, 90 minutes; H_2O, 100 minutes. Light, NaAc, 110 minutes; H_2O, 120 minutes. Water controls contained no measurable Na. The ratios of Na/Ca scintillations for NaAc-treated pulvinules were: ventral—red, dark, 0.14; far-red, dark, 0.22; light, 0.30; and dorsal-red, dark 0.13; far-red, dark, 0.19; light, 0.32. (A) Leaflet movement; (B) ventral pulvinules; (C) dorsal pulvinules.

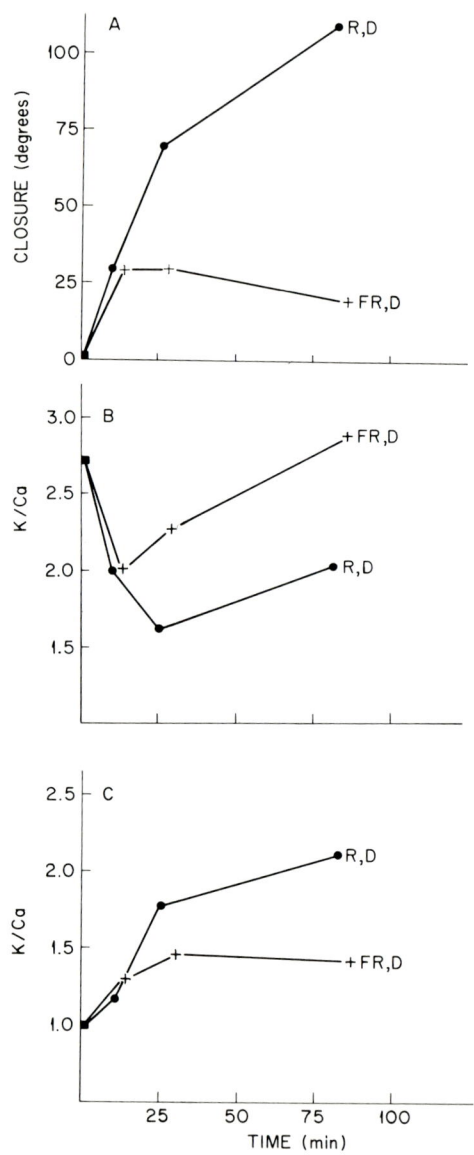

Fig. 13. A kinetic study of potassium flux and leaflet movement during the dark period following red and far-red irradiation. Excised open pinnae with their cut ends in water were irradiated with red or far-red light and then darkened. During the closure period, pulvinules were removed and frozen in preparation for potassium analysis. +, far-red, dark; ●, red, dark; ■, light. (A) Leaflet movement; (B) ventral pulvinules; (C) dorsal pulvinules.

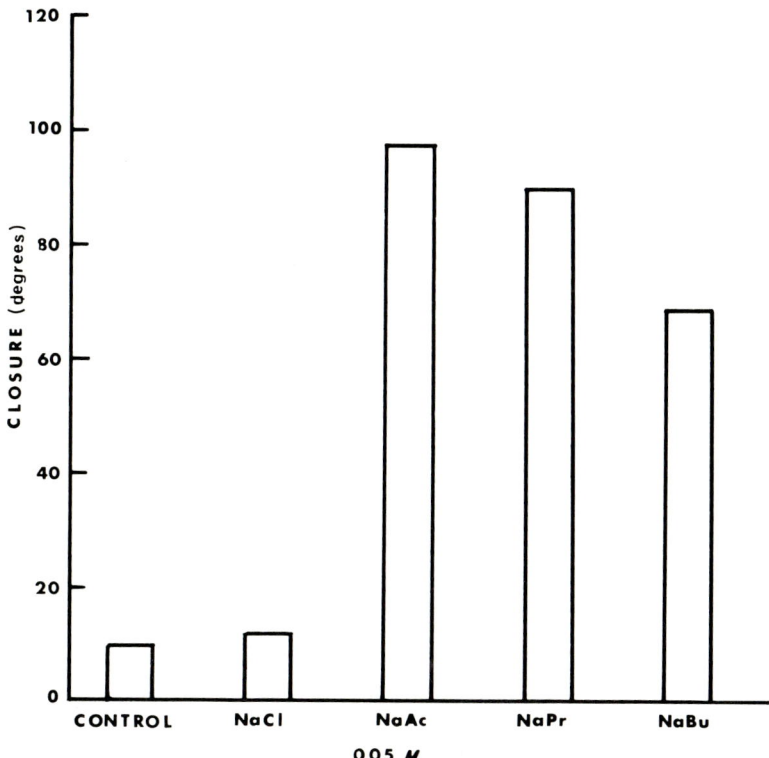

FIG. 14. Closure in the light of pinnule pairs floating for 2 hours 20 minutes on sodium salts of organic acids. From Satter et al. (1970b), by courtesy of the *American Journal of Botany*.

Discussion

We conclude that leaflets close when potassium ions move across the membranes of key pulvinule cells or through their plasmodesmata. Such movement may be controlled by phytochrome, endogenous rhythms, or various organic anions. During the first few hours of the light period, when phytochrome control predominates, there are several requirements for optimal potassium movement out of ventral cells: (1) Phytochrome must be in the P_{fr} form. (2) ATP or some other aerobically generated energy source must be available. (3) Leaflets must be darkened. (4) Electrical neutrality must be preserved, either by movement of a proton into the

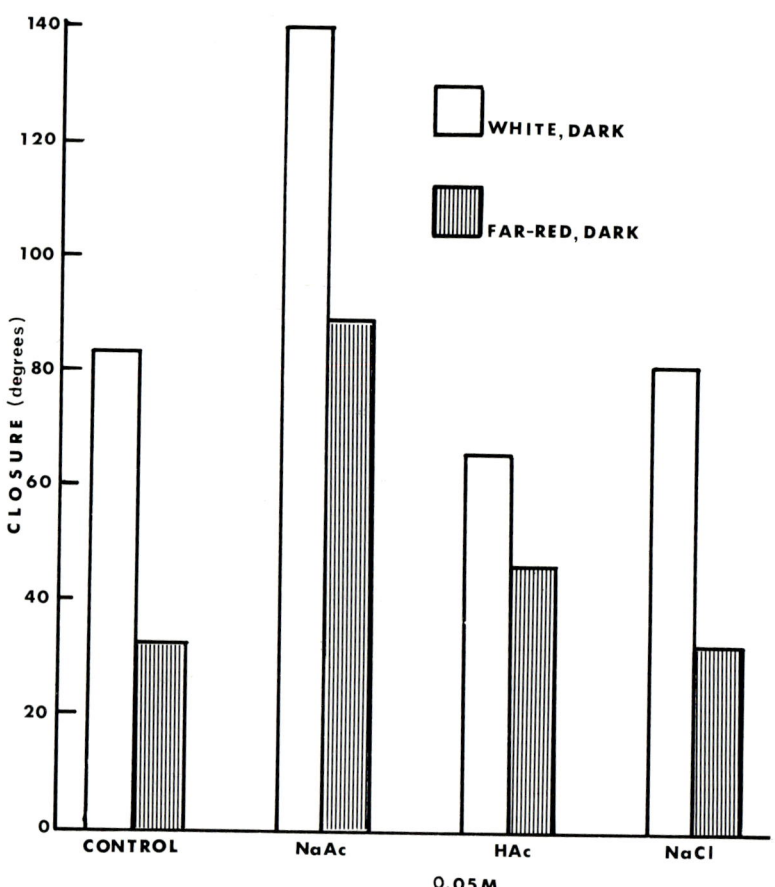

Fig. 15. Dark closure (during 90 minutes) of leaflets floating on sodium acetate or acetic acid. From Satter *et al.* (1970b), by courtesy of the *American Journal of Botany*.

cell in exchange for potassium or by the availability of an anion to accompany potassium out of the cell.

One possible interpretation of our data is, as already suggested by others (Hendricks and Borthwick, 1967; Smith, 1970), that phytochrome is located in the plasma membrane, where it controls the availability of certain permease loci. We can envision several possibilities in our system: Such loci, on motor cell membranes, might represent potassium-activatable ATPases. Under conditions where potassium ions cannot reach these loci

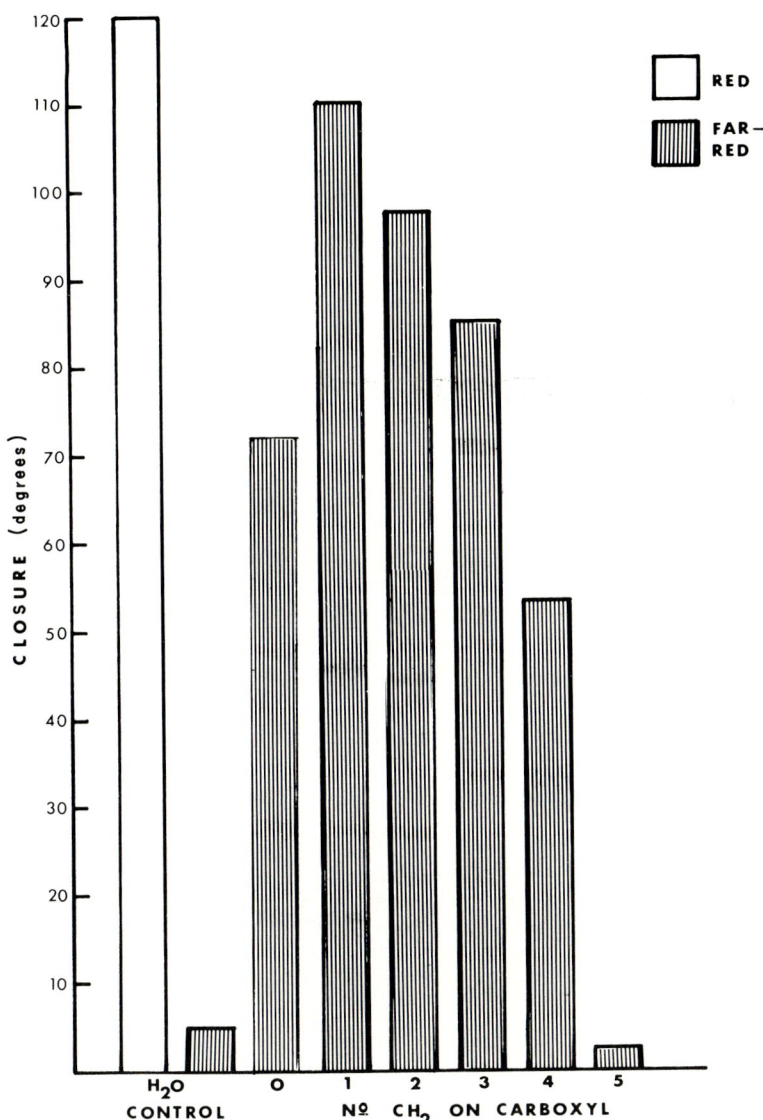

Fig. 16. Dark closure (during 45 minutes) of leaflets floating on 0.05 M sodium salts of alkylmonocarboxylic acids of varying chain length. All pinnule pairs except a water control were preirradiated with far-red light.

(i.e., when phytochrome occurs as P_r) potassium efflux occurs at only minimal rates through alternate pathways. When phytochrome is transformed from P_r to P_{fr}, the permease loci become exposed and, in the resulting action, ATP is used to move potassium across the membrane.

Alternatively such loci could control an inwardly directed proton pump that would carry H^+ into the cell in exchange for K^+. Evidence in some systems suggests that proton movement is the primary energetic process and ionic flux secondary (reviewed by Packer et al., 1970). It is also possible that organic anions, such as glycolate, formed photosynthetically in the light and sequestered in discrete cytoplasmic compartments (glyoxysomes) (Tolbert et al., 1968) might become available in the dark due to leakage from such bodies or to their total lysis, if phytochrome is in the P_{fr} form.

It is also possible that the supply of ATP in the motor cells might be controlled, directly or indirectly, by phytochrome. If ATP is required in stoichiometric amounts to move potassium across membranes, we wonder whether enough ATP is generated within the motor cells to provide energy for potassium flux of the magnitude depicted in Table 2. The motor cells have a normal but not unusually large number of mitochondria and chloroplasts, but most of the photosynthetic tissue in the pulvinule is in the inner cortical cells (Satter et al., 1970a). Thus the ultimate energy sources for the ATP required for potassium secretion from ventral motor cells might be supplied by other cells in the pulvinule or lamina. Phytochrome could control the transport of such energy substrates from these cells to the motor cells, or could influence the supply of ATP by controlling the rate of dark respiration.

If, in fact, phytochrome is localized on the membrane and controls its permeability, must we regard this effect as being specific for potassium? We think not. Previous experiments from this laboratory have demonstrated, for example (Goren and Galston, 1966), that the state of phytochrome in etiolated pea epicotyls controls the rate at which sucrose reaches the distant apical bud when it is applied basally to excised epicotyls. This effect might be a consequence of altered rate of movement of sucrose across cell membranes.

The complex photoperiodic requirements for flowering, in which phytochrome may also play a decisive role, involve interactions between light and endogenous rhythms (Cumming, 1967), perhaps acting on the same basic process, such as membrane permeability. Thus, the synthesis of floral hormone in the leaf may not occur until alterations in membrane permeability make precursors available. Alternatively, already synthesized hormone may have no consequences for bud differentiation until basic alterations in the transport mechanism occur through appropriate phytochrome conversion. A verification of this hypothesis would involve either an isolation of membrane fragments containing phytochrome activity or the addition of phytochrome to synthetic lipid membranes *in vitro* and noting accompanying changes in membrane characteristics following selective irradiation with red or far-red light.

Extracellular acetate or other effective organic acids can facilitate closure in leaflets exposed to far-red light before darkness, thereby circumventing the P_r barrier to potassium efflux. These same anions also promote light-dependent potassium uptake in pea leaf fragments (Nobel, 1970). They might promote closure of *Albizzia* leaflets either by binding to a carrier molecule in the membrane as suggested by Nobel (1970), by transforming phytochrome, or by facilitating H^+ for K^+ exchange. The last hypothesis is tenable only if the extracellular H^+ ion concentration is high and acetate moves into ventral motor cells as the undissociated acid.

In considering the relevance of the *Albizzia* mechanisms for other phenomena in plant physiology, we have been struck by the fact that stomatal guard cells, which also depend on light-triggered movement of potassium ions to expedite turgor changes and consequent movements (Sawhney and Zelitch, 1969) also have multiple vacuoles (Thomson and de Journett, 1970). Also, salt gland cells of *Tamarix*, which secrete large quantities of salts, have numerous spherosome-like bodies and vacuoles containing tannins (Thomson *et al.*, 1969).

ACKNOWLEDGMENT

Aided by grants from the National Science Foundation and the Herman Frasch Foundation. We thank Dr. D. D. Sabnis for the electron microscopic studies, Dr. Horace Winchell of the Yale Geology Department for the use of the electron microprobe, and Messrs. Philip Marinoff, Rolf Drivdahl, and Malcolm McConnell for technical assistance.

REFERENCES

Anderson, C. A. 1967. An introduction to the electron probe microanalyzer and its application to biochemistry. *Methods Biochem. Anal.* **15**:147–270.

Attridge, T. H., and H. Smith. 1967. A phytochrome-mediated increase in the level of phenylalanine ammonia-lyase activity in the terminal buds of *Pisum sativum*. *Biochim. Biophys. Acta* **148**:805–807.

Borthwick, H. A., S. B. Hendricks, M. W. Parker, E. H. Toole, and V. K. Toole. 1952. A reversible photoreaction controlling seed germination. *Proc. Nat. Acad. Sci. U.S.* **38**:662–666.

Briggs, W. R., H. V. Rice, G. Gardner, and C. S. Pike. 1972. Physiological and biochemical properties of the plant chromoprotein phytochrome. *In* "Recent Advances in Phytochemistry" (V. C. Runeckles, ed.), Vol. 5, pp. 35–49. Academic Press, New York, N.Y.

Cumming, B. C. 1967. Circadian rhythmic flowering responses in *Chenopodium rubrum* L.: effects of glucose and sucrose. *Can. J. Bot.* **45**:2173–2193.

Fondéville, J. C., H. A. Borthwick, and S. B. Hendricks. 1966. Leaflet movement of *Mimosa pudica* L. indicative of phytochrome action. *Planta* **69**:357–364.

Galston, A. W. 1968. Microspectrophotometric evidence for phytochrome in plant nuclei. *Proc. Nat. Acad. Sci. U.S.* **61**:454–460.

Goren, R., and A. W. Galston. 1966. Control by phytochrome of ^{14}C-sucrose incorporation into buds of etiolated pea seedlings. *Plant Physiol.* **41**:1055–1064.

Haupt, W. 1965. Perception of environmental stimuli orienting growth and movement in lower plants. *Annu. Rev. Plant Physiol.* **16**:267–290.

Haupt, W., G. Mörtel, and I. Winkelnkemper. 1969. Demonstration of different dichroic orientation of phytochrome P_r and P_{fr}. *Planta.* **88**:183–186.

Hendricks, S. B., and H. A. Borthwick. 1967. The function of phytochrome in regulation of plant growth. *Proc. Nat. Acad. Sci. U.S.* **58**:2125–2130.

Hillman, W. S., and W. L. Koukkari. 1967. Phytochrome effects in the nyctinastic leaf movements of *Albizzia julibrissin* and some other legumes. *Plant Physiol.* **42**:1413–1418.

Jaffe, M. J. 1968. Phytochrome-mediated bioelectric potentials in mung bean seedlings. *Science* **162**:1016–1017.

Jaffe, M. J. 1970. Evidence for the regulation of phytochrome mediated processes in bean roots by the neurohumor, acetylcholine. *Plant Physiol.* **46**: 768–777

Jaffe, M. J., and A. W. Galston. 1967. Phytochrome control of rapid nyctinastic movements and membrane permeability in *Albizzia julibrissin*. *Planta* **77**:135–141.

Koukkari, W. L., and W. S. Hillman. 1968. Pulvini as the photoreceptors in the phytochrome effect on nyctinasty in *Albizzia julibrissin*. *Plant Physiol.* **43**:698–704.

Lange, H., and H. Mohr. 1965. The inhibitory effect of actinomycin D and puromycin on the phytochrome induced synthesis of anthocyanin. *Planta* **67**:107–121.

Mohr, H. 1969. Photomorphogenesis. *In* "An Introduction to Photobiology" (C. P. Swanson, ed.), p. 99-141. Prentice-Hall, Englewood Cliffs, New Jersey.

Nobel, P. S. 1970. Relation of the light-dependent potassium uptake by pea leaf fragments to the pK of the accompanying organic acid. *Plant Physiol.* **46**:491–493.

Packer, L., D. W. Deamer, and A. R. Crofts. 1967. Conformational changes in chloroplasts. *Brookhaven Symp. Biol.* **19**:281–302.

Packer, L., S. Murakami, and C. W. Mehard. 1970. Ion transport in chloroplasts and plant mitochondria. *Annu. Rev. Plant Physiol.* **21**:271–304.

Parker, M. W., S. B. Hendricks, and H. A. Borthwick. 1950. Action spectrum for the photoperiodic control of floral initiation of the long-day plant, *Hyoscyamus niger*. *Bot. Gaz. (Chicago)* **111**:242–252.

Satter, R. L., D. D. Sabnis, and A. W. Galston. 1970a. Phytochrome controlled nyctinasty in *Albizzia julibrissin*. I. Anatomy and fine structure of the pulvinule. *Amer. J. Bot.* **57**(4):374–381.

Satter, R. L., P. Marinoff, and A. W. Galston. 1970b. Phytochrome controlled nyctinasty in *Albizzia julibrissin*. II. Potassium flux as a basis for leaflet movement. *Amer. J. Bot.* **57**(8):916–926.

Sawhney, B. L., and I. Zelitch. 1969. Direct determination of potassium ion accumulation in guard cells in relation to stomatal opening in light. *Plant Physiol.* **44**:1350–1354.

Smith, H. 1970. Phytochrome and photomorphogenesis in plants. *Nature (London)* **227**:665–668.

Tanada, T. 1967. A rapid photoreversible response of barley root tips in the presence of 3-indoleacetic acid. *Proc. Nat. Acad. Sci. U.S.* **58**:376–380.

Thomson, W. W., and R. de Journett. 1970. Studies on the ultrastructure of the guard cells of *Opuntia*. *Amer. J. Bot.* **57**:309–316.

Thomson, W. W., W. L. Berry, and L. L. Liu. 1969. Localization and secretion of salt by the salt glands of *Tamarix aphylla*. *Proc. Nat. Acad. Sci. U.S.* **63**:310–317.

Tolbert, N. E., A. Oeser, T. Kisaki, R. H. Hageman, and R. K. Yamazaki. 1968. Peroxisomes from spinach leaves containing enzymes related to glycolate metabolism. *J. Biol. Chem.* **243**:5179–5184.

ACETYLCHOLINE AS A NATIVE METABOLIC REGULATOR OF PHYTOCHROME-MEDIATED PROCESSES IN BEAN ROOTS

M. J. JAFFE

Department of Botany, Ohio University, Athens, Ohio

Introduction.	81
Methods.	85
Identification and Measurement of Acetylcholine in Bean Roots.	88
The Role of Calcium.	94
Interactions with Respiration.	94
Synthesis.	101
References.	102

Introduction

Phytochrome is the pigment responsible for photoperiodic responses in green plants and, depending on its physical state, is capable of absorbing red (660 nm) light or far-red (730 nm) light. The pigment itself is believed to undergo a configurational change due to irradiation, and it is expected that this change becomes translated to conformational changes in the protein part of the holochrome (Hopkins and Butler, 1970). Although permeability changes and development of bioelectric potentials (Jaffe and Galston, 1967; Jaffe, 1968) have been shown to be rapidly associated with red or far-red irradiation, none of the immediate biochemical events have heretofore been measured. The general way in which the phytochrome

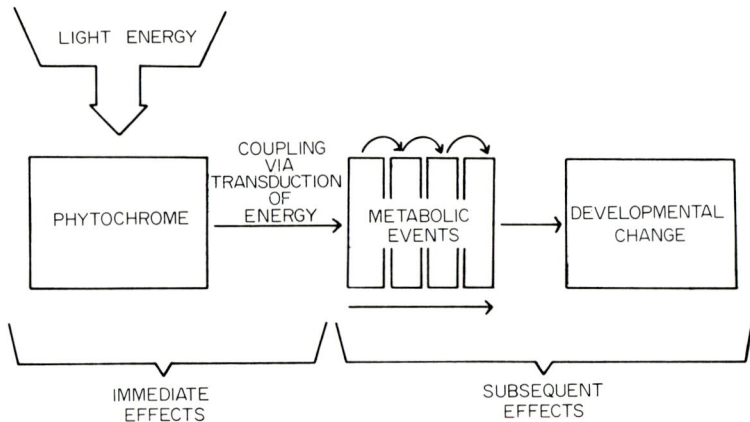

Fig. 1. Graphic representation of the possible course of events, which begin with absorption of the stimulus light energy and find expression in a developmental or behavioral change.

may regulate morphogenesis is shown in Fig. 1. In this scheme, phytochrome absorbs the energy from the primary signal, which is red light, and becomes physically transformed by it. In order to link the initial phytochrome transformation to the ultimate photomorphogenic change, numerous metabolic changes have been documented (Goren and Galston, 1966; Kohler et al., 1968; Bottomly et al., 1966; Attridge and Smith, 1967; Mohr et al., 1967; Steiner, 1968; Engelsma, 1966; Jaffe, 1969), all of which have been measured 0.5 hour to many hours after absorption of the primary stimulus. The means of energy transduction by which photoconversion of the phytochrome is coupled to these metabolic events is not known. However, two mechanisms are found in nature that transmit sensory information: propagated bioelectric potentials and hormones. Bioelectric potentials may be vectorial, as in the case of a nervous system or in the sensitive plant, *Mimosa pudica* (Umrath, 1934; Sibaoka, 1966), or radiating from a point source in both animal and plant systems (Sibaoka, 1966). Recently, this laboratory has demonstrated a positive bioelectric potential correlated with red irradiation of bean root tips and a negative one with far red (Jaffe, 1968). Of the plant responses mediated by the phytochrome system, we may make a distinction between the responses involving growth and development, which take a relatively long time to be observed, and the relatively rapid behavioral responses. These latter usually involve a dramatic change in the behavior of the test material which can be observed within seconds or minutes of irradiation. Thus,

Fondéville et al. (1967) have shown with *M. pudica*, and Hillman and Koukkari (1967) and Jaffe and Galston (1967) have shown with *Albizzia julibrissin*, that rapid nyctinasty is under phytochrome control; Forward (1970) has demonstrated a very rapid phototactic response in the dinoflagellate *Gyrodinium dorsum*; and rapid phototactic responses of isolated bean roots have been demonstrated by Tanada (1968a,b) and by Yunghans and Jaffe (1970). In *Mimosa*, phytochrome-controlled nyctinasty has been shown to be kinetically related, and possibly caused by the conformational changes of contractile vacuoles in the motor cells (Setty and Jaffe, 1970) which, via intercellular calcium fluxes, may control the intercellular potassium fluxes which seem to be responsible for the turgor move-

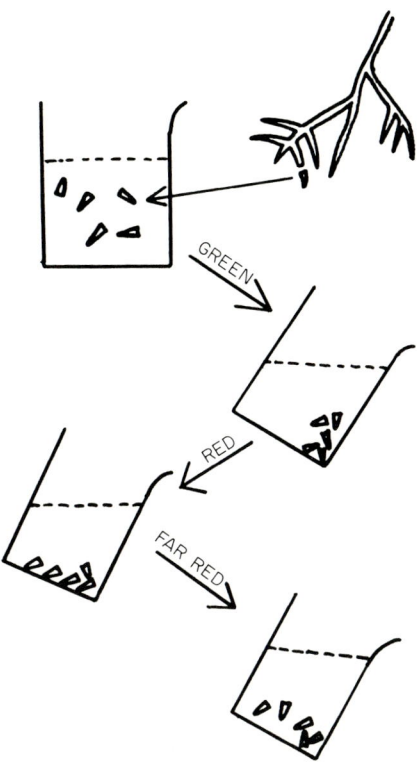

Fig. 2. The Tanada effect. Excised tips of secondary bean roots are placed in a special solution in a beaker whose interior surface has been given a negative charge. Under a green safelight, root tips slide to the side when the beaker is tipped. However, when they are exposed to red irradiation for a number of seconds, the root apices adhere to the glass. Far-red light causes them to release, and the effect is repeatedly photoreversible.

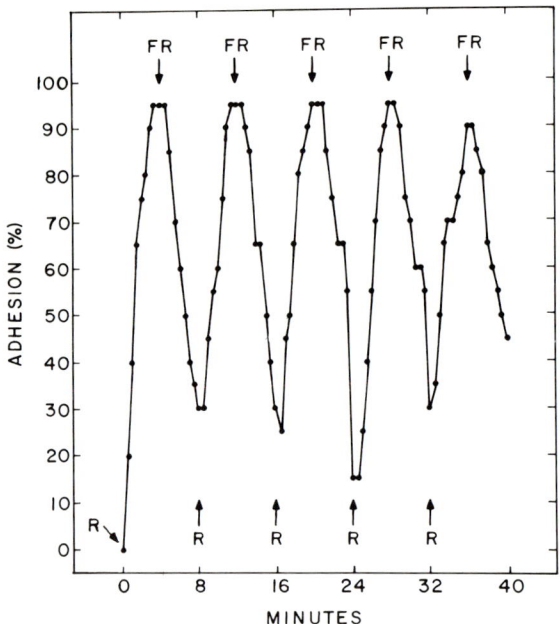

Fig. 3. The effect of 5 cycles of red (R) and far-red (FR) light on the adhesion of excised secondary root tips to a negatively charged glass surface. Each cycle consisted of 4 minutes of red light followed by 4 minutes of far-red light. From Jaffe (1970).

ments of the motor tissues in *Mimosa* and *Albizzia* (Toriyama, 1955; Satter and Galston, 1970).

Since animal bioelectric potentials are associated with ion fluxes, which in turn are often induced by neurohumors, it seemed appropriate to investigate the possibility that such a hormone might be intervening between the photochromic and morphogenic events caused by the primary stimulus. To test this possibility, the phytochrome-mediated effect first described by Tanada (1968a) was used. In this system, excised 2-mm secondary root tips of light- or dark-grown seedlings of the mung bean (*Phaseolus aureus*) are caused to adhere in red light and to release in far-red light from a negatively charged glass surface, when suspended in a carefully designed solution (Fig. 2). These adhesion and release responses are very rapid, are repeatedly photoreversible (Fig. 3), and their kinetics seem to be closely parallel to those of the bioelectric potentials that seem to cause them (Jaffe, 1968). This system was therefore chosen because it was thought that any biochemical changes that were found to occur would probably be close enough to the photodynamic change in the phytochrome molecule, and that they would probably be part of the transductional coupling mechanism outlined in Fig. 1. This paper describes the presence and activity of the neurohumor acetylcholine (ACh) as a hormone mediating phytochrome responses in bean roots.

Methods

Mung bean seedlings were grown in the dark or in the light as previously described (Yunghans and Jaffe, 1970). Two-millimeter tips of the secondary roots of dark grown plants were used for experiments involving red light-induced adhesion and far red light-induced release from a negatively charged glass surface. With this system, various addenda, such as neurohumors or inhibitors, could be added to the bathing solution.

ACh was studied with the ventricular beat of the marine clam *Mercenaria mercenaria* used as a bioassay (Greenberg, 1965). The preparation and apparatus are shown in Fig. 4. The ventricle was surgically removed and suspended by 2 threads in seawater in an aerated perfusion chamber modified after Florey (1967). One end of one thread was fastened inside the perfusion chamber. The other thread led from the ventricle to a mechanical strain transducer in circuit with a multiple gain recorder (Jaffe, 1970). The extraction method was modified after that of Fowler and Lewis (1958). One hundred to 500 mg of plant tissue was appropriately treated and quickly frozen. The tissue was then boiled in seawater for 10 minutes, homogenized in glass, and centrifuged at about 2000 g; the supernatant fluid was recovered. The ventricle was challenged with this extract or known concentrations of ACh by introducing 1 volume of test solution to 9 volumes of perfusion bath. The normal response to ACh was inhibition of the amplitude and frequency of the ventricular beat (Fig. 5). Therefore, the ventricular beat in the test solution was compared to that of a seawater control immediately preceding it, and the amount of ACh was computed with the use of a standard curve drawn from the inhibition due to known concentrations of authentic ACh (Fig. 5, inset). Because of the variation between ventricles, a separate standard curve was prepared for each one.

To irradiate root tips as well as to measure H^+ efflux from them, a special procedure was followed (Jaffe, 1970). Thirty secondary roots were laid parallel to one another on a glass microscope slide with their tips about 1 mm from one end. Another slide was placed over the roots 2 mm behind the tips and the slides were held together with a rubber band. Two such "sandwiches" were placed at 45° angles in a shallow depression cut into a

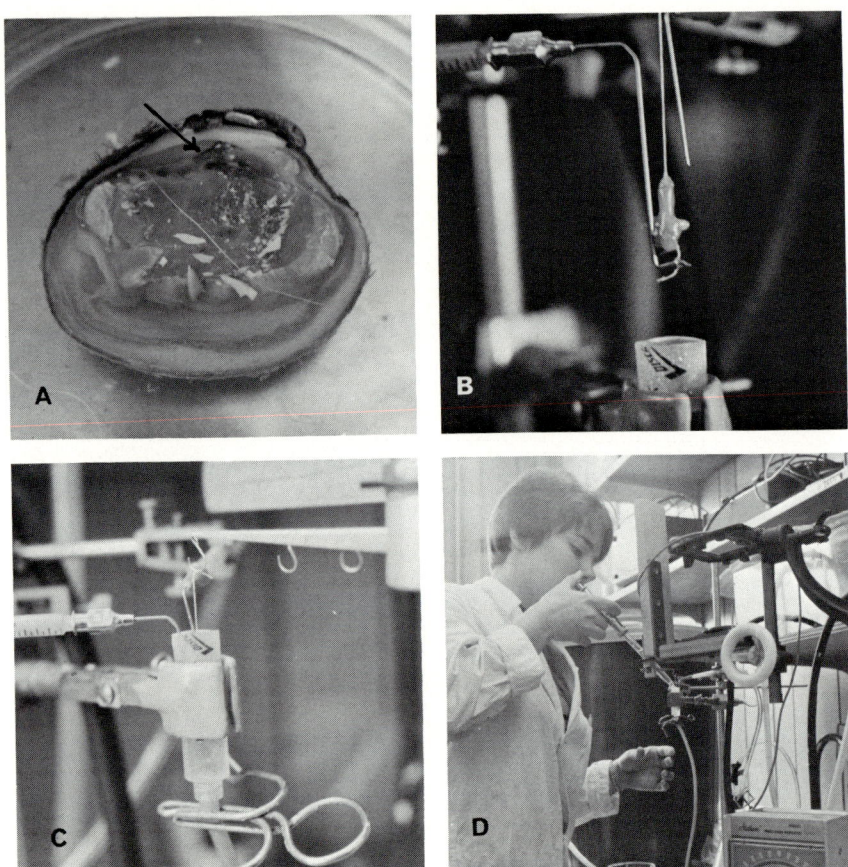

Fig. 4. Preparation for the clam ventricle bioassay for acetylcholine (ACh). (A) Appearance of the animal with one half of the shell removed. The location of the excised ventricle is indicated by the arrow. (B) Ligated and isolated ventricle. The ligatures are attached to the ventricular aspects of the auricular–ventricle junctions. (C) Perfusion chamber with the isolated ventricle in place. One cotton ligature is tied to a hook at the end of a lever in circuit with an electronic strain gauge. The lower part of the perfusion chamber is fitted with a clamped drain for flushing the ventricle. (D) The strain gauge was attached to a rack and pinion rider so that the cotton thread could be gently tightened. The strain gauge was in circuit with the voltage regulator, from which the signal went to a multiple gain recorder (not shown). In the picture, fresh seawater is being added to the perfusion chamber after flushing.

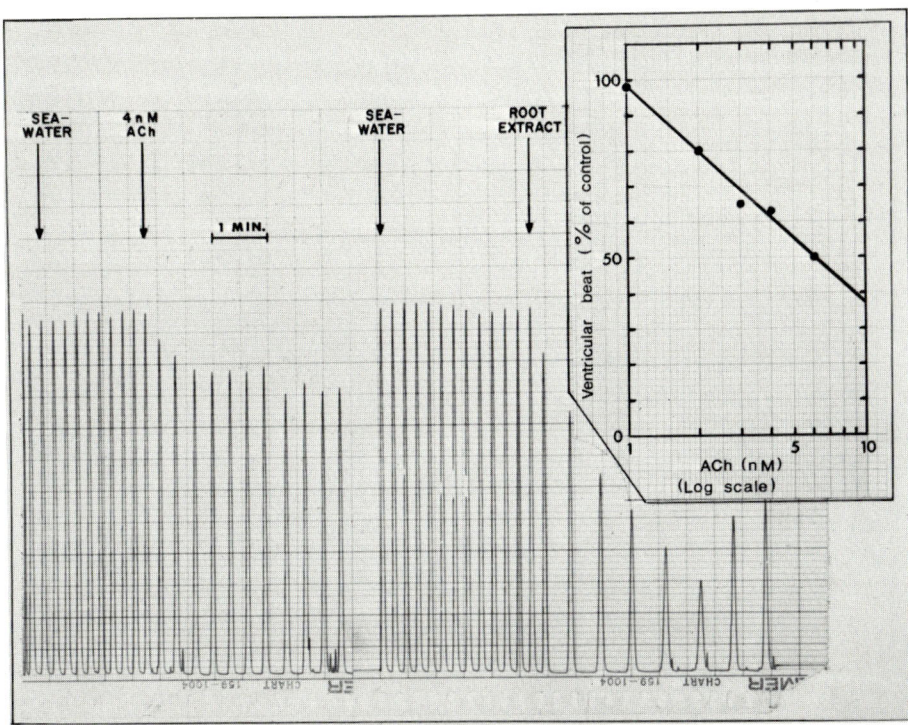

Fig. 5. Recording of the response of the clam ventricle to authentic acetylcholine (ACh) and to extracts of secondary bean roots. For extraction, the roots were frozen, boiled in seawater for 10 minutes, and homogenized in glass. The homogenate was centrifuged, and the supernatant fluid was used to challenge the ventricle. In the inset is shown a typical dose response curve of the ventricular beat to authentic ACh.

Lucite block which held 3.0 ml of bathing solution (Yunghans and Jaffe, 1970) and a tiny spin bar. The whole assembly was placed on a wooden shelf over a magnetic stirrer inside an irradiation chamber (Yunghans and Jaffe, 1970). The bathing solution as well as the 60 root tips were assayed for ACh. Red irradiation was at 50 and far-red at 150 $\mu W/cm^2/$sec/nm. The light treatments were: 4 minutes darkness (Dk); 4 minutes Dk + 4 minutes red light (R); 4 minutes Dk + 4 minutes R + 4 minutes far-red light (FR). In some experiments, an additional treatment of 4 minutes of 100 μM ACh in the dark (ACh) was also given. For measurements of H^+ concentration, a small combination pH electrode was positioned in the bathing solution between the secondary root sandwiches.

In order to localize and measure intracellular calcium, the cytochemical technique of Toriyama and Jaffe (1971) was coupled with microspectrophotometry. In this procedure, killed, fixed, and sectioned root tips, were stained with Alizarin Red and the absorbance of the resulting pink complex was measured with a Leitz Ortholux microscope fitted with an MPV microspectrophotometric apparatus. The electron micrographs were obtained by the procedures of Green et al. (1968) and Harris et al. (1969), and the various respiratory experiments were done as previously described by Yunghans (1970).

Identification and Measurement of Acetylcholine in Bean Roots

The extraction and assay methods described above were adequate for the identification and, at least, semiquantification of the very low levels of ACh present in mung bean tissue. There was an inhibition of the ventricular beat by extracts of secondary roots (Fig. 5). Table 1 shows that if extracts of these roots were treated with acetylcholinesterase (AChE), their inhibitory effect was reduced. Treatment of the ventricle with atropine sulfate or benzoquinonium chloride (Mytolon) decreased its sensitivity to the extract, and with eserine sulfate increased it. These agonists and antagonists are all quite specific for ACh (Florey, 1967; MacIntosh and Perry, 1950). The chromatographic migration of the ex-

TABLE 1

Effect of Various Treatments on the Response of the Isolated and Perfused Ventricle of *Mercenaria mercenaria* to Extracts of Mung Bean Tissue

Treatment	Percent change[a]
None; the same extract was tested twice on the same ventricle, or on separate ventricles	$+12 \pm 7$
The ventricle was perfused for ca. 4 minutes with 100 μM atropine sulfate between challenges with the extract	-32 ± 17
The ventricle was perfused for 30 minutes with 1 μM benzoquinonium chloride between challenges with the extract	-99 ± 1
The ventricle was perfused with 5 μM eserine sulfate for 30 minutes between challenges with the extract	$+59 \pm 20$
The extract was treated with $10^{-3}\%$ AChE. The reaction mixture was boiled for 10 minutes, and then used to rechallenge the ventricle.	-60 ± 9

[a] Comparison of the effect on the ventricular beat of the fresh extract to the extract after the appropriate treatment. Each datum is followed by its standard error.

TABLE 2

MIGRATION OF AUTHENTIC ACETYLCHOLINE (ACh) AND ACTIVITY EXTRACTED FROM SECONDARY ROOTS ON PAPER CHROMATOGRAMS USING TWO SOLVENT SYSTEMS[a]

Solvent system	Applied at origin	R_f exhibiting the greatest inhibition of the ventricular beat
Propanol–benzyl alcohol–H_2O (5:2:2)	ACh	0.6
	Plant extract	0.6
Butanol saturated with H_2O	ACh	0.2
	Plant extract	0.1

[a] After chromatography on Whatman No. 1, the dried chromatograms were cut into 10 equal strips which were eluted in seawater. The eluates were used to challenge the ventricle.

tract was quite similar to that of authentic ACh in 2-solvent systems (Table 2) as well as to published R_f values (Florey, 1967). Because of the relative specificity of the clam ventricle (Florey, 1967), AChE (Hestrin, 1949), the sensitization by eserine, and the desensitization by atropine and Mytolon, as well as the chromatographic evidence, it seems very probable that the active principle in the root extracts is ACh.

TABLE 3

AMOUNT OF EXTRACTABLE ACETYLCHOLINE IN BEAN SEEDLINGS[a]

Organ	11-Day-old light grown plants	6-Day-old dark grown plants
Leaf	34.2 ± 7.6	—
Bud	106.5 ± 29.2	79.8 ± 18.5
Hypocotyl hook	—	49.7 ± 24.4
Top internode	25.4 ± 9.3	—
Bottom internode	32.6 ± 12.9	—
Top half of hypocotyl	—	27.2 ± 8.9
Middle of primary root	32.7 ± 10.2	23.8 ± 3.8
Whole secondary root	97.2 ± 37.3	135.3 ± 25.1
Apical half of secondary root	—	203.9 ± 35.2
Basal half of secondary root	—	89.9 ± 32.3

[a] For the method of extraction and assay, see text. Each datum is followed by its standard error.

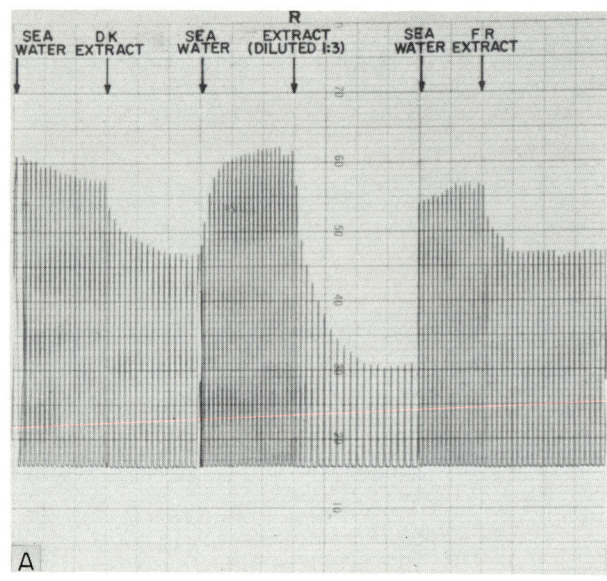

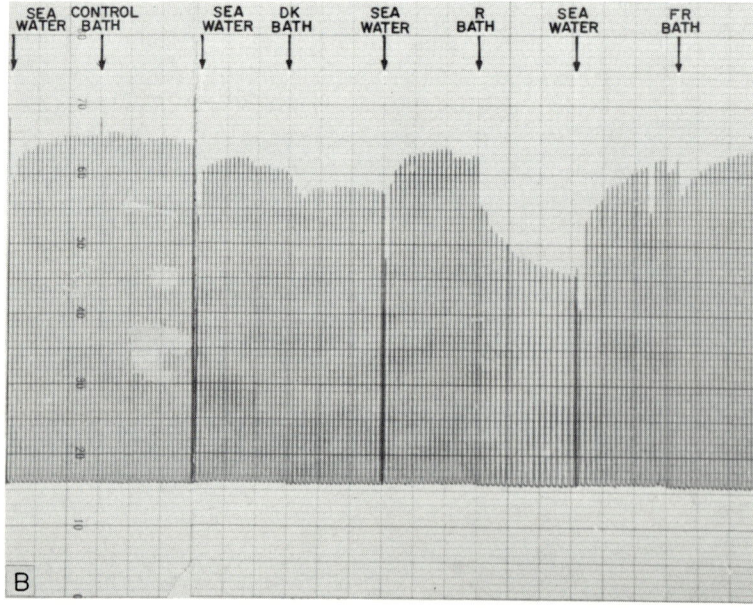

TABLE 4

Effects of Red and Far-Red Light, and the Photomimetic Effect of Acetylcholine (ACh), on Various Processes in Bean Roots

Response	4 min Dk[a] + 4 min R	4 Min Dk + 4 min FR	4 Min Dk + 4 min R + 4 min FR	4 Min ACh in the Dk	Average SE
Secondary root formation from day 3 to day 6 in the dark (No. of new 2° roots)	13	6	10	5	1.5
Adhesion of excised root tips to a negatively charged glass surface (% adhesion)	0	85	10	82	3.5
Net change in bioelectric potential of excised root tips (mV)	0	+1.16	−0.43	—	0.28
Net change of the hydrogen ion concentration in the root bathing solution. ($[H^+]$)	−4.8	+3.6	−6.2	+17.7	1.9

[a] Dk, dark; R, red; FR, far-red.

The topographical distribution of ACh in both light- and dark-grown seedlings is presented in Table 3. It is of special interest to note that the actively dividing tissues, contained in the secondary root tips and in the shoot buds, have the greatest endogenous titer of ACh. It may also be significant that the distribution of ACh in the etiolated plant follows a similar pattern to the distribution of phytochrome reported by Furuya and Hillman (1964).

Red light increased the endogenous titer of ACh in root tips from etiolated plants, and far-red light reduced it (Fig. 6). The computed average values of ACh for Dk, R, and FR were 27.4, 64.9, and 13.7 ng per 60 root tips. The average measurements of the root bathing solution were 1.0, 2.0, and 1.5 ng of ACh per 60 root tips, respectively. If red light were merely causing release of ACh, the endogenous ACh titer in the dark would be expected to be similar to that in red light. Thus it seems that red light immediately causes the synthesis, and far-red light the destruc-

Fig. 6. Response of the clam ventricle to extracts of roots (A) and exudates of roots (B) exposed to 4-minute consecutive intervals darkness, darkness followed by red light, or darkness followed by red light followed by far-red light.

TABLE 5

Interaction of Various Treatments with Darkness (Dk), Red Light (R), and Far-Red Light (FR) in the Adhesion of Excised Bean Root Tips to a Negatively Charged Glass Surface[a]

Addendum	Adhesion (%)		
	Dk	R	FR
None	0 ± 0	83 ± 4	4 ± 3
40 μM ACh	27 ± 10	—	—
100 μM ACh[b]	80 ± 0	73 ± 7	63 ± 3
100 μM Atropine	22 ± 4	40 ± 9	6 ± 4
5 μM Eserine	0 ± 0	73 ± 10	21 ± 12
1 μM Ecothiopate iodide	0 ± 0	23 ± 10	0 ± 0
1 μM Paraoxon	0 ± 0	23 ± 10	0 ± 0
100 μM Pyridine-2-aldoxime methiodide	0 ± 0	31 ± 7	0 ± 0
100 μM AMO-1618	40 ± 0	50 ± 10	17 ± 3
100 μM CCC	0	50	0
30 min of AChE, then observe	0 ± 0	25 ± 5	0 ± 0
15 min of 5 μM eserine, then 40 μM ACh	53 ± 10	—	—
15 min of 100 μM atropine, then 100 μM ACh	0 ± 0	—	—

[a] The light treatments were 4 minutes each of Dk, Dk + R, or Dk + R + FR, in consecutive order. The acetylcholine (ACh) treatment for root inhibition was for 4 minutes in the dark. The adhesion measurements were made during the 4 minutes of irradiation, with or without a pretreatment. Each datum represents the average of 3–5 experiments using 10 root tips per experiment, and is followed by its standard error.

[b] In this experiment, the dark treatment was separate from, and not consecutive with, the red and far-red treatments.

tion, of a substance having the properties of ACh. Since red light causes ACh efflux, it seems reasonable to conclude that ACh satisfies the hormone requirement of stimulable mobility. ACh also satisfies the causal requirement, since it is photomimetic for red light in at least 4 processes in the bean roots (Table 4). This table also indicates that both red light and ACh are able to stimulate what appears to be an increase in H^+ efflux from the roots. Since such an efflux may cause positive bioelectric potentials in roots, the phytochrome and ACh-mediated efflux may account for the red light-induced adhesion to a negatively charged glass surface (Tanada effect).

In regard to the Tanada effect, Table 5 presents data that support the hypothesis that ACh mediates this phytochrome response. Thus, ACh is able to substitute for red light; atropine, ecothiopate iodide, and paraoxon (the latter two compounds are widely used as insecticides and are competitive inhibitors of ACh) inhibit adhesion in red light; eserine, which is an inhibitor of AChE, inhibits release in far-red light, as well as increasing the response to ACh and to red light (Jaffe, 1970); and AMO-1618 and CCC, both plant growth retardants which are known, respectively, to inhibit and promote AChE (Newhall, personal communication), have similar effects on red light-induced adhesion.

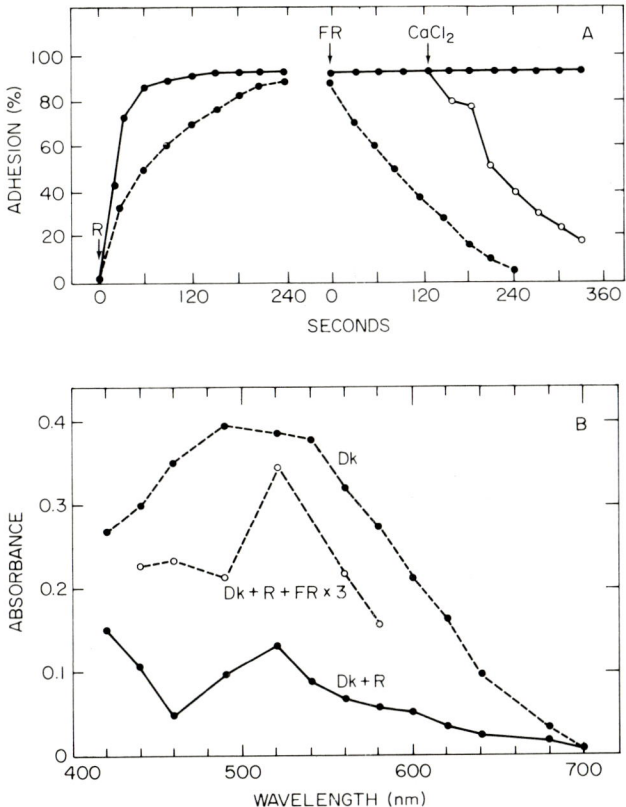

Fig. 7. The results of experiments dealing with the role of calcium in phytochrome-mediated processes in bean roots. (A) The effect of 20 mM CaCl$_2$ on release in far-red light from an oxalate-charged glass surface. ●---●, Control, using a glass beaker charged with a saturated aqueous phosphate solution; ●——●, beaker charged with a saturated oxalate solution, without excess CaCl$_2$ added; ○——○, beaker charged with a saturated oxalate solution, with the addition of excess CaCl$_2$. The average standard error is ± 0.83. (B) Effect of light on the presence of calcium on the nuclear envelope. The roots were given the standard light treatments and fixed in Lillie's neutral buffered formalin and stained with alizarin red. The absorption spectra of the colored complex formed by alizarin red and Ca^{2+} were measured with a Leitz microspectrophotometer and are shown in the figure. R, red light; FR, far-red light; Dk, dark.

The Role of Calcium

When the anion used to give the beaker its negative charge is glutamate, citrate, or oxalate, the root tips do not release in far-red light (Yunghans and Jaffe, 1970). This is apparently due to the chelating action of these anions upon calcium, for when an excess of $CaCl_2$ is added to the bathing solution, the root tips do release (Fig. 7). It is known from the literature dealing with animal systems that calcium ions seem to be mobilized by ACh and that the calcium then acts as a secondary hormone (Rasmussen, 1970), possibly responsible for monovalent ion fluxes. Since calcium is necessary for the Tanada effect, we studied the intracellular movement of calcium during irradiation. The lower part of Fig. 7 shows that in cells of dark-grown roots there is a certain amount of ionizable calcium on the nucleus, probably on or in the nuclear envelope. When the roots are irradiated with red light, the complex formed between alizarin red and Ca^{2+} is no longer to be found on the nucleus, whereas subsequent irradiation with far-red light causes it to reappear. It is not yet possible to make a complete interpretation of these data, but they seem to support the theory that when Ca^{2+} ions are bound to a membrane, it is in the resting state, whereas when they are released, the membrane becomes "active" (Rasmussen, 1970). Furthermore, these data are not inconsistent with the report by Galston (1968) that the phytochrome type of photoreversibility seems to be localized in the nuclear envelopes of root cells.

Interactions with Respiration

Since the Tanada effect has a specific requirement for ATP (Tanada, 1968b; Yunghans, 1970), and because there are no chloroplasts in the

TABLE 6

Oxygen Consumption of Excised Root Tips from Etiolated Plants[a]

Treatment	Oxygen uptake (μmoles/100 2-mm root tips)
4 Minutes dark	0.016 ± 0.003
4 Minutes dark + 4 minutes red	0.040 ± 0.005
4 Minutes dark + 4 minutes red + 4 minutes far red	0.026 ± 0.006
4 Minutes dark in the presence of 100 μM ACh	0.056 ± 0.005

[a] Oxygen uptake was measured with a Warburg-type respirometer.

TABLE 7

Oxygen Consumption of Excised Light-Grown Roots and of Mitochondrial Suspensions Isolated from Them[a]

ACh concentration (M)	Oxygen consumption (% change)		
	Forty light-grown root tips		Isolated mitochondria
	Tested in the dark	Tested in the light	
0	0	0	−4
8×10^{-8}	−16	−6	—
3×10^{-7}	—	—	−26
8×10^{-7}	−8	−13	—
3×10^{-6}	—	—	−5
8×10^{-6}	−23	+10	—
3×10^{-5}	—	—	+23
8×10^{-5}	+18	+16	—
3×10^{-4}	—	—	+14
8×10^{-4}	+1	−13	—
3×10^{-3}	—	—	−9
Average standard error:	8	8	11

[a] Excised 5-mm root tips were placed in the reaction flask of a Yellow Springs respirometer, and oxygen uptake was measured with a Clark O_2 electrode in circuit with a multiple gain recorder. For a measurement, O_2 uptake was measured for approximately 4 minutes and then the appropriate acetylcholine (ACh) solution was added, and measurements were recorded again. The data are presented as the percentage change of the control, and each experiment was done at least 3 times.

TABLE 8

ATP, P_i Levels, and ATPase Activity Extracted from 2 mm Long Root Tips Given the Standard Light and Acetylcholine (ACh) Treatments[a]

Treatment	ATP (nmoles/4 min)	ATPase activity (OD units)	P_i (μmoles/4 min)
4 Minutes dark	14.0 ± 0.1	0.035 ± 0.03	0.55 ± 0.01
4 Minutes dark + 4 minutes red	1.2 ± 0.2	0.035 ± 0.04	7.95 ± 0.01
4 Minutes dark + 4 minutes red + 4 minutes far red	10.0 ± 0.4	0.035 ± 0.02	0.50 ± 0.02
4 Minutes dark in the presence of 100 μM ACh	1.7 ± 0.5	0.037 ± 0.06	7.50 ± 0.03

[a] For details of the procedures, see the text. Each datum is followed by its standard error.

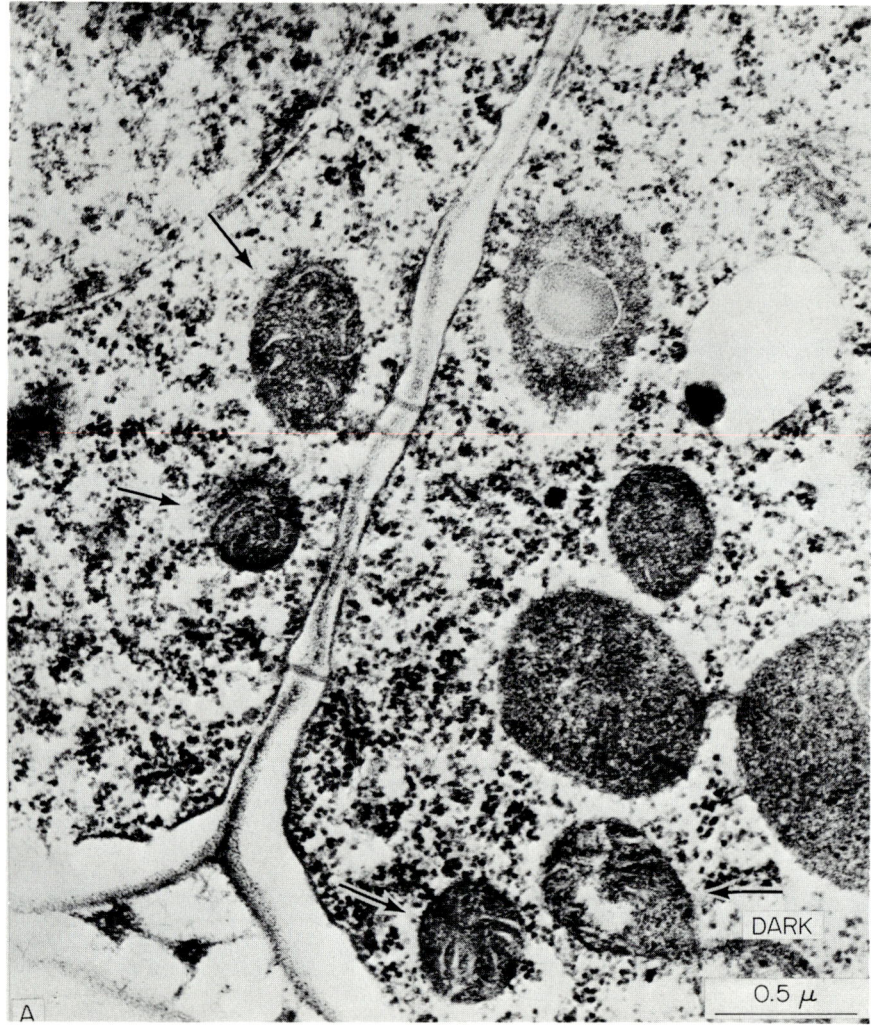

Fig. 8. Electron micrographs of variously irradiated root cell mitochondria, observed *in situ*. See the text for details of preparation. (A) Four minutes dark; (B) 4 minutes dark + 4 minutes red; (C) 4 minutes dark + 4 minutes red + 4 minutes far-red.

dark-grown roots, we thought it appropriate to see whether the energetics of respiration were affected by ACh and light. Accordingly, oxygen uptake was measured by Warburg respirometry of root tips given the standard treatments. Table 6 shows that both red light and ACh had a promotive effect on O_2 consumption and that far-red light reversed the effect of red light. In order to make our short-term measurements more accurate, we turned to the use of an electronic means of measuring respiration. For these studies, a Clark oxygen electrode used with a Yellow Springs respirometer coupled to a multiple gain recorder was used to measure O_2 uptake in excised root tips as well as in suspensions of isolated mitochondria (Table 7). It is clear from Table 7 that, when added in the same concentration range as caused responses in other root phenomena (see Table 4), ACh produced a significant increase in respiratory oxygen consumption in the excised root tips. The mitochondria responded in a similar manner; however, they required a little less ACh to produce similar promotion of O_2 uptake, an observation that is probably related to the difficulty of ACh uptake by the roots. The only comparable studies that a search of the literature has revealed are those by Vdovichenko and Demin (1965; Vdovichenko, 1965, 1966). Working with mitochondria isolated from the

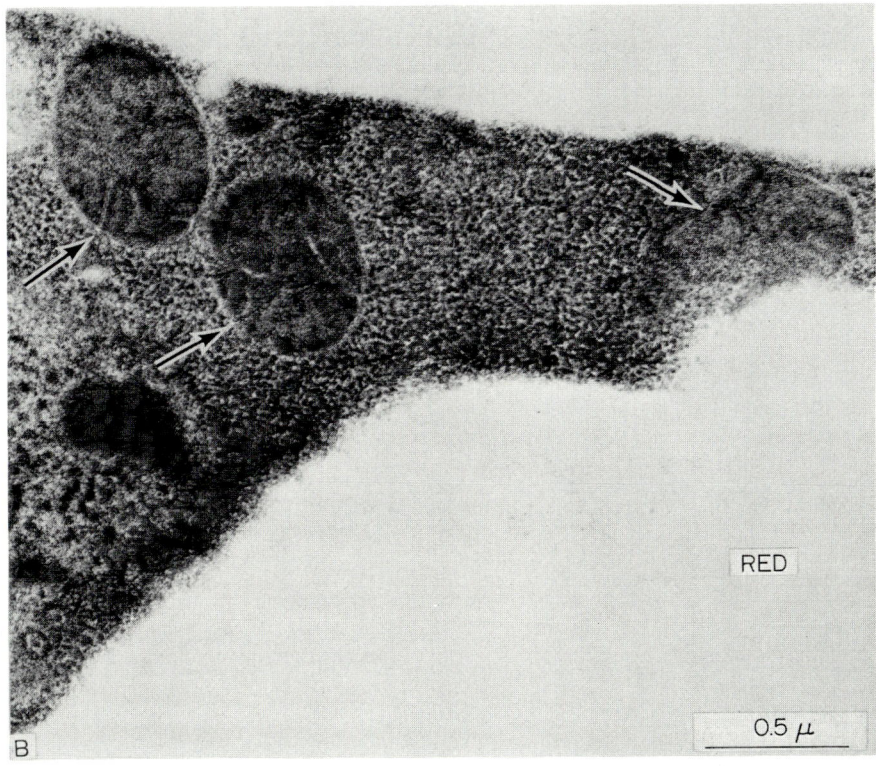

Fig. 8B. See caption under Fig. 8A.

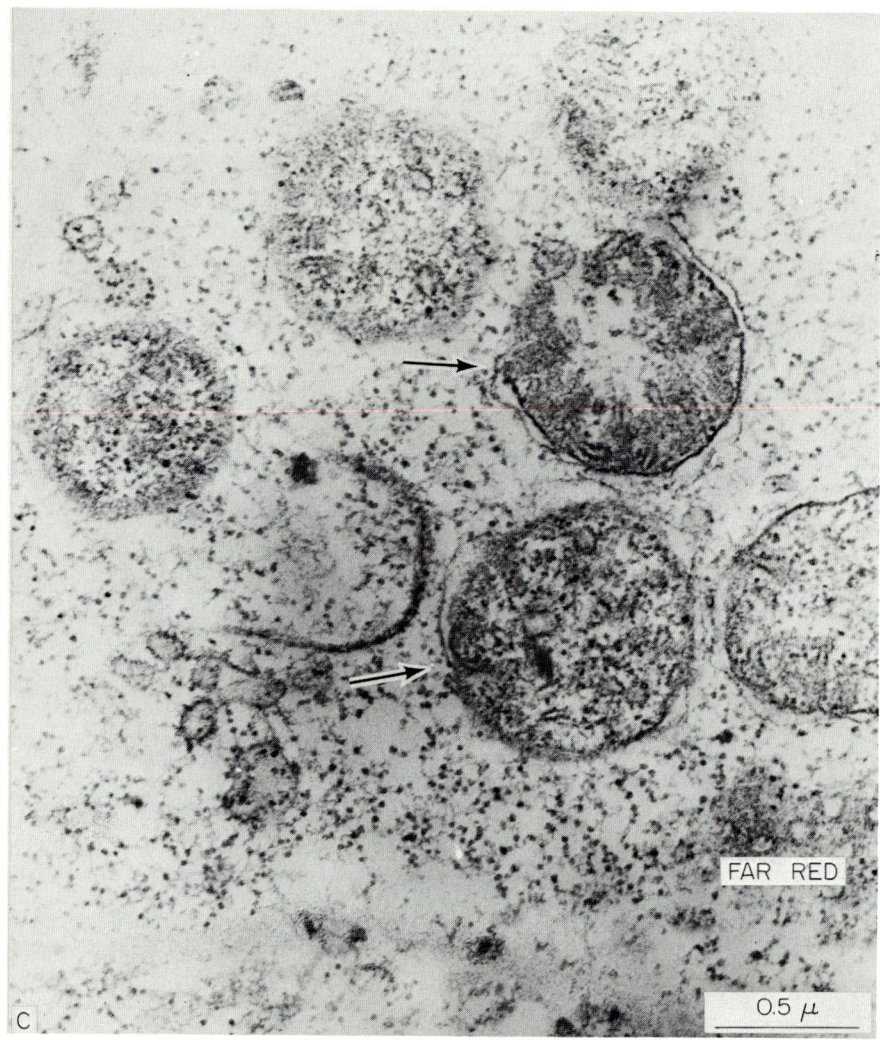

Fig. 8C. See caption under Fig. 8A.

brains and livers of rats and rabbits, they reported essentially the same kind of dose response to ACh as are shown herein. It may be, therefore, that the mitochondria are target sites of ACh in all eukaryotic cells. Since there are impressive arrays of mitochondria specially related to the ACh-containing vesicles in neuronal presynaptic regions, the function of mitochondria as target sites for ACh may prove to play a role in transmission of the nerve impulse across the synapse. When ATP and inorganic phosphate levels were measured under the standard treatments, it was found that both red light and ACh decreased the ATP level about 10-fold, and increased the P_i titer by about 14-fold (Table 8). The amount and changes of inorganic phosphate were, however, from 2 to 3 orders of magnitude greater than those of ATP, indicating a much greater pool size. The decrease in ATP is probably not due to an increase in ATPase activity, since such an enhancement due to red light or ACh could not be demonstrated (Table 8).

From the above data, it is possible to speculate about the effect of red light and of ACh on respiratory energetics and the possible relationship of this to the Tanada effect. Since ACh increases oxygen consumption of both root tips and their mitochondria, it seems possible that some rate-limiting step is being removed. Since ATP synthesis is such a rate-limiting step, and since treatment with ACh and red light cause a rapid decrease in the endogenous ATP titer, it may be that these agents are able to uncouple oxidative phosphorylation from respiratory electron transport, or that ATP is being used up more rapidly. We have evidence that bears on this question. Figure 8 shows electron micrographs of mitochondria in cells of variously treated root tips. Using the terminology of Green and his co-workers (Asai *et al.*, 1969; Green *et al.*, 1968; Harris *et al.*, 1969), mito-

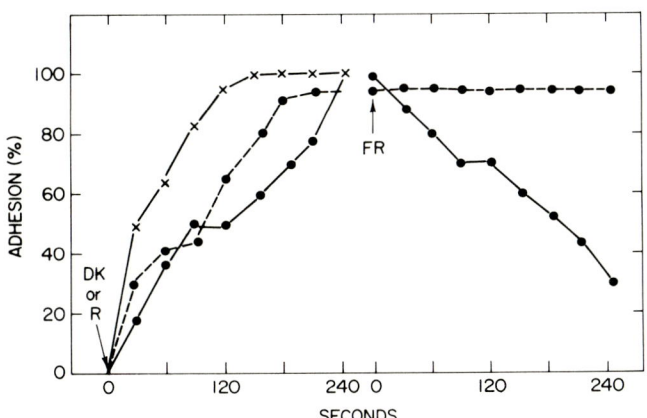

FIG. 9. The effects of 10 μM gramicidin on adhesion and release of bean root tips to negatively charged glass. Dark + gramicidin (×——×); Control, adhesion and release by red and far-red light (●——●); red and far-red light + gramicidin (●---●). The average standard error is ± 1.2. From Yunghans and Jaffe (1972).

chondria in the dark are in the so-called "energized state," a state that has been correlated with the ability to actively synthesize ATP. After 4 minutes of red light, however, the conformation of the inner mitochondrial membranes has changed, and they now resemble the so-called "nonenergized" state, and are unable to synthesize ATP. This condition seems to be partially reversed by far-red light, in which we see somewhat swollen mitochondria, in a partially energized state. One further kind of evidence seems to support this contention. It is well known that certain antibiotics, as well as other agents, affect oxidative phosphorylation (Lehninger, 1967). We have tried a number of these uncoupling agents (e.g., 10 μM gramicidin, 25 μM oligomycin, 1 mM atractyloside, 1 mM digitoxin, and 10 μM valinomycin). All of these produced similar results on the Tanada effect; the responses to gramicidin are shown in Fig. 9. The figure indicates that the uncoupler is able to substitute for red light, has no effect on red light-induced adhesion, but inhibits release under far-red light. These are essentially similar to the responses that we have been able to obtain by adding ACh to the bathing solution, the difference being that ACh is a compound that occurs naturally in the bean root, whereas the other agents presumably do not.

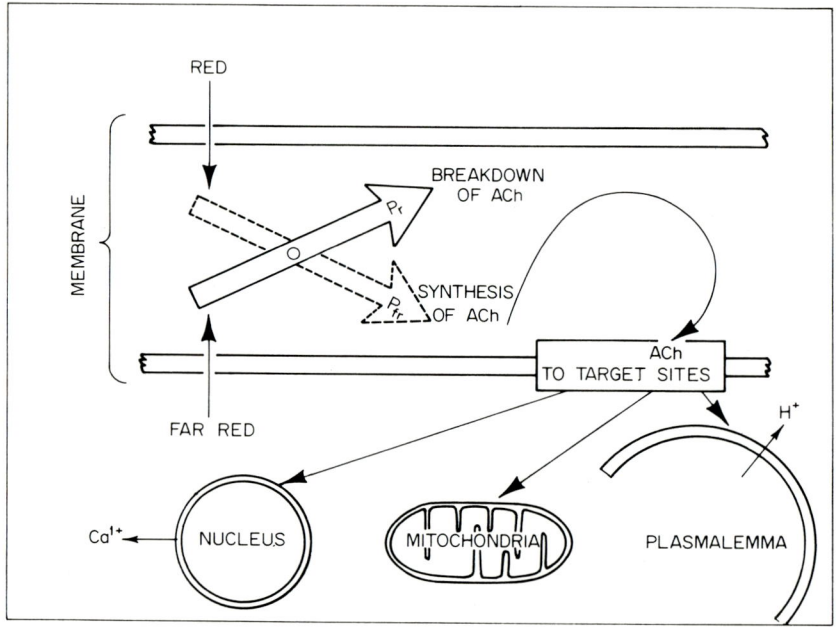

Fig. 10. Aspects of the proposed regulation of phytochrome-mediated responses in bean roots by acetylcholine.

Synthesis

Although ACh has been tentatively identified in several plant organs (Appel and Werle, 1959; Lin, 1957; Marquardt and Falk, 1957; Saxena et al., 1966; Tulus et al., 1961), no function has been assigned to it. The present data indicate that its hormonal function in regulating ion fluxes and bioelectric potentials in animal systems (Nachmanson, 1970) may now be extended to plants as well. If ACh does function in this way in bean root cells, it may be considered evidence for the commonality of fundamental membrane processes in eukaryotic cells and provide additional evidence for the membrane theory of phytochrome action (Hendricks and Borthwick, 1967; Jaffe and Galston, 1967; Jaffe, 1968). Since ACh is thought to exert its primary effect on the transport of ions across animal cell membranes, it would be of interest to see whether it also does so in plants. Certainly, our observations concerning H^+ and Ca^{2+} would seem to support this view.

From these data, it is possible to formulate an explanation of ACh-regulated phytochrome control in bean roots. Such a model is shown schematically in Fig. 10. First, since red light increases the ACh titer and ACh is photomimetic for red light, red irradiation probably induces the synthesis of ACh. The animal ACh-synthesizing system uses choline and acetyl-CoA in the presence of the enzyme, choline acetyl transferase (formerly called choline acetylase) to make ACh. To the best of my knowledge, this enzyme has not yet been reported in plants, and this laboratory is currently beginning a search for its presence. Second, far-red light induces a decrease of the endogenous ACh level, and eserine, which inhibits AChE activity in animal systems, inhibits far red-induced release from a negatively charged glass surface. Thus, far-red light may function by activating a cholinesterase in some way. True AChE activity has already been reported by Dettbarn (1962) in *Nitella* and by Schwartz (1967) in pea roots, and there is no reason to expect that it will not be present in the roots of the mung bean plant. In this connection, it is interesting to note, that Newhall (1971) has found a significant positive correlation between growth retardant activity of various compounds, and their ability to inhibit blood serum cholinesterase. Since some of these compounds are choline derivatives, it is possible that they function in plants by affecting the ACh–AChE system. Thus, it may be that the light-induced conformational change in the membrane-bound phytochrome protein is directly coupled to the systems that synthesize and break down acetylcholine. According to this hypothesis, it does not matter just where in the cell the ACh is made, since it then diffuses randomly throughout the cell, unless it is broken down by AChE. By way of this random diffusion, it may then be expected to reach various target sites, such as the nucleus, where it can be seen to cause a release of Ca^{2+}, and the plasmalemma where a "proton pump" may be stimulated. Another target site apparently is the mitochondrion, where the electron energy is diverted to the facilitation of transmembrane movement of monovalent cations. If all of these functions are also involved in the ultimate photormorphogenesis, the ACh system should be coupled to the metabolic changes that are known to occur in phytochrome-mediated systems (see Fig. 1), and I shall be looking for such relationships.

Acetylcholine is a natural metabolite of the bean roots and its mobility and titer is promoted by red light. This, together with the fact that ACh is able to substitute for red light in the long-term photomorphogenic effect on inhibition of rooting, leads to the possibility that ACh is a hormone, new to plants, which couples the photoconversion of the phytochrome with the later metabolic and morphogenic effects that have been so well documented.

ACKNOWLEDGMENTS

I thank Dr. J. Vail and Dr. R. Harris for the electron micrographs, and Dr. H. Toriyama for the cytochemical preparations in the calcium localization study. I am grateful to Miss Linda Thoma for her skilled and devoted technical assistance. The ecothiopate iodide was a gift of the Ayerst Laboratories, New York, New York, and the paraoxon was a gift of the American Cyanimid Co., Princeton, New Jersey. The benzoquinonium chloride was a gift of the Sterling Winthrop Research Institute, Rensselaer, New York. This research was supported by NSF grants GB-7609 and GB-20474.

REFERENCES

Appel, W., and E. Werle. 1959. Nachweis von Histamin, N-Dimethylhistamin, N-Acetylhistamin und Acetylcholin in *Spinacia oleracea. Arzneim. Forsch.* **9**:22–26.

Asai, J., G. A. Blondin, W. J. Vail, and D. E. Green. 1969. The mechanism of mitochondrial swelling: IV. Configurational changes during swelling of beef heart mitochondria. *Arch. Biochem. Biophys.* **132**:524–544.

Attridge, T. H., and H. Smith. 1967. A phytochrome-mediated increase in the level of phenylalanine ammonialyase activity in the terminal buds of *Pisum sativum. Biochim. Biophys. Acta* **148**:805–807.

Bottomley, W., H. Smith, and A. W. Galston. 1966. Flavonoid complexes in *Pisum sativum*. III. The effect of light on the synthesis of kaempferol and quercetin complexes. *Phytochemistry* **5**:117–123.

Dettbarn, W. D. 1962. Acetylcholinesterase activity in *Nitella. Nature (London)* **194**: 1175–1176.

Engelsma, G. 1966. The influence of light of different spectral regions on the synthesis of phenolic compounds in gherkin seedlings in relation to Photomorphogenesis. III. Hydroxylation of cinnamic acid. *Acta Bot. Neer.* **15**:394–405.

Florey, E. 1967. The clam-heart bioassay for acetylcholine. *Comp. Biochem. Physiol.* **20**:365–377.

Fondéville, J. C., M. J. Schneider, H. A. Borthwick, and S. B. Hendricks. 1967. Photocontrol of *Mimosa pudica* L. leaf movement. *Planta* **75**:228–238.

Forward, R. B. 1970. Change in the photoresponse action spectrum of the dinoflagellate Gyrodinium dorsum Kofoid by red and far-red light. *Planta* **92**:248–258.

Fowler, K. S., and S. E. Lewis. 1958. The extraction of acetylcholine from frozen insect tissue. *J. Physiol. (London)* **142**:165–172.

Furuya, M., and W. S. Hillman. 1964. Observations on spectrophotometrically assayable phytochrome *in vivo* in etiolated *Pisum* seedlings. *Planta* **63**:31–42.

Galston, A. W. 1968. Microspectrophotometric evidence for phytochrome in plant nuclei. *Proc. Nat. Acad. Sci. U.S.* **61**:454–460.

Goren, R., and A. W. Galston. 1966. Control by phytochrome of ^{14}C-sucrose incorporation into buds of etiolated pea seedlings. *Plant Physiol.* **41**:1055–1064.

Green, D., J. Asai, R. Harris, and J. Penniston. 1968. Conformational changes in the mitochondrial inner membrane induced by changes in functional states. *Arch. Biochem. Biophys.* **125**:684–705.

Greenberg, M. J. 1965. A compendium of responses of bivalve hearts to acetylcholine. *Comp. Biochem. Physiol.* **14**:513–539.

Harris, R. A., C. H. Williams, M. Caldwell, D. E. Green, and E. Valdivia. 1969. Energized configurations of heart mitochondria *in situ*. *Science* **165**:700–703.

Hendricks, S. B., and H. A. Borthwick. 1967. The function of phytochrome in regulation of plant growth. *Proc. Nat. Acad. Sci. U.S.* **58**:2125–2130.

Hestrin, S. 1949. The reaction of acetylcholine and other carboxylic acid derivatives with hydroxylamine, and its analytical application. *J. Biol. Chem.* **180**:249–261.

Hillman, W. S., and W. L. Koukkari. 1967. Phytochrome effects in the nyctinastic leaf movements of *Albizzia julibrissin* and some other legumes. *Plant Physiol.* **42**:1413–1418.

Hopkins, D. W., and W. L. Butler. 1970. Immunochemical and spectroscopic evidence for protein conformational changes in phytochrome transformations. *Plant Physiol.* **45**:567–570.

Jaffe, M. J. 1968. Phytochrome-mediated bioelectric potentials in mung bean seedlings. *Science* **162**:1016–1017.

Jaffe, M. J. 1969. Phytochrome controlled RNA changes in terminal buds of dark grown peas (*Pisum sativum* cv. Alaska). *Physiol. Plant.* **22**:1033–1037.

Jaffe, M. J. 1970. Evidence for the regulation of phytochrome mediated processes in bean roots by the neurohumor, acetylcholine. *Plant Physiol.* **46**:768–777.

Jaffe, M. J., and A. W. Galston. 1967. Phytochrome control of rapid nyctinastic movements and membrane permeability in *Albizzia julibrissin*. *Planta* **77**:135–141.

Kohler, D., K. V. Willert, and U. Luttge. 1968. Phytochromabhangige Veranderungen des Wachstums und der Ionenaufnahme etiolierter Erbsenkeimlinge. *Planta* **83**:35–48.

Lehninger, A. L. 1967. Cell organelles: The mitochondrion. *In* "The Neurosciences: A Study Program" (G. C. Quarton, T. Melnechuk, and F. O. Schmitt, eds.), pp. 91–100. Rockefeller Univ. Press, New York.

Lin, R. C. Y. 1957. Distribution of acetylcholine in the Malayan Jack-fruit plant *Artocarpus integra*. *Brit. J. Pharmacol. Chemother.* **12**:265–269.

MacIntosh, F. C., and W. L. M. Perry. 1950. Biological estimation of acetylcholine. *Methods Med. Res.* **3**:78–94.

Marquardt, P., and H. Falk. 1957. Vorkommen und Synthese von Acetylcholin in Pflanzen und Bakterien. *Arzneim.-Forsch.* **7**:203–211.

Mohr, H., C. Holderied, W. Link, and K. Roth. 1967. Protein- und RNS-Gehalt des

Hypokotyle beim stationaren Wachstum im Dunkeln und unter dem Einfluss von Phytochrom (Keimlinge von Sinapis alba L.). *Planta* **76**:348–358.

Nachmansohn, D. 1970. Proteins in excitable membranes. *Science* **168**:1059–1066.

Newhall, W. F. 1971. Effect of phenyl substituents in benzyl quaternary ammonium derivatives of (+)-limonene on plant growth-retardant activity. *J. Agr. Food Chem.* **19**:294–297.

Rasmussen, H. 1970. Cell communication, calcium ion, and cyclic adenosine monophosphate. *Science* **170**:404–412.

Satter, R. L., and A. W. Galston. 1970. Phytochrome controlled nyctinasty in *Albizzia julibrissin*. II. K^+ flux as a basis for leaflet closure. *Plant Physiol.* **46**:2. (Abstr. Suppl.)

Saxena, P. R., K. K. Tangri, and K. P. Bhargava. 1966. Identification of acetylcholine, histamine, and 5-hydroxytryptamine in *Girardinia heterophylla*. *Can. J. Physiol. Pharmacol.* **44**:621–627.

Schwartz, O. 1967. The preliminary characterization of cholinesterase activity from etiolated pea seedlings. 45 pp. M.S. Thesis, North Carolina State Univ., Raleigh, North Carolina.

Setty, S. L., and M. J. Jaffe. 1970. Phytochrome controlled contraction and recovery of contractile vacuoles in the motor cells of *Mimosa pudica* L. *Amer. J. Bot.* **57**:762. (Abstr. Suppl.)

Sibaoka, T. 1966. Action potentials in plant organs. *Symp. Soc. Exp. Biol.* **20**:49–74.

Steiner, A. M. 1968. Rash ablaufende Anderunsen im Gehalt an loslichen Zuckern und Zellwandkohlenhydraten beider phytochrominduzierten Photomorphogenese des Senfkeimlings (*Sinapis alba* L.). *Planta* **82**:223–234.

Tanada, T. 1968a. A rapid photoreversible response of barley root tips in the presence of 3-indole-acetic acid. *Proc. Nat. Acad. Sci. U.S.* **59**:376–380.

Tanada, T. 1968b. Substances essential for a red, far-red light reversible attachment of mung bean root tips to glass. *Plant Physiol.* **43**:2070–2071.

Toriyama, H. 1955. Observational and experimental studies of sensitive plants. VI. The migration of potassium in the primary pulvinus. *Cytologia* **20**:367–377.

Toriyama, H., and M. J. Jaffe. 1971. The migration of calcium and its role in the regulation of seismonasty in the motor cell of *Mimosa pudica* L. *Plant Physiol.* In press.

Tulus, R., A. Uleubelen, and F. Ozer. 1961. Choline and acetylcholine in the leaves of *Digitalis ferruginia*. *Arch. Pharm. (Weinheim)* **294**:11–17.

Umrath, K. 1934. Uber die elektrischen Erscheinungen bei thigmischer Reizung der Ranken von *Cucumis melo*. *Planta* **23**:47–50.

Vdovichenko, L. M. 1965. Effect of acetylcholine on swelling and respiration in liver mitochondria. *Tsitologiya* **7**:756–759.

Vdovichenko, L. M. 1966. Acetylcholine and oxidative processes in mitochondria. *Ukr. Biokhim. Z.* **38**(2):185–193.

Vdovichenko, L. M., and N. N. Demin. 1965. Acetylcholine and the respiration of mitochondria of brain cells. *Dokl. Akad. Nauk SSSR* **162**:1434–1436.

Yunghans, H. O. 1970. A study of the early bioenergetic events of phytochrome mediated photomorphogenesis. M.S. Thesis, Ohio Univ., Athens, Ohio.

Yunghans, H. O., and M. J. Jaffe. 1970. Phytochrome controlled adhesion of mung bean root tips to glass: A detailed characterization of the phenomenon. *Physiol. Plant.* **23**:1004–1016.

Yunghans, H. O., and M. J. Jaffe. 1972. Rapid respiratory changes due to red light or acetylcholine during the early events of phytochrome mediated photomorphogenesis. *Plant Physiol.* In press.

THE BETALAINS: STRUCTURE, FUNCTION, AND BIOGENESIS, AND THE PLANT ORDER CENTROSPERMAE*

TOM J. MABRY, LINDA KIMLER, and CHRISTINA CHANG

*The Cell Research Institute and Department of Botany,
The University of Texas at Austin, Austin, Texas*

Introduction.	106
Summary of Reports Pertaining to the Betalains (Tables 1–7)	107
DNA-DNA and DNA-RNA Hybridization Studies in the Centrospermae.	107
Present and Proposed Research on the Betalains and the Plant Order Centrospermae.	115
Structure Determinations and Distributional Studies.	115
Biogenetic Studies.	121
Total Syntheses of Betalains.	130
Genetics of Betalain Biogenesis.	130
The Function of Betalains.	131
Ultrastructural Research on Sieve-Tube Plastids in the Centrospermae.	131
References.	132

---

* Some of the research described herein was supported by the National Institutes of Health (Grant HD-04488), the National Science Foundation (Grant GB-29576X), and the Potts and Sibley Foundation.

Introduction

The betalains* represent a class of red and yellow alkaloids which have long intrigued not only natural products chemists, but plant systematists as well. The early research performed on the betalains was reviewed by Dreiding (1961) and Mabry (1964); in addition, several chapters (Mabry, 1966, 1970; Mabry and Dreiding, 1968) have dealt with the structural, systematic, and biogenetic investigations carried out on the betalains in

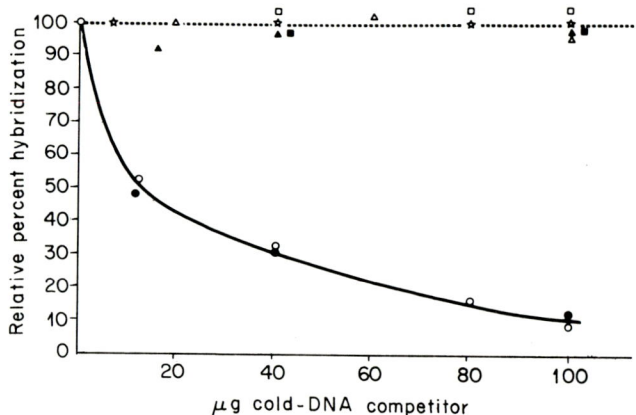

Fig. 1. The data show the extent of competition in the hybridization reaction of ^{32}P-labeled, sheared sugar beet DNA with long-stranded sugar beet DNA filters in the presence of increasing amounts of unlabeled DNA competitor. In each experiment, two filters each containing 10 μg of bound sugar beet DNA (long-stranded) were incubated in 0.175 ml of 1 SSC containing 2 μg of sheared ^{32}P-labeled sugar beet DNA (2450 cpm/μg) and increasing amounts of various sheared, unlabeled DNA competitors. After 14 hours of incubation at 60°C, the filters were washed twice with 5 ml of 1 SSC for 5 minutes: 6.4% of the input ^{32}P sugar beet DNA bound to filters (thus, the value represents 100% hybridization in this figure) in the presence or in the absence of competitor DNA in all experiments with the exception of the experiments that employed as competitors unlabeled sheared DNA from the sugar and red beet varieties of *Beta vulgaris*. All values have been corrected for background binding using calf thymus DNA filters.

Source of competitor DNA: Betalain-producing species—○, *Beta vulgaris* var. sugar beet (Chenopodiaceae); ●, *Beta vulgaris* var. red beet (Chenopodiaceae); □, *Chenopodium* sp. (Chenopodiaceae); △, *Celosia argentea* (Amaranthaceae); ▲, *Alternanthera crassa* (Amaranthaceae). Anthocyanin-producing species—☆, *Dianthus caryophyllus* (Caryophyllaceae); ■, *Pisum sativum* var. Alaska (Leguminosae).

* The expression "betalains" was introduced in April, 1966, to include both the red betacyanins and yellow betaxanthins; see Mabry and Dreiding (1968) for the relevant publication.

the past ten years.* Therefore, this review will not discuss previously reviewed material but will present most of these early data in tabular form. Thus in the second section of this chapter, Tables 1–5 (pp. 108–110, 112–114, 116–120) present the known structures for the betacyanins and betaxanthins and reports relating to the biogenesis and synthesis of the betalains. In Table 6 (pp. 122–128), we have summarized the flavonoid reports for the ten betalain-producing families which, according to Mabry et al. (1963), constitute the Order Centrospermae. For comparison, a summary of the flavonoids reported from the closely related anthocyanin-containing families Caryophyllaceae, Molluginaceae, and Polygonaceae is included. Also, the single flavonoid report for the Batidaceae is recorded. This latter family, for which there are as yet no reports of either anthocyanins or betalains, has also been placed in or near the Centrospermae by various systematists.

Table 7 (p. 129) includes recent reports of the occurrence of betalains in Centrospermae plants.

Finally, the last two main sections of this review present comments on a number of research projects pertaining to the betalains which are either currently in progress or are proposed for future study.

Summary of Reports Pertaining to the Betalains

These reports are summarized in tabular form (Tables 1–7).

DNA-DNA and DNA-RNA Hybridization Studies in The Centrospermae†

DNA-DNA and DNA-RNA hybridization studies have been successfully employed for elucidating the evolutionary relationships among animals (see, e.g., Bolton et al., 1964) and a few plant taxa (Bendich and Bolton, 1967; Bendich and McCarthy, 1970). We have now explored the application of these techniques for determining the extent of genetic homology among species that belong to various betalain-producing plant families of the Centrospermae and species that are members of the anthocyanin-producing but otherwise Centrospermae-like family, the Caryophyllaceae.

---

* Although the work was not published until later (Mabry et al., 1962), evidence that the betalains were not structurally related to the common plant pigments, the anthocyanins, was available in February, 1961.

† This is the first report of these investigations, which were carried out in our laboratory by Christina Chang; the data will be published in detail later in collaboration with Dr. Charles Laird, Department of Zoology, The University of Texas at Austin, Austin, Texas.

TABLE 1
The Betacyanins

Compound	Compound	Substitution pattern for both series	References
Betanidin	Isobetanidin	R, R' = H	Wyler et al. (1963; see also Mabry et al., 1962, 1967)
2-Decarboxybetanidin	—	R, R' = H; compound is missing the COOH at C-2	Piattelli and Impellizzeri (1970)
Betanin	Isobetanin	R = β-D-glucosyl; R' = H	Piattelli et al. (1964a); Wilcox et al. (1965)
Amarantin	Isoamarantin	R = 2'-O-(β-D-glucosyluronic acid)-β-D-glucosyl; R' = H	Piattelli et al. (1964b)
Prebetanin	Isoprebetanin	R = 6'-O-(SO$_3$H)-β-D-glucosyl; R' = H	Wyler et al. (1967)
Phyllocactin	Isophyllocactin	R = 6'-O-(malonyl)-β-D-glucosyl; R' = H	Minale et al. (1966)

Name	Structure	Reference
Celosianin	R = 2'-O-[O-(trans-p-feruloyl)-β-D-glucosyluronic acid]-O-trans-p-coumaroyl)-β-D-glucosyl; R' = H	Minale et al. (1966)
Isocelosianin		
Iresinin-I	R = 2'-O-(β-D-glucosyluronic acid)-6'-O-(3-hydroxy-3-methylglutaryl)-β-D-glucosyl; R' = H	Minale et al. (1966)
Isoiresinin-I		
Lampranthin-I	R = (feruloyl-p-coumaroyl-)-β-D-glucosyl; R' = H	Piattelli and Impellizzeri (1969)
Isolampranthin-I		
Lampranthin-II	R = O-(diferuloyl-p-coumaroyl-β-D-glucosyl; R' = H	Piattelli and Impellizzeri (1969)
Isolampranthin-II		
Bougainvillein-r-I	R = β-sophoroside; R' = H	Piattelli and Imperato (1970a)
Isobougainvillein-r-I		
Gomphrenin-I	R = H; R' = β-D-glucosyl	Minale et al. (1967)
Gomphrenin-II		
Gomphrenin-III[a]	R = H; R' = 6'-O-(cis-p-coumaroyl)-β-D-glucosyl	Minale et al. (1967)
Gomphrenin-V	R = H; R' = 6'-O-(trans-p-feruloyl)-β-D-glucosyl	Minale et al. (1967)
Gomphrenin-VI		
Bougainvillein-V (DP3)	R = H; R' = β-sophoroside plus acyl groups presumably attached to R'	Piattelli and Imperato (1970b)
Isobougainvillein-V (DP4)		
Bougainvillein-V (DP1)	R = H; R' = 6'-O-rhamnosyl-β-sophoroside plus acyl groups presumably attached to R'	Piattelli and Imperato (1970b)
Isobougainvillein-V (DP4)		

[a] It was not established unambiguously that gomphrenin-III was a member of the betanidin series of betacyanins; the material used in the investigation may have been a mixture of the C_{15} epimers (Minale et al., 1967).

The methods employed in our investigations are modifications of those previously described (Bonner et al., 1967; Laird and McCarthy, 1969; Gillespie and Spiegelman, 1965). In general, the experiments are carried out as follows: single-stranded high molecular weight DNA is immobilized on nitrocellulose filters which are then incubated in a standard saline citrate (SSC) solution containing homologous (from the same plant) but sheared and labeled single-stranded DNA or labeled 16 S rRNA. The

TABLE 2

THE BETAXANTHINS

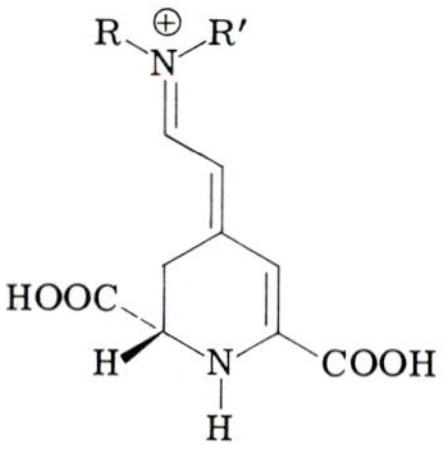

Compound	Substitution pattern	References
Indicaxanthin	R, R' together form a proline group	Piattelli et al. (1964c)
Vulgaxanthin-I	R = H; R' = $-\text{CHCH}_2\text{CH}_2\text{COOH}$ with COOH on first C (glutamic acid group)	Piattelli et al. (1965a)
Vulgaxanthin-II	R = H; R' = $\text{CHCH}_2\text{CH}_2\text{CNH}_2$ with COOH on first C and =O on last C (glutamine group)	Piattelli et al. (1965a)
Miraxanthin-I	R = H; R' = $\text{CHCH}_2\text{CH}_2\text{SCH}_3$ with COOH on first C and =O on S (methionine sulfoxide group)	Piattelli et al. (1965b)
Miraxanthin-II	R = H; R' = CHCH_2COOH with COOH on first C (aspartic acid group)	Piattelli et al. (1965b)
Miraxanthin-III	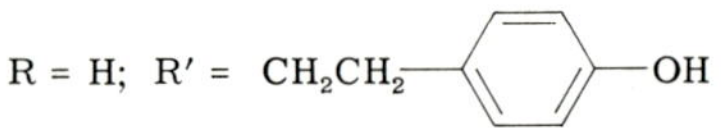 R = H; R' = CH_2CH_2–C$_6$H$_4$–OH (Tyramine group)	Piattelli et al. (1965b)
Miraxanthin-V	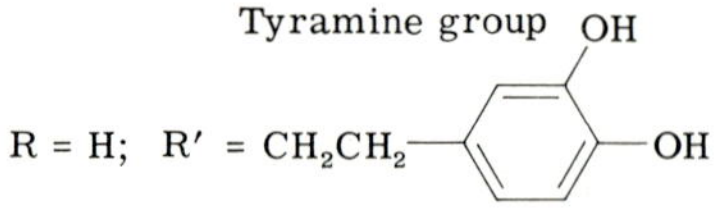 R = H; R' = CH_2CH_2–C$_6$H$_3$(OH)–OH (Dopamine group)	Piattelli et al. (1965b)

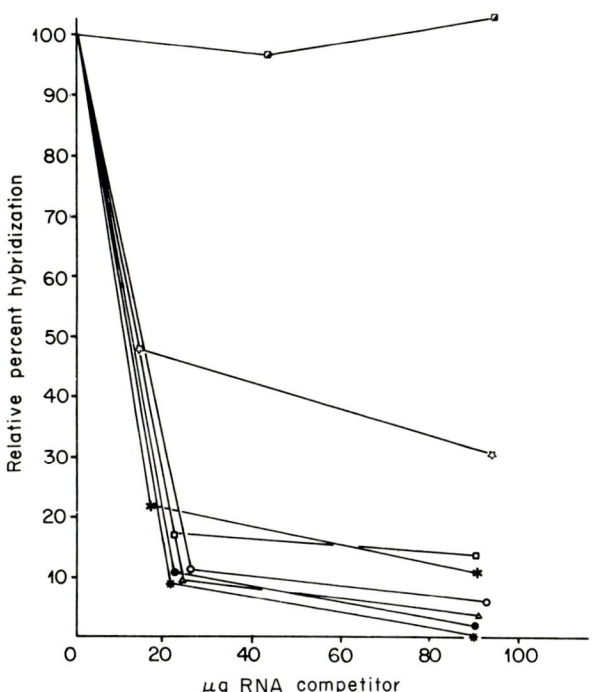

Fig. 2. In all experiments, 0.6 μg of ³H-labeled spinach 16 S rRNA (4000 cpm/μg) were incubated with 12 μg of spinach DNA bound on a nitrocellulose filter in 0.1 ml of 2 SSC containing 50% formamide at 40°C, in the presence of increasing amounts of rRNA from other taxa. After 38 hours, each filter was washed with 2 ml of 50% formamide and then 2 SSC for 5 minutes at 40°C then 2 SSC for another 5 minutes. Each point in this figure presents the average value obtained from three determinations. The ratio of labeled rRNA to DNA in the hydrid in the absence of competitor rRNA was 0.33%, representing 7% binding of the input labeled rRNA. All values have been corrected for background binding using calf thymus DNA filters.

Source of competitor rRNA: Betalain-producing species—★, *Spinacia oleracea* (Chenopodiaceae); ●, *Beta vulgaris* var. sugar beet (Chenopodiaceae); △, *Alternanthera crassa* (Amaranthaceae); ○, *Rivina humilis* (Phytolaccaceae). Anthocyanin-producing species—□, *Stellaria sp.* (Caryophyllaceae); ✻, *Dianthus caryophyllus* (Caryophyllaceae); ☆, *Pisum sativum* var. Alaska (Leguminosae); ◢, Yeast (Saccharomycetaceae).

incubation is carried out in the presence of increasing amounts of a short single-stranded nucleic acid selected as a competitor. If the unlabeled short-stranded competitor nucleic acid (DNA or rRNA) contains similar base sequences to those which occur in the DNA bound on the filter, then the competitor strands compete with the labeled short strands or homologous nucleic acid for hybridization sites on the filter-bound DNA.

The ³²P-labeled DNA is obtained from seedlings grown in the presence

TABLE 3: Reports Relating to the Biogenesis of the Betalains

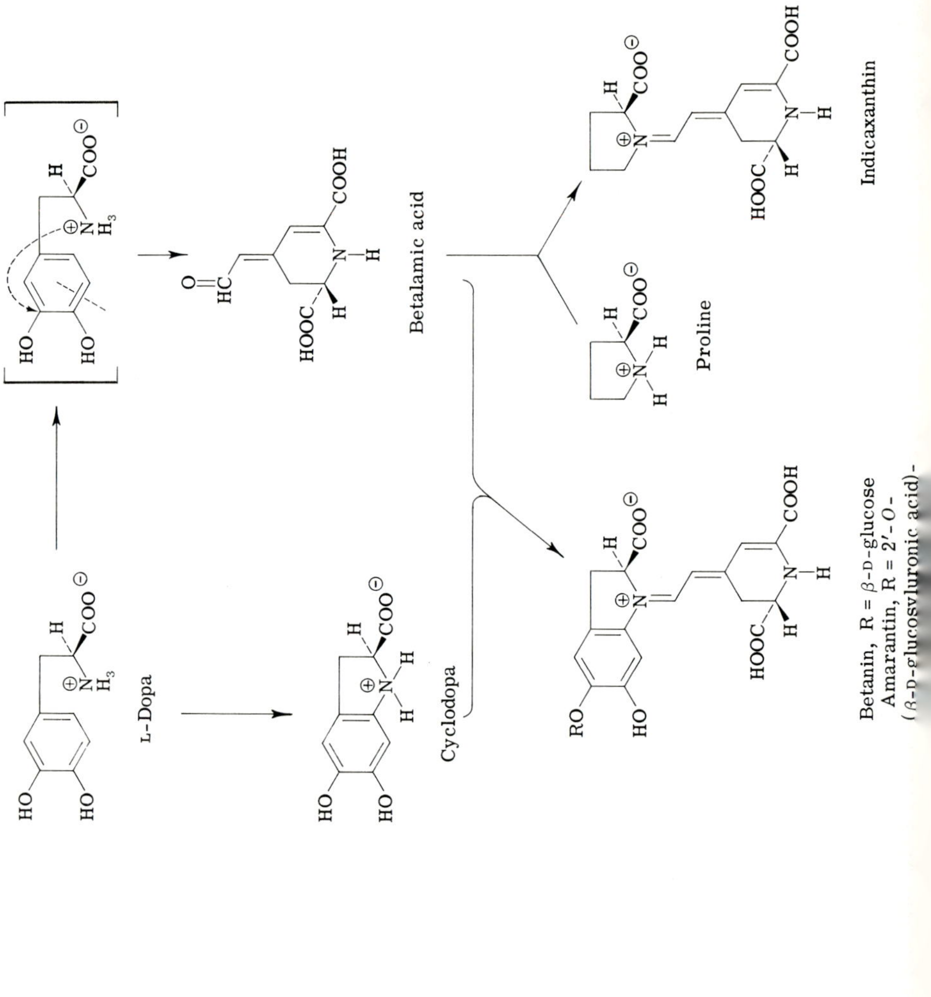

Species	Substrate	Results	Reference
Beta vulgaris	DL-Dopa-2-^{14}C	Pathway from dopa to the betalains first proposed	Wyler et al. (1963); see also Mabry, 1964)
	DL-Dopa-2-^{14}C	Beet seedlings grown on a medium containing DL-dopa-2-^{14}C were reported to synthesize labeled betanin; however, the conclusion was based on the radioactivity measured for small quantities of noncrystalline electrophoretically purified betanin. It is now known that radioactive impurities accompany labeled betanin during electrophoresis	Hörhammer et al. (1964)
Opuntia ficus-indica	Proline-^{14}C (randomly labeled); DL-dopa-2-^{14}C	When the homogenized pulp of Opuntia ficus-indica fruits was incubated with L-proline-^{14}C, the indicaxanthin which was produced contained label only in the proline moiety. A similar experiment using DL-dopa-2-^{14}C produced labeled betanin and labeled indicaxanthin. The results supported the scheme shown above but were inconclusive since a labeled compound derived from the dihydropyridine ring of the betalains was not isolated and the experiments were performed in vitro	Minale et al. (1965)
Amaranthus, Molten Fire	Dopa-U-^{14}C; tyrosine-U-^{14}C; L-phenylalanine-U-^{14}C	L-Tyrosine was shown to be a good precursor of amarantin; however, the conclusion was based upon noncrystalline material that had been purified by paper chromatography	Garay and Towers (1965)
Opuntia decumbens Opuntia bergeriana	DL-Dopa-1-^{14}C; DL-Dopa-2-^{14}C	Dopa was incorporated into betanin in high yield (more than 10% in some cases) using fruits on Opuntia plants; the betanin was obtained radiopure after several purification steps including crys-	Miller et al. (1968)

(Continued)

TABLE 3—Continued
Reports Relating to the Biogenesis of the Betalains

Species	Substrate	Results	Reference
Beta vulgaris	DL-Dopa-1-^{14}C; DL-dopa-2-^{14}C; DL-tyrosine-1-^{14}C	tallization. Reaction of the radiopure betanin with proline afforded indicaxanthin containing more than 90% of the radioactivity. Thus, in this system dopa is a more efficient precursor for the dihydropyridine moiety than for the cyclodopa group. Decarboxylation studies established that the ^{14}C had been incorporated without randomization. Under the same conditions, mevalonic acid, aspartic acid, and phenylalanine showed low, random incorporation. These data established conclusively for the first time that dopa serves as a precursor for the dihydropyridine moiety of the betalains	Miller et. al. (1968)
Beta vulgaris	Tyrosine-1-^{14}C-^{15}N; tyrosine-2-^{14}C-^{15}N; phenylalanine-^{14}C	Results similar to those for the Opuntia fruits were obtained; however, the overall incorporation was considerably less, and the relative percent of incorporation into the cyclodopa moiety was higher Labeled tyrosine was incorporated intact into both the cyclodopa and dihydropyridine portions of betanin (2/1 ratio). Phenylalanine was not incorporated	Liebisch et al. (1969)

of phosphate-^{32}P, and the tritium-labeled 16 S rRNA is obtained by a controlled *in vitro* methylation of the total RNA which is extracted from the plant; the tritium-labeled 16 S and 25 S components of the rRNA are then separated out by sucrose gradient ultracentrifugation. The methylation is carried out with ^{3}H-labeled dimethyl sulfate.

The DNA-DNA hybridization results shown in Fig. 1 (p. 106) are somewhat surprising in that only among varieties of the same species (*Beta vulgaris*, the red and sugar beets) could competition be observed. Therefore, the DNA-DNA hybridization at these reaction conditions apparently cannot be used to distinguish even genera in the Centrospermae, and thus certainly not families. As a result of these DNA-DNA studies, we turned our attention to the possibility of using ribosomal RNA for hybridization with the DNA since it is well known that the cistrons for rRNA are relatively conservative. The preliminary results for these DNA-RNA hybridization studies are summarized in Fig. 2: Excess rRNA from the distantly related yeast did not prevent the labeled 16 S rRNA from the betalain-producing spinach (*Spinacia oleracea*, Family Chenopodiaceae) from hybridizing with the filter-bound DNA from the same plant (top curve in Fig. 2, p. 111); on the other hand, excess rRNA from the still somewhat distantly related species, *Pisum sativum*, a member of the anthocyanin-producing Leguminosae, reduced the degree of hybridization between spinach DNA and spinach rRNA to 30 percent (second curve from the top, Fig. 2). It was of particular interest to observe that excess rRNA from either of two taxa (*Stellaria* sp., *Dianthus caryophyllus*) from the anthocyanin-producing Caryophyllaceae reduced the spinach-DNA to spinach rRNA hybridization to about 10-15 percent. Finally, rRNA from other betalain-producing families reduced the hybridization of spinach DNA to spinach rRNA to less than 5 percent. Thus, these preliminary results support a close phylogenetic relationship of the betalain-producing families and indicate that the Caryophyllaceae while probably distinct is nevertheless evolutionarily close to the Order Centrospermae.

Present and Proposed Research on the Betalains and the Plant Order Centrospermae

Structure Determinations and Distributional Studies

In the future, new structural reports for naturally occurring betalains will, for the most part, result directly from the increasing interest in distributional aspects of these pigments within the Centrospermae. In order to further establish the reliability of the betalains as criteria for circumscribing the Centrospermae, more extensive surveys for these plant pigments are required; and in the course of these phylogenetically oriented

TABLE 4
Reports Relating to the Photoinduction of Betalain Biogenesis

Species	Results	References
Amaranthus, Molten Fire; *Amaranthus*, Love Lies Bleeding; *Amaranthus*, Joseph's Coat; *Amaranthus caudatus*; *Amaranthus paniculatus*	No betalain synthesis was observed when seedlings of the listed species and varieties were grown in the dark; however, when the seedlings were placed in the light for a short time and then returned to darkness, amarantin was synthesized in the dark	Garay and Towers (1965)
Amaranthus tricolor	1. In continuous darkness, *A. tricolor* seedlings produced only a small amount of amarantin. When seedlings were irradiated for 2 hours with fluorescent light and returned to darkness, amarantin synthesis increased after 4 hours. 2. Photoinduction of amarantin synthesis could be reversed by exposure of the seedlings to far-red light (>700/nm) for 5 minutes, but could be reinitiated by exposure of the seedlings to red light (600–700 nm). These results are typical for the phytochrome response and are similar to those reported for anthocyanin biogenesis in many non-Centrospermae plants. 3. Two photo reactions were found to be involved in betalain biogenesis; however, only the low energy one was dependent upon the phytochrome system. Phytochrome (P_{fr}) which promotes betalain biogenesis can be converted to phytochrome (P_r) if irradiated with far-red light; the latter system inhibits betalain biogenesis; a 3–5 minute irradiation period was found to be sufficient to cause this reversible change. The high energy reaction was not di-	Piattelli *et al.* (1969)

Species	Notes	Reference
	rectly dependent upon the phytochrome system; instead, a different photoreceptor appeared to promote betalain biogenesis under conditions involving exposure of the seedlings to either red or far-red light for long periods of time	
Amaranthus, Molten Fire; Amaranthus caudatus L.; Beta vulgaris; Celosia, Cristate nana; Celosia, Golden Feather; Celosia, Pride of Castle Gould; Celosia, Flaming Fire; Chenopodium nuttallae; Gomphrena globosa; Mirabilis, Four O'Clocks; Portulaca (Double Flowered Pink); Portulaca (Double Flowered Yellow)	All species examined with the exception of Amaranthus, Molten Fire, Celosia, Golden Feather, and Mirabilis, Four O'Clock produced betalains when the seeds germinated in the absence of light. However, under the conditions of the experiments the amount of betalains synthesized in each case was always less when the seedlings were grown in the dark relative to seedlings grown under the same conditions except exposed to continuous light	Wohlpart and Mabry (1968a)
Amaranthus caudatus Chenopodium rubrum	Photoinduction was required for betalain biogenesis 1. Betalain biogenesis was shown to be under phytochrome control involving both low- and high-energy reactions. 2. Rhythmicity in rate of betalain synthesis was shown to be circadian	Liebisch et al. (1969) Wagner and Cummings (1970)
Amaranthus tricolor	The photoreaction response for the synthesis of amarantin was reduced considerably by actinomycin D (an inhibitor of DNA-dependent RNA polymerase) and puromycin (an inhibitor of protein synthesis). The authors concluded that the synthesis of betalains is controlled by P 730 (phytochrome) as a result of the activation of potentially active genes	Piattelli et al. (1969)

TABLE 5
Reports Relating to the Synthesis of the Betalains

Structure synthesized	Comments	References
I R = H; R′ = CH$_3$ II R = CH$_3$; R′ = H (indoline with RO, R′O substituents and N-H, COOH)	I and II were prepared from nitrobenzene derivatives in connection with the determination of the position of glucose in betanin	Piattelli et al. (1964a)
III R = CH$_3$ IV R = C$_2$H$_5$ V R = CH$_3$ VI R = C$_2$H$_5$ (N-alkyl styryl pyridines)	III, IV, V, and VI were synthesized by the condensation of N-alkyl-N-formylindolines with γ-picoline or phenylacetaldehyde	Omote et al. (1967)
VII R = H VIII R = CH$_3$ (dihydroxyindoline-COOR)	Cyclodopa (VIIa) and the methyl ester (VIII) derivatives were prepared by oxidation of dopa methyl ester to dopachrome methyl ester, which was then reduced	Wyler and Chiovini (1968)

Neobetenamine (IX) was synthesized by condensation of *N*-formylindoline with γ-picoline and also by exchange between indoline and the product of γ-picoline and *N*-methylformanilide. 5,6-Di-*O*-methylneobetanidin trimethyl ester (XII) was synthesized from an exchange reaction of (S)-cyclodopa methyl ester (X) with compound XI. The synthesis of XI was also reported. The synthetic pentamethylneobetanidin (XII) was identical with authentic material obtained by diazomethane treatment of betanidin (see Mabry *et al.*, 1967)

Badgett *et al.* (1970; see also Parikh, 1966)

TABLE 5—Continued

REPORTS RELATING TO THE SYNTHESIS OF THE BETALAINS

Structure synthesized	Comments	References
HO─⟨benzene ring fused to pyrrolidine⟩─N─H, HO─ XIII	5,6-Dihydroxy-2,3-dihydroindole (X) was prepared by the oxidative cyclization of 3-hydroxytyramine	Piattelli and Impellizzeri (1970)

studies, the number of fully characterized betalains can be expected to increase markedly. Nevertheless, new betacyanins will probably represent minor modifications of the betanidin and isobetanidin skeletons and new betaxanthins will probably be derived from the common free amino acids and related amines which occur in plants belonging to the Centrospermae.

The pigment content and evolutionary relationship of the following plant families, which have been placed at one time or another in or near the ten betalain-producing families in the Centrospermae, (Willis, 1966), remain to be determined:

1. Achatocarpaceae, Agdestidaceae (both found in the tropical Americas), and Gyrostemonaceae (endemic to Australia) are taxa that remain to be examined for their pigments; all these taxa have been raised to familial level by some systematists although they are also frequently placed in the betalain-producing Phytolaccaceae.

2. The Batidaceae is often placed in the Centrospermae, and while neither betalains nor anthocyanins were detected in *Batis maritima* (one of the two species of this family) Van Royen, who described a second species *B. argillicola* (from New Guinea), has privately informed us that the plant contained purple pigmentation; however, we have as yet been unable to obtain any pigmented samples for chemical analysis.*

3. Halophytaceae, which has some of the morphological features of Chenopodiaceae, Phytolaccaceae, Basellaceae, and Batidaceae, contains the single genus *Halophytum* and occurs in temperate South America.

4. Sphenocleaceae is a tropical monogeneric (*Sphenoclea*) family often placed near the Phytolaccaceae.

5. Theligonaceae contains only one genus, *Theligonum*, and is found in the Mediterranean, China, and Japan; the group has been allied with Portulacaceae.

6. Dysphaniaceae is an Australian monogeneric family (*Dysphania*); it is of special interest because it is considered by some to be an intermediate taxon between the Chenopodiaceae and Caryophyllaceae.

7. Barbeuiaceae, which contains the single genus *Barbeuia*, occurs in Madagascar and remains to be examined for the presence or absence of betalains.

Biogenetic Studies

Investigations of the enzymology and metabolism of betalain synthesis and utilization are presently being initiated in a number of laboratories. These studies represent a logical consequence of our present state of knowl-

* We examined the voucher specimens of this taxa which Van Royen deposited in the Ryksherbarium, Leiden, Netherlands, but did not observe any purple pigmentation.

TABLE 6
FLAVONOID COMPOUNDS IN FAMILIES BELONGING TO OR CLOSELY ALLIED WITH THE CENTROSPERMAE

Species	Flavonoid	Reference
Aizoaceae		
Aptenia cordifolia Schwantes	Quercetin	Bate-Smith (1962)
	Kaempferol	Bate-Smith (1962)
Carpobrotus acinaciformis L. Bolus	Leucocyanidin	Bate-Smith (1954)
C. edulis (L.) N.E. Br.	Leucocyanidin	Kimler *et al.* (1970)
Conophytum (37 species)	Kaempferol glycosides	Reznik (1957)
Faucaria duncani	Kaempferol glycosides	Reznik (1957)
Fenestraria aurantiaca N.E. Br.	Kaempferol glycosides	Reznik (1957)
Gibbaeum gibbosum	Kaempferol glycosides	Reznik (1957)
Lampranthus zeyheri	Kaempferol glycosides	Reznik (1957)
Lithops kuibisensis	Kaempferol glycosides	Reznik (1957)
Maliphora mollis (W. Aiton) N.E. Br.	Kaempferol glycosides	Reznik (1957)
Pleiospilos herrei	Kaempferol glycosides	Reznik (1957)
Rhombophyllum dolabriforme	Kaempferol glycosides	Reznik (1957)
Ruschia nonimpressa L. Bolus	Quercetin	Bate-Smith (1962)
	Kaempferol	Bate-Smith (1962)
	A leucoanthocyanin	Bate-Smith (1962)
Amaranthaceae		
Amaranthus blitum L.	Quercetin	Bate-Smith (1962)
A. caudatus L.	Quercetin 3-rutinoside	Reznik (1957)
A. paniculatus L.	Quercetin 3-rutinoside	Reznik (1957)
	Quercetin 3-glycoside	Reznik (1957)
A. retroflexus L.	Quercetin 3-rutinoside	Reznik (1957)
Bosea yervamora L.	Quercetin	Bech (1966)
Celosia thompsonii	Kaempferol 3-glycoside	Bate-Smith (1962)
	Quercetin 3-glycoside	Reznik (1957)
Iresine celosia L.	Tlattlancuayin (an isoflavone)	Crabbé *et al.* (1958)

The Betalains

Basellaceae		
Basella alba L.	Quercetin	Bate-Smith (1962)
Cactaceae		
Ariocarpus retusus Scheidweiler	Quercetin tetramethyl ether (retusine)	Dominguez et al. (1968)
Cactus grandiflora	Isorhamnetin 3-galactoside	Harborne (1967, p. 339)
	Isorhamnetin 3-rutinoside	Harborne (1967, p. 339)
Carnegiea gigantea Britton & Rose	Quercetin	Steelink et al. (1967)
Cereus grandiflorus (L.) Miller	Isorhamnetin 3-glucoside	Harborne (1967, p. 135)
Epiphyllum angulifer G. Don & Dayton	Quercetin	Bate-Smith (1962)
Mammillaria sp.	Quercetin 3-rutinoside	Reznik (1957)
Notocactus scopa Berger	Quercetin 7-glycoside	Reznik (1957)
Opuntia sp.	Isorhamnetin 3-rhamnoglucoside	Harborne (1967, p. 67)
	Isorhamnetin 3-rhamnogalactoside	Harborne (1967, p. 67)
O. dillenii Haw.	Quercetin glucoside	Nair and Subramanian (1961)
	Isorhamnetin glucoside	Nair and Subramanian (1961)
O. elatior Mill.	Isorhamnetin 3-glucoside	Shabbir and Zaman (1968)
O. ficus-indica Mill.	Isorhamnetin	Harborne (1967, p. 135)
	Isorhamnetin glucoside	Hegnauer (1964, p. 333)
O. humifusa Rafin.	Quercetin	Bate-Smith (1962)
	Kaempferol	Bate-Smith (1962)
O. lindheimeri	Quercetin 3-galactoside	Hairborne (1967, p. 135)
	Isorhamnetin 3-galactoside	Harborne (1967, p. 135)
	Isorhamnetin 3-rutinoside	Harborne (1967, p. 135)
	Isorhamnetin 3-rutinosylgalactoside	Harborne (1967, p. 135)
O. vulgaris Mill.	Isorhamnetin glucoside (rutoside)	Paris (1951)
Rebutia sp.	Quercetin 7-rutinoside	Reznik (1957)
Chenopodiaceae		
Atriplex hortensis L.	Quercetin	Bate-Smith (1962)
	Kaempferol	Bate-Smith (1962)
Axyris amaranthoides L.	Quercetin	Bate-Smith (1962)
	Kaempferol	Bate-Smith (1962)

(Continued)

TABLE 6—Continued

Flavonoid Compounds in Families Belonging to or Closely Allied with the Centrospermae

Species	Flavonoid	Reference
Beta vulgaris L.	Vitexin-xyloside	Gardner et al. (1967)
	Quercetin glycoside	Gardner et al. (1967)
	Quercetin	Bate-Smith (1962)
	Kaempferol	Bate-Smith (1962)
Chenopodium amaranticolor Coste & Reyn	Quercetin 3-rutinoside	Reznik (1957)
	Quercetin 3-glycoside	Reznik (1957)
	Kaempferol 3-glycoside	Reznik (1957)
C. bonus-henricus L.	Kaempferol	Bate-Smith (1962)
Hablitzia tamnoides Bieb.	Kaempferol	Bate-Smith (1962)
	Quercetin	Bate-Smith (1962)
Spinacia oleracea L.	Quercetagetin 6,3'-dimethyl ether (spinacetin)	Zane and Wender (1961)
	Quercetagetin-6-methyl ether (patuletin)	Zane and Wender (1961)
	Quercetin 3-rutinoside	Medovar (1968)
	Kaempferol	Bate-Smith (1962)
Suaeda fruiticosa Forsk.	Kaempferol	Bate-Smith (1962)
	Quercetin	Bate-Smith (1962)
Didieraceae	No report given	
Nyctaginaceae		
Bougainvillea glabra Choisy	Quercetin	Bate-Smith (1962)
	Kaempferol	Bate-Smith (1962)
Mirabilis jalapa L.	Quercetin	Bate-Smith (1962)
	Kaempferol	Bate-Smith (1962)
Pisonia cuspidata Himerl.	Quercetin	Bate-Smith (1962)
	Kaempferol	Bate-Smith (1962)
	Leucocyanidin	Bate-Smith (1962)
P. eggersiana Himerl.	Quercetin	Bate-Smith (1962)
	Kaempferol	Bate-Smith (1962)

P. salicifolia Himerl.	Kaempferol	Bate-Smith (1962)
	Leucocyanidin	Bate-Smith (1962)
Phytolaccaceae		
Phytolacca sp.	Isorhamnetin glucoside (rutoside)	Plouvier (1967a,b)
P. clavigera W. W. Smith	Kaempferol	Bate-Smith (1962)
P. decandra L.	Isoquercetin	Hegnauer (1969, p. 307)
	Quercetin 3-glucoside	Hegnauer (1969, p. 307)
	Kaempferol 3-glucoside	Hegnauer (1969, p. 307)
P. diocia L.	Quercetin 3-rutinoside	Hegnauer (1969, p. 307)
	7,4'-Dimethyl-quercetin 3-β-Rutinoside (ombuoside)	Hörhammer et al. (1968)
Rivina humulis L.	Kaempferol	Bate-Smith (1962)
Portulacaceae		
Anacampseros rufescens Sweet	Quercetin 3-rutinoside	Reznik (1957)
Calandrinia grandiflora Lindl.	Quercetin glycoside	Reznik (1957)
	Kaempferol glycoside	Reznik (1957)
Portulacaria afra Jacq.	Quercetin glycoside	Reznik (1957)
	Kaempferol glycoside	Reznik (1957)
	Quercetin	Bate-Smith (1962)
	Kaempferol	Bate-Smith (1962)
	Leucocyanidin	Bate-Smith (1962)
Stegnospermaceae	No report given	

Flavonoid Reports for Some Centrospermae-like Families

Batidaceae		
Batis maritima L.	Isorhamnetin 3-rutinoside	Kagan and Mabry (1969)
Caryophyllaceae[a]		
Agrostemma githago L.	Quercetin	Bate-Smith (1962)
	Kaempferol	Bate-Smith (1962)

[a] Glycoflavonoids have been found in leaves of *Arenaria*, *Minuarta*, *Cerastium*, and *Stellaria* (unpublished results from SiSi, Ph.D. Thesis, Univ. of Liverpool, 1970, communicated to us by J. B. Harborne).

(Continued)

TABLE 6—Continued

FLAVONOID COMPOUNDS IN FAMILIES BELONGING TO OR CLOSELY ALLIED WITH THE CENTROSPERMAE

Species	Flavonoid	Reference
Alsinodendron trinerve H. Mann	Cyanidin	Beck *et al.* (1962)
Arenaria graminiflora	A leucoanthocyanin	Bate-Smith and Lerner (1954)
A. purpurascens	An anthocyanin (cyanidin or delphinidin)	Beck *et al.* (1962)
Corrigiola littoralis L.	Quercetin	Bate-Smith (1962)
	Kaempferol	Bate-Smith (1962)
Dianthus sp.	Pelargonidin 3-glycoside	Harborne (1967, p. 127)
	Cyanidin 3-glycoside	Harborne (1967, p. 127)
	Isosalipurposide (a 2′-glucoside of 4-hydroxyflavone)	Harborne (1967, p. 127)
	Kaempferol 3-glycoside	Reznik (1957)
D. alpinus L.	Quercetin	Bate-Smith (1962)
Gypsophila sp.	Saponaretin 4-β-glucoside	Krivenchuk *et al.* (1968)
G. paniculata L.	Vitexin	Darmograi *et al.* (1968)
	Saponaretin	Darmograi *et al.* (1968)
	Isosaponarin	Darmograi *et al.* (1968)
	Adonivernitol	Darmograi *et al.* (1968)
	Orientin	Darmograi *et al.* (1968, 1969)
	Isoorientin	Darmograi *et al.* (1968, 1969)
Herniaria ciliata Bab., non Clairv.	Quercetin	Bate-Smith (1962)
	Kaempferol	Bate-Smith (1962)
H. glabra L.	Quercetin 3-rutinoside	Hegnauer (1964, p. 388)
	Umbelliferone (a coumarin)	Hegnauer (1964, p. 388)
	Herniarin (a coumarin)	Hegnauer (1964, p. 388)
	Quercetin	Bate-Smith (1962)
	Quercetin arabinoside	Borkowski and Pasich (1958)
	Quercetin galactoside	Borkowski and Pasich (1958)

The Betalains

Genera	Compound reported (Number of species reported from)	
H. hirsuta L.	Narcissin (isorhamnetin rhamnoglucoside)	Hegnauer (1964, p. 388)
	Isorhamnetin 3-rutinosylglucoside	Harborne (1967, p. 135)
	Kaempferol	Bate-Smith (1962)
	Herniarin (a coumarin)	Borkowski and Pasich (1958)
Paronychia kapela A. Kern	Quercetin	Bate-Smith (1962)
	Kaempferol	Bate-Smith (1962)
	Quercetin	Bate-Smith (1962)
	Kaempferol	Bate-Smith (1962)
Saponaria officinalis L.	Saponarin (isovitexin 7-glucoside)	Harborne (1967, p. 145)
	Saponaretin (isovitexin)	Krivenchuk *et al.* (1968)
	Saponaretin 4-β-glucoside	
Silene schafta	Schaftoside (a C-glycoside)	Plouvier (1967a,b)
Spergularia salina J. & C. Presl.	An anthocyanin (probably malvidin)	Beck *et al.* (1962)
Molluginaceae		
Mollugo nudicaulis Lamk.	Flavonol 3-xyloside	Sosa (1959, 1962)

Genera	Compound reported (Number of species reported from)
Polygonaceae[b]	
Atraphaxis	Quercetin glycosides (1); a leucoanthocyanin (1)
Calligonum	Quercetin glycosides (4)
Coccoloba	Quercetin (3); kaempferol (1); cyanidin (3)
Emex	Quercetin 3-rhamnoside (2); quercetin 3-rhamnoglucoside (2)
Fagopyrum	Quercetin (1); quercetin 3-galactoside (3); quercetin 3-rhamnoside (3); quercetin 3-rhamnoglucoside (5); vitexin (1); orientin (1); isoorientin (1); saponarin (1); saponaretin (1); leucocyanidin (2); cyanidin (1)
Oxyria	Quercetin 3-galatoside (2)

[b] For specific references for many flavonoid reports from the Polygonaceae see Hegnauer (1969, pp. 361–382); others are available upon request.

(Continued)

TABLE 6—Continued

FLAVONOID COMPOUNDS IN FAMILIES BELONGING TO OR CLOSELY ALLIED WITH THE CENTROSPERMAE

Genera	Compound reported (Number of species reported from)
Polygonum	Quercetin (7); quercetin 3-glucoside (4); quercetin 3-galactoside (15); quercetin 3-rhamnoside (9); quercetin 3-diglucoside (2); quercetin 3-rhamnoglucoside (2); quercetin 3-xyloside (1); quercetin 3-S-arabinoside (3); quercetin glycosides (5); kaempferol (3); kaempferol glycosides (1); myricetin (1); myricitin glycosides (1); isorhamnetin (1); vitexin (1); orientin (2); isoorientin (1); saponaretin (1); isorhamnetin-3-KSO$_3$ (2); isorhamnetin 3-KSO$_3$ 7-methyl ether (1); leucocyanidin (1); leucodelphindin (1); a leucoanthocyanin (1); cyanidin (1); cyanidin hexoside (5)
Rheum	Quercetin (1); quercetin 3-glucoside (10); quercetin 3-galactoside (16); quercetin 3-rhamnoside (9); quercetin 3-diglucoside (1); quercetin 3-rhamnoglucoside (18); leucocyanidin (2); cyanidin 3-glucoside (2); cyanidin 3-rutinoside (1); cyanidin 3,5-diglucoside (1)
Rumex	Quercetin (1); quercetin 3-galactoside (17); quercetin 3-rhamnoside (15); quercetin 3-rhamnoglucoside (16); quercetin rhamnoside (1); quercetin rhamnoglucoside (1); quercetin galactoside (1); quercetin 3-arabinogalactoside (1); quercetin glycosides (1); vitexin (1); leucocyanidin (2); leucodelphinidin (1); a leucoanthocyanin (3); cyanidin (2)

TABLE 7

NEW REPORTS OF BETACYANINS IN CENTROSPERMAE PLANTS[a]

Amaranthaceae
  Acnida tamariscina[3][b]
  Amaranthus fimbriatus,[3] *A. tuberculatus*[4]
  Chamissoa altissima[1]
  Froelichia floridana,[3] *F. gracilis*[3]
  Gomphrena officinalis,[3] *G. nitida*,[3] *G. dispersa*[4]
  Iresine celosia,[3] *I. schaffneri*[4]
Cactaceae
  Aporocactus flagelliformis[2]
  Borzicactus sepium[2]
  Cereus comarapanus,[2] *C. stenogonus*[2]
  Cleistocactus parviflorus,[2] *C. smaragdiflorus*,[2] *C. strauski*[2]
  Epicactus, cultivated varieties[2]
  Eriocereus guelichii[2]
  Haageocereus acranthus[2]
  Mammillaria centricirrha,[2] *M. magnimamma*,[2] *M. neumannian*,[2] *M. sietziana*,[2] *M. zuccariniana*[2]
  Nopalea dejecta[2]
  Opuntia engelmanii,[2] *O. guatemalersis*,[2] *O. monacantha*,[2] *O. paraguayensis*,[2] *O. polycantha*,[2] *O. ritteri*,[2] *O. soehrensii*,[2] *O. streptacantha*,[2] *O. tomentella*,[2] *O. tomentosa*[2]
  Pilosocereus glaucescens,[2] *P. nobilis*,[2] *P. leucocephalus*[2]
  Soehronsia bruchii[2]
  Stenocereus stellatus[2]
Chenopodiaceae
  Atriplex argentia,[3] *A. pusilla*,[3] *A. wolfii*[4]
  Chenopodium atrevirens,[4] *C. capitatum*,[3] *C. hybridum*,[3] *C. leptophyllum*,[3] *C. rubrum*[3]
  Grayia spinosa[3]
  Monolepis nuttalliana,[3] *M. pusilla*[3]
Nyctaginaceae
  Abronia choisyi[4]
  Boerhaavia purpurescens[4]

[a] For a listing of all other reports of the occurrence of betalains, see Wohlpart and Mabry (1968b).

[b] Superscript numbers refer to the following reports: 1 = Seeligmann and Kraponickas (1969). 2 = Piattelli and Imperato (1969). 3 = Kimler *et al.* (1970). 4 = Reported here for the first time. Voucher specimens are deposited in the Herbarium, The University of Texas at Austin, Austin, Texas.

edge with regard to the intermediates in the biogenetic pathway leading to the betalains. In this connection, we (Kimler *et al.*, 1971) have characterized betalamic acid both as a natural product in betalain-producing species and as an alkaline cleavage product from betanin.

In connection with the enzymes that may be responsible in the betalain-producing plants for the cleavage of dihydroxyphenylalanine to yield, at least theoretically, betalamic acid, it should be mentioned that enzymes have been isolated from microorganisms which cleave dihydroxyphenylalanine both between the two hydroxyl groups [protocatechuate-3,4-dioxygenase from *Pseudomonas aeruginosa* (Fujasawa and Hayaishi, 1968)] and at an extra diol position [metapyrocatechase from *P. arvilla* (Nozaki *et al.*, 1963, 1968)]. A search for similar enzymes in the betalain-containing plants is presently underway.

Finally, the biogenesis of betalains, at least in some plants, appears to be related to a photoresponse which is mediated by the phytochrome system (see Table 4); however, the detailed manner in which light is involved in betalain synthesis has not yet been determined.

Total Syntheses of Betalains

None of the betalains have been totally synthesized. However, the recent reports of the synthesis of pentamethylneobetanidin (Badgett *et al.*, 1970) and cyclodopa (Wyler and Chiovini, 1968) indicate the direction of research in this area (see Table 5). The major difficulty in the synthesis of a naturally occurring betalain is, of course, the synthesis of an appropriate dihydropyridine derivative such as betalamic acid or the reduction of the pyridine moiety in a neobetalain.

Genetics of Betalain Biogenesis

Since only the few preliminary studies mentioned below have been carried out on the genetics of betalain biogenesis, this area of study will no doubt be pursued in the coming decade.

Portulaca grandiflora. Ootani and Hagiwara (1969) determined the coloration genetics for the common *Portulaca;* in general, purple coloration (betacyanins) was found to be dominant to the yellow-orange coloration (betaxanthins) in F_1 hybrids. The white forms were found to be recessive to all colored forms.

Sugar beet (Beta vulgaris). Because of the economic importance of the sugar beet, some preliminary studies have been performed on the genetics of betalain biogenesis in these plants (see, e.g., Shevtsov and Trukhanov, 1969). However, much of the work was directed toward increasing the sugar content in the beets; as a result, little information was obtained with regard to the genetics of pigment formation. Butterfass (1968) established that gene R for hypocotyl color was located on chromosome II.

Mirabilis jalapa. Melcher (1966) reported the development of a new color strain in *Mirabilis jalapa* but did not elaborate on the genetics of the system.

The Function of Betalains

As with most so-called "secondary" compounds, the question of the function, utilization, and possible degradation of betalains has not been explored in detail.

In non-Centrospermae families, several functions have been attributed to the common anthocyanin pigments aside from their role in floral coloration to attract animal vectors which aid in reproduction; these additional functions include participation in biological oxidations (Lebeden and Litvinenko, 1968; Nagornaya, 1968) and providing resistance to microbial infection (Mandre, 1968; Yarwood et al., 1969).

In the betalain-producing families, the betalains appear to have replaced the anthocyanins as pollination factors. However, like anthocyanins, betalains probably have multiple functions. Like the widespread anthocyanins, betalains are water soluble, occurring naturally as salts in the cell vacuoles of the flowers, fruits, and leaves. Tronchet (1968) noted that both betalains and anthocyanins are often present in the epidermal layers of the plant tissue. Moreover, betalains often accumulate in the stalk, and in the red beet they are found in high concentrations in underground parts.* It has also been observed that betalains frequently accumulate at wound or injury sites in the plants that synthesize them normally.

Ultrastructural Research on Sieve-Tube Plastids in the Centrospermae†

Investigations on the ultrastructure of Centrospermae and Centrospermae-like plants (Behnke, 1969) have revealed structural variation which could aid in the taxonomic grouping of these families. Sieve-tube plastids of the Centrospermae are characterized by ringlike inclusions composed of proteinaceous filaments. These structures have not been observed in most other dicot families (most sieve-tube plastids in dicots contain starch but no filaments).

Twenty-nine species belonging to the following 12 families were found to contain these unique inclusions: Phytolaccaceae, Nyctaginaceae, Didieraceae, Amaranthaceae, Chenopodiaceae, Aizoaceae, Molluginaceae (often included in Aizoaceae), Cactaceae, Portulacaceae, Basellaceae, Caryophyllaceae, and Dysphaniaceae (this latter taxon remains to be investigated for its pigments). These data support a close evolutionary

---

* The high level of betalains present in the underground part of the red beet (*Beta vulgaris*) apparently resulted from selection processes carried out by farmers over several hundred years in Northern Europe along the Baltic Sea.

† We wish to thank Dr. H. -D. Behnke, University of Heidelberg, for making some of these data available to us prior to their publication elsewhere.

relationship of the anthocyanin-producing Caryophyllaceae to the Order Centrospermae.

It is also interesting that the Centrospermae-like families, the Polygonaceae (anthocyanin-containing) and Batidaceae (neither anthocyanins nor betalains yet reported) did not contain the proteinaceous inclusions in their sieve-tube plastids.

REFERENCES

Badgett, B., I. Parikh, and A. S. Dreiding. 1970. *Helv. Chim. Acta* **53**:433.
Bate-Smith, E. C. 1954. *Biochem. J.* **58**:122.
Bate-Smith, E. C. 1962. *J. Linn. Soc. London, Bot.* **58**:95.
Bate-Smith, E. C., and N. H. Lerner. 1954. *Biochem. J.* **58**:126.
Bech, T. D. 1966. *Farm. Zh. (Kiev)* **20**:38. (*Biol. Abstr.* **48**:82535.)
Beck, E., H. Merxmüller, and H. Wagner. 1962. *Planta* **58**:220.
Behnke, H. D. 1969. *Planta* **89**: 275.
Bendich, A., and E. Bolton. 1967. *Plant Physiol.* **42**:959.
Bendich, A., and B. J. McCarthy. 1970. *Proc. Nat. Acad. Sci. U.S.* **65**:349.
Bolton, E. T., B. H. Hoyer, and B. J. McCarthy. 1964. *Science* **144**:959.
Bonner, J., G. Kung, and I. Bekhor. 1967. *Biochemistry* **6**:3650.
Borkowski, B., and B. Pasich. 1958. *Planta Med.* **6**:225. (*Biol. Abstr.* **53**:22272.)
Butterfass, T. 1968. *Theor. Appl. Genet.* **38**:348. (*Biol. Abstr.* **50**:90123.)
Crabbé, P., P. R. Leeming, and C. Djerassi. 1958. *J. Amer. Chem. Soc.* **80**:5258.
Darmograi, N. N., V. I. Litvinenko, and P. E. Krivenchuk. 1968. *Khim. Prir. Soedin.* **4**:248. (*Biol. Abstr.* **51**: 39989.)
Darmograi, N. N., P. E. Krivenchuk, and V. I. Litvinenko. 1969. *Farmatsiya (Moscow)* **18**:30. (*Biol. Abstr.* **50**:72549.)
Dominguez, X. A., R. H. Ramirez, O. L. Ugas, D. J. Garcia, and R. Ketcham. 1968. *Planta Med.* **16**:182. (*Biol. Abstr.* **50**:10264.)
Dreiding, A. S. 1961. *In* "Recent Developments in the Chemistry of Natural Phenolic Compounds" (W. D. Ollis, ed.), p. 194. Pergamon, Oxford.
Fujasawa, H., and O. Hayaishi. 1968. *J. Biol. Chem.* **243**:2673.
Garay, A. S., and G. H. N. Towers. 1965. *Can. J. Bot.* **44**:231.
Gardner, R., A. F. Kerst, D. M. Wilson, and M. G. Payne. 1967. *Phytochemistry* **6**:417.
Gillespie, D., and S. Spiegelman. 1965. *J. Mol. Biol.* **12**:829.
Harborne, J. B. 1967. "Comparative Biochemistry of the Flavonoids." Academic Press, New York.
Haynes, L. J., and K. E. Magnus. 1961. *Nature (London)* **191**:1108.
Hegnauer, R. 1964. "Chemotaxonome der Pflanzen." Vol. III. Birkhäuser, Basel.
Hegnauer, R. 1969. "Chemotaxonome der Pflanzen," Vol. V. Birkhäuser, Basel.
Hörhammer, L., H. Wagner, and W. Fritzsche. 1964. *Biochem. Z.* **399**:398.
Hörhammer, L., H. Wagner, H. Arndt, G. Hitzler, and L. Farkas. 1968. *Chem. Ber.* **101**:1183.
Kagan, J., and T. J. Mabry. 1969. *Phytochemistry* **8**:325.
Kimler, L., J. Mears, H. Rösler, and T. J. Mabry. 1970. *Taxon* **19**:875.
Kimler, L., R. A. Larson, L. Messenger, J. B. Moore, and T. J. Mabry. 1971. *Chem. Commun.* 1329.
Krivenchuk, P. E., V. I. Litvinenko, and V. M. Darmograi. 1968. *Farm. Zh. (Kiev)* **23**:62. (*Chem. Abstr.* **71**:839.)

Laird, C. D., and B. J. McCarthy. 1969. *Biochemistry* **8**:3289.
Lebeden, S. I., and L. G. Litvinenko. 1968. *Nauk. Pr. Ukr. Akad. Sil's'kogospod. Nauk* **4**:55. (*Chem. Abstr.* **71**:88491.)
Liebisch, H. W., B. Matschiner, and H. R. Schuette. 1969. *Z. Pflanzenphysiol.* **61**:269.
Mabry, T. J. 1964. *In* "Taxonomic Biochemistry and Serology" (C. A. Leone, ed.), p. 239. Ronald Press, New York.
Mabry, T. J. 1966. *In* "Comparative Phytochemistry" (T. Swain, ed.), p. 231. Academic Press, New York.
Mabry, T. J. 1970. *In* "Chemistry of the Alkaloids" (S. W. Pelletier, ed.), p. 367. Van Nostrand-Reinhold, Princeton, New Jersey.
Mabry, T. J., and A. S. Dreiding. 1968. *In* "Recent Advances in Phytochemistry" (T. J. Mabry, R. E. Alston, and V. C. Runeckles, eds.), p. 145. Appleton, New York.
Mabry, T. J., A. Taylor, and B. L. Turner. 1963. *Phytochemistry* **2**:61.
Mabry, T. J., H. Wyler, G. Sassu, M. Mercier, I. Parikh, and A. S. Dreiding. 1962. *Helv. Chim. Acta* **45**:640.
Mabry, T. J., H. Wyler, I. Parikh, and A. S. Dreiding. 1967. *Tetrahedron* **23**:3111.
Mandre, M. 1968. *Eesti NSV Tead. Akad. Toim., Biol.* **17**:229. (*Biol. Abstr.* **50**:27530.)
Medovar, B. Y. 1968. *Vop. Ratsion. Pitan.* **4**:151. (*Chem. Abstr.* **71**:79918.)
Melcher, R. 1966. *Z. Pflanzensuecht.* **55**:383.
Miller, H. E., H. Rosler, A. Wohlpart, H. Wyler, M. E. Wilcox, H. Frohofer, T. J. Mabry, and A. S. Dreiding. 1968. *Helv. Chim. Acta* **51**:1470.
Minale, L., M. Piattelli, and R. A. Nicolaus. 1965. *Phytochemistry* **4**:593.
Minale, L., M. Piattelli, S. DeStefano, and R. A. Nicolaus. 1966. *Phytochemistry* **5**:1037.
Minale, L., M. Piattelli, and S. DeStefano. 1967. *Phytochemistry* **6**:703.
Nagornaya, R. V. 1968. *Fiziol.-Biokhim. Osn. Pitan. Rast.* **4**:203. (*Chem. Abstr* **71**:46735.)
Nair, A. G. R. and S. S. Sabramanian. 1961. *J. Sci. Ind. Res., Sect. B* **20**:507. (*Chem. Abstr.* **56**:6372.)
Nozaki, M., H. Kagiyama, and O. Hayaishi. 1963. *Biochem. Z.* **338**:582.
Nozaki, M., T. Nakazawa, H. Fujasawa, S. Kotani, Y. Kojima, and O. Hayaishi. 1968. *Advan. Chem. Ser.* **77**:242.
Omote, Y., K.-T. Kuo, N. Fukada, M. Matsuo, and N. Sugiyama. 1967. *Bull. Chem. Soc. Jap.* **40**:234.
Ootani, S., and T. Hagiwara. 1969. *Jap. J. Genet.* **44**:65. (*Biol. Abstr.* **50**:112675.)
Parikh, I. 1966. Thesis, Univ. of Zurich, Zurich.
Paris, R. 1951. *C. R. Acad. Sci., Ser. D* **233**:90.
Piattelli, M., and G. Impellizzeri. 1969. *Phytochemistry* **8**:1595.
Piattelli, M., and F. Impellizzeri. 1970. *Phytochemistry* **9**:2553.
Piattelli, M., and F. Imperato. 1969. *Phytochemistry* **8**:1503.
Piattelli, M., and F. Imperato. 1970a. *Phytochemistry* **9**:455.
Piattelli, M., and F. Imperato. 1970b. *Phytochemistry* **9**:2557.
Piattelli, M., L. Minale, and G. Prota. 1964a. *Ann. Chim. (Rome)* **54**:955.
Piattelli, M., L. Minale, and G. Prota. 1964b. *Ann. Chim. (Rome)* **54**:963.
Piattelli, M., L. Minale, and G. Prota. 1964c. *Tetrahedron* **20**:2325.
Piattelli, M., L. Minale, and G. Prota. 1965a. *Phytochemistry* **4**:121.
Piattelli, M., M. Minale, and R. A. Nicolaus. 1965b. *Phytochemistry* **4**:817.
Piattelli, M., M. Giudicide Nicola, and V. Castrogiovanni. 1969. *Phytochemistry* **8**:731.
Plouvier, V. 1967a. *C. R. Acad. Sci., Ser. D* **264**:145. (*Biol. Abstr.* **50**:55356.)
Plouvier, V. 1967b. *C. R. Acad. Sci., Ser. D* **265**:516. (*Biol. Abstr.* **50**:16225.)
Reznik, H. 1957. *Planta* **49**:406.
Seeligmann, P., and A. Kraponickas. 1969. *Chemical Plant Taxonomy Newsletter* **13**:8.

Shabbir, M., and A. Zaman. 1968. *J. Indian Chem. Soc.* **45**:81.
Shevtsov, I. A., and V. A. Trukhanov. 1969. *Tsitol. Genet.* **3**:164. (*Biol. Abstr.* **50**:95842.)
Sosa, A. 1959. *C. R. Acad. Sci., Ser. D* **248**:1699.
Sosa, A. 1962. *Ann. Pharm. Fr.* **20**:257.
Steelink, C., M. Yeung, and R. L. Caldwell. 1967. *Phytochemistry* **6**:1435.
Tronchet, J. 1968. *Ann. Sci. Univ. Bescancon Bot.* **5**:9. (*Chem. Abstr.* **71**:109761.)
Wagner, E., and B. G. Cummings. 1970. *Can. J. Bot.* **48**:1.
Wilcox, M. E., H. Wyler, T. J. Mabry, and A. S. Dreiding. 1965. *Helv. Chim. Acta* **48**:252.
Willis, J. C. 1966. "A Dictionary of the Flowering Plants and Ferns," 7th Ed. Cambridge Univ. Press, London and New York.
Wohlpart, A., and T. J. Mabry. 1968a. *Plant Physiol.* **43**:457.
Wohlpart, A., and T. J. Mabry. 1968b. *Taxon* **17**:148.
Wyler, H., and J. Chiovini. 1968. *Helv. Chim. Acta* **51**:1476.
Wyler, H., T. J. Mabry, and A. S. Dreiding. 1963. *Helv. Chim. Acta* **46**:1745.
Wyler, H., H. Rosler, M. Mercier, and A. S. Dreiding. 1967. *Helv. Chim. Acta* **50**:545.
Yarwood, C. E., H. I. Kegami, and K. K. Batra. 1969. *Phytopathology* **5**:596.
Zane, A. and S. H. Wender. 1961. *J. Org. Chem.* **26**:4718.

RECENT PROGRESS IN THE CHEMISTRY OF FLAVYLIUM SALTS

LEONARD JURD

Western Regional Research Laboratory, United States Department of Agriculture, Agricultural Research Service, Albany, California

Introduction	135
Hydrolytic Reactions of Flavylium Salts	138
Peroxide Oxidation of Flavylium Salts	148
Flavylium Salt–Polyphenol Condensation Reactions	151
Synthesis and Some Reactions of Flavenes	155
Formation of Xanthylium Salts From Proanthrocyanidins	160
References	162

Introduction

The water-soluble, red anthocyanin pigments of plants are polyphenolic flavylium salts. Since the work of Willstätter and Robinson, it has been recognized that these natural pigments are derived chiefly from three basic nuclei, namely, pelargonidin (I), cyanidin (II), and delphinidin (III), by glycosidation of the hydroxyl group located at position 3. Hydroxyl groups at position 3' and 5' may be methylated, as in peonidin and malvidin, and a second sugar is often combined with the hydroxyl at position 5. The sugars present in anthocyanins include monosaccharides, disaccharides, and so forth, and their hydroxyl groups may be acylated

I, II, III

by one of a variety of aliphatic acids, phenolic benzoic acids, or phenolic cinnamic acids. It is apparent from this brief summary that a very considerable number of different anthocyanins may be derived from the three anthocyanidins (I–III), and with the development of modern chromatographic and spectral techniques numerous studies on the separation and identification of individual anthocyanin constituents of plant extracts have been reported during the last fifteen years. This aspect of anthocyanin chemistry has been extensively reviewed by Harborne (1967).

In view of the wide occurrence of related flavones and flavanones, e.g., apigenin and naringenin, it is of interest that 3-deoxyanthocyanidins, such as apigeninidin (IV), and luteolinidin (V), are rare, although these compounds have recently been detected in *Sorghum vulgare* (Stafford, 1965), in ferns (Harborne, 1965), and in mosses (Bendz *et al.*, 1962; Nilsson, 1967). Unusual 3-deoxyanthocyanidins have been detected in some plants in the form of their quinonoidal anhydro bases, and it is probable that more compounds of these types will be identified in future studies. Carajurin, (VI) (Ponniah and Seshadri, 1953) and dracorubin (VII), (Robertson *et al.*, 1950), the complex pigment from the palm *Dracaena draco*, are natural 3-deoxyanthocyanidin anhydro bases.

Being electron deficient, the flavylium nucleus of an anthocyanin is highly susceptible to attack by nucleophilic reagents. The instability of anthocyanins in food products presents a problem of considerable economic importance, since the attractive red pigments of many fruits frequently undergo undesirable structural and color changes under the variety of conditions used in their processing. A number of kinetic investigations have been undertaken in an attempt to determine the factors that may contribute to accelerated pigment breakdown in processed strawberries, raspberries, and other fruits (Lukton *et al.*, 1956; Decareau *et al.*, 1956; Markakis *et al.*, 1957; Daravingas and Cain, 1968). These studies have shown that the rate of pigment loss is increased by higher temperatures, by high pH, and by the presence of such normal, co-occurring cellular constituents as ascorbic acid, amino acids, sugars and sugar breakdown products (Meschter, 1953; Tinsley and Bockian, 1960; Starr and Francis, 1968; Sondheimer and Kertesz, 1948, see also 1952). Although some suggestions have been made to account for the kinetic observations, it is clear that the actual reactions involved in different types of pigment degradation are still largely obscure. For this reason extensive chemical studies, primarily employing model phenolic flavylium salts, have been initiated in the last few years to provide precise structural information on the types of hydrolytic, oxidative, reductive, and condensation reactions which anthocyanins may be expected to undergo.

As previously mentioned, flavylium salts are decolorized because they are electron deficient and, therefore, readily attacked by water, peroxides, sulfur dioxide, and other nucleophiles. In attempting to formulate reasonable mechanisms for flavylium degradation reactions, it should be recognized that, although these compounds are commonly represented as oxonium salts of types (I) to (V), the positive charge is actually delocalized over the nucleus. Flavylium salts, in fact, may be regarded as resonance hybrids of oxonium ions, e.g., (VIII), carbonium ions of types (IX) and (X), and of other charged species such as (XI).

Hydrolytic Reactions of Flavylium Salts

Phenolic flavylium salts are stable only in strongly acid media, and it has long been known that at higher pH these compounds lose a proton to form highly colored, unstable quinonoidal anhydro bases, which rapidly hydrate to yield colorless chromenols. Recent spectral and polarographic studies have further clarified the structural response to changes in pH when position 3 of the flavylium nucleus is (1) unsubstituted, as in (IV) or (V); (2) glycosylated, as in natural anthocyanins, or (3) hydroxylated, as in anthocyanidins (I) to (III) (Jurd, 1963a; Jurd and Geissman, 1963; Sperling et al., 1966; Timberlake and Bridle, 1966a,b; Harper and Chandler, 1967).

The effects of pH changes on flavylium salts of type (1) are clearly indicated by measurements of the spectra of 3'-methoxy-4', 7-dihydroxyflavylium chloride (XII), at equilibrium in aqueous solutions of pH 1 to pH 4 (Fig. 1). The intensity of absorption at 468 nm, the λ_{max} of the flavylium cation, markedly decreases as the pH is raised. Simultaneously

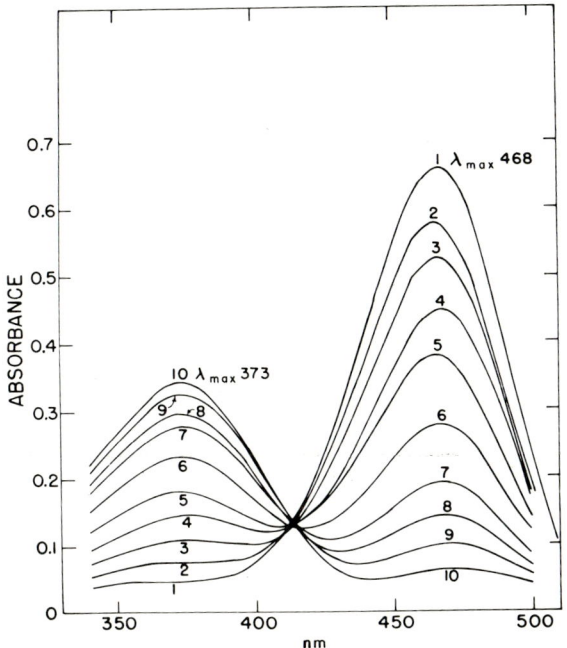

FIG. 1. Effect of pH on the visible spectra of aqueous solutions of 3'-methoxy-4',7-dihydroxyflavylium chloride (XII) (5.2×10^{-3} grams/liter) after standing for 2 hours. pH values: curve 1 = 0.63; 2 = 2.30; 3 = 2.65; 4 = 2.94; 5 = 3.11; 6 = 3.40; 7 = 3.67; 8 = 3.81; 9 = 4.10; 10 = 4.36 (Reprinted from L. Jurd, 1963. *J. Org. Chem.* **28**, 987. Copyright 1963 by the American Chemical Society. Reprinted by permission of the copyright owner.)

the absorption at 373 nm progressively increases due to formation of the 2-hydroxychalcone (XVI). These spectra show that at pH 4 the flavylium salts exists largely (about 90%) in the form of the corresponding chalcone. Identification of the equilibrium product, λ_{max} 373 nm, as the *trans*-chalcone (XVI), has been unequivocally established by the isolation and synthesis of a number of the crystalline equilibrium products from similar flavylium salts, and by nuclear magnetic resonance (NMR) spectra in which the ethylenic protons of the chalcone appear as well-defined doublets with J = 16.0 Hz. The magnitude of this coupling establishes the trans configuration of the chalcone. To summarize the results of numerous spectral measurements of the above type, it is now believed that flavylium salts, which lack a 3-substituent, form equilibrium mixtures with the corresponding *trans*-2-hydroxychalcones. These chalcones are formed from intermediate anhydro bases such as (XIII) via the highly unstable chromenol (XIV) and, presumably, a *cis*-chalcone (XV).

Sperling *et al.* (1966) and Timberlake and Bridle (1966a,b) have observed that the above flavylium–chalcone equilibrium is markedly affected by light. Light shifts the equilibrium in the direction of the flavylium cation form. Thus, for example, a standard solution of 4′, 5, 7-trimethoxyflavylium perchlorate (XVII), in 1 percent HCl (λ_{max} 469 nm, ϵ_{max} 37,200), diluted with aqueous buffer, pH 5.60, yields *trans*-2-hydroxy-4, 6, 4′-trimethoxychalcone (XVIII). In the absence of light at this pH, conversion to the chalcone (λ_{max} 378 nm, ϵ_{max} 22,500) is almost complete

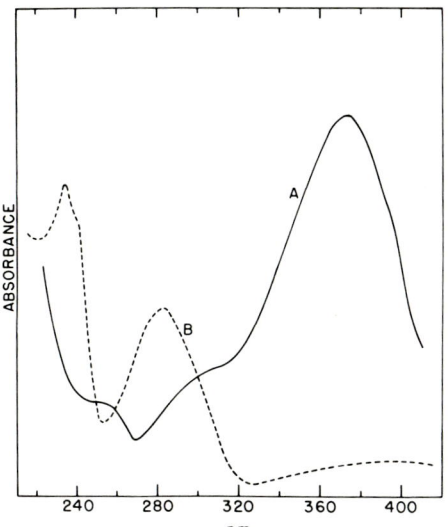

Fig. 2. Ultraviolet spectrum in ethanol of (A) *trans*-2-hydroxy-4,6,4'-trimethoxy chalcone (XVIII); (B) the *trans*-chalcone exposed to sunlight for 30 minutes. (From Jurd, 1969a, by permission of Pergamon, Oxford and New York.)

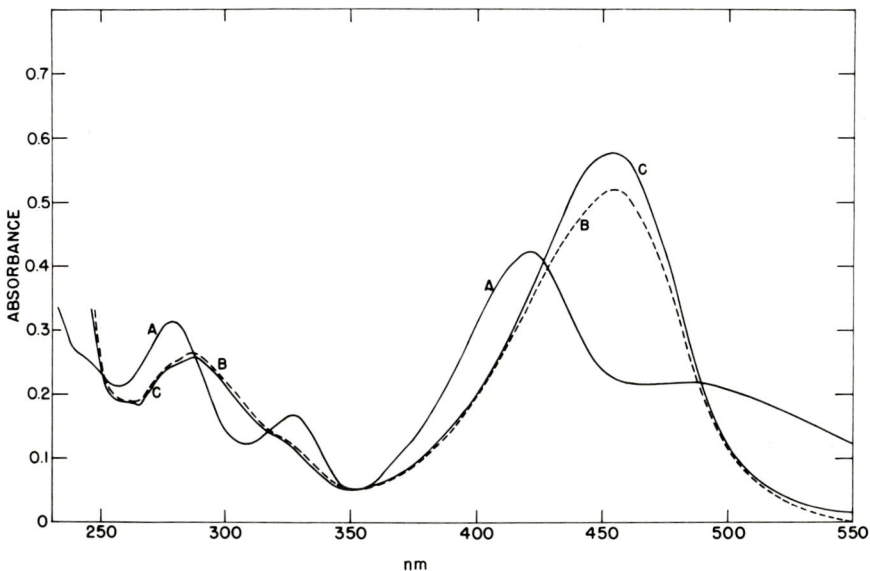

Fig. 3. Isomerization of 5,8-dihydroxyflavylium salts. Curve A: Spectrum of 5,7,8,4'-tetrahydroxyflavylium iodide (XIX) (7.2×10^{-3} grams/liter) at pH 2.6 taken immediately on mixing (2 minutes). Curve B: Spectrum of A after 24 hours. Curve C: Spectrum of 5,6,7,4'-tetrahydroxyflavylium iodide (XX) (7.04×10^{-3} grams/liter) at pH 2.6 (stable). (From Jurd, 1962, by permission.)

within 30 minutes. In light, however, an equilibrium is rapidly reached, containing approximately 60 percent flavylium salt and 40 percent chalcone. Exposure of a solution of the chalcone at pH 5.6 (obtained in the dark) to light results in its rapid recyclization to the flavylium salt and formation of the same 60:40 equilibrium mixture. It has been shown that it is the ease with which a photochemical and/or acid-catalyzed transformation of the *trans*-chalone to the *cis*-chalcone can occur that governs the position of the flavylium–chalcone equilibrium. As indicated in Fig. 2, these *trans*-chalcones undergo an extremely facile photoisomerization in light to *cis*-chalcones, which then rapidly cyclize (in acid solutions) to flavylium salts. In the absence of light, on the other hand, the necessary initial trans to cis isomerization is acid-catalyzed, and this process is much slower than the photochemical process (Jurd, 1969a).

It is of interest to note that because of the ease of reversible ring opening of these flavylium salts, 5,7,8-trihydroxyflavylium salts of type (XIX) isomerize at room temperature in aqueous acid solutions to the more stable 5,6,7-trihydroxyflavylium salts of type (XX) (Fig. 3). This flavylium rearrangement (Jurd, 1962) represents an unusually facile variant of the well-known Wessely-Moser rearrangement of 5,8- to 5,6-dihydroxyflavones.

Anthocyanins, flavylium salts of type (2), form very unstable anhydro bases at higher pH, and these rapidly decolorize with the formation of colorless chromenols, e.g., (XXI) → (XXII) → (XXIII). The chromenols formed from anthocyanins are quite stable at room temperature (although they do undergo a slow, irreversible decomposition on long standing), and they instantly regenerate the anthocyanin upon acidification. There is no spectral evidence of chalcone formation in these anthocyanin equilibria. Cyanidin 3-rhamnoglucoside, (XXI), for example, has λ_{max} 510 nm (log ϵ 4.47) in strongly acid solutions. At higher pH the intensity of absorption at this λ_{max} decreases *without* an increase in absorption in the 350–400 nm (chalcone) region (Fig. 4).

In contrast to the stability of chromenols (XXIII), from anthocyanins, the intermediate chromenols derived from anthocyanidins, 3-hydroxyflavylium salts of type (3), are unstable and readily undergo ring fission to yield α-diketones. Recent spectral and polarographic studies by Harper (1968) indicate that in the range of pH 2–5 pelargonidin (I), is in equilibrium with the chromenol (XXIV) and α-diketone (XXV).

It has often been reported that free anthocyanidins are highly unstable in aqueous solutions and rapidly undergo "spontaneous" irreversible de-

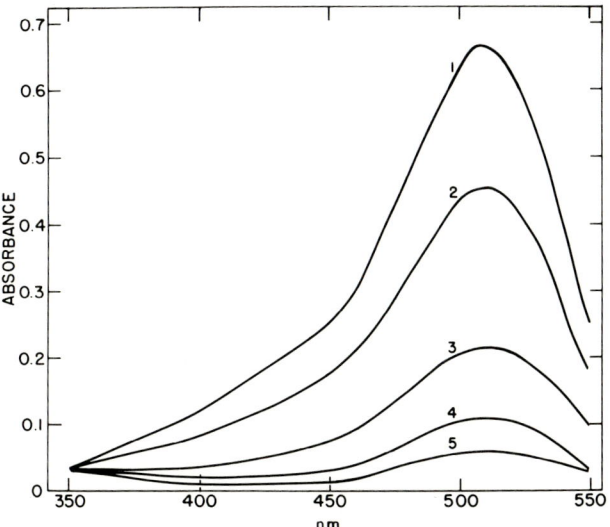

FIG. 4. Spectra of cyanidin 3-rhamnoglucoside, (XXI) (1.6 × 10^{-2} grams/liter) at equilibrium in buffered aqueous solutions after 2 hours. pH values: curve 1 = 0.71; 2 = 2.53; 3 = 3.31; 4 = 3.70; 5 = 4.02. γ_{max} = 510 nm; log ϵ = 4.47. (From Jurd, 1964b. Copyright © 1964 by Institute of Food Technologists.)

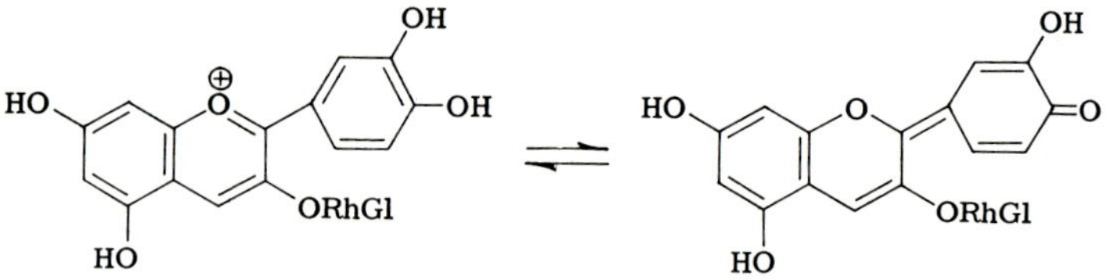

XXI ⇌ XXII 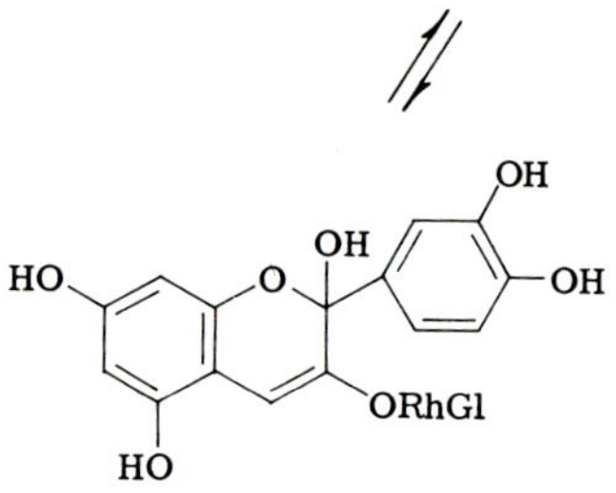
⇌
XXIII

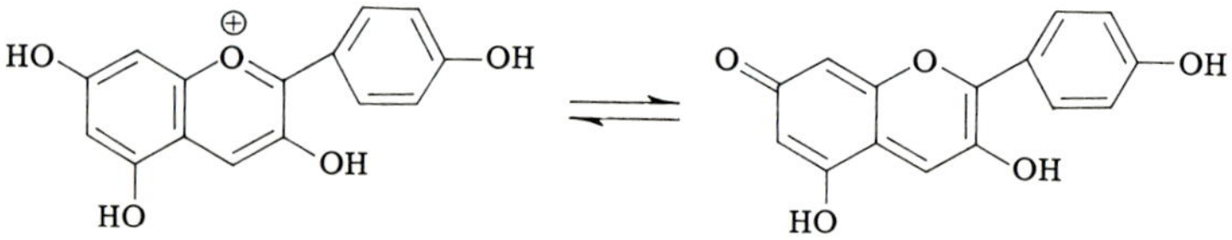

I ⇌ (quinonoid)
⇌ ⇌
XXIV
⇌
⇌ XXV

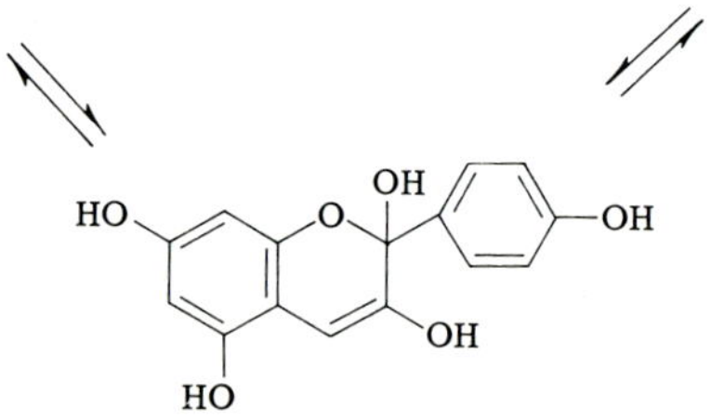

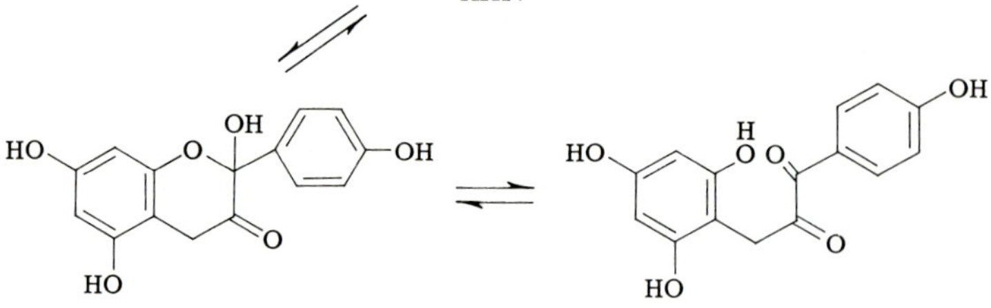

composition. The author has noted that aqueous solutions of cyanidin at pH 2 to pH 5 decompose with the liberation of 3,4-dihydroxybenzoic acid. Harper's demonstration of the existence of α-diketones in aqueous anthocyanidin solutions adequately accounts for the author's observation, since it is known that α-diketones are easily hydrolyzed to acids.

The above spectral studies on flavylium salt equilibria are closely involved with similar studies on sulfite bleaching of flavylium salts, on the stability of anthocyanin–metal complexes, and on the structural modi-

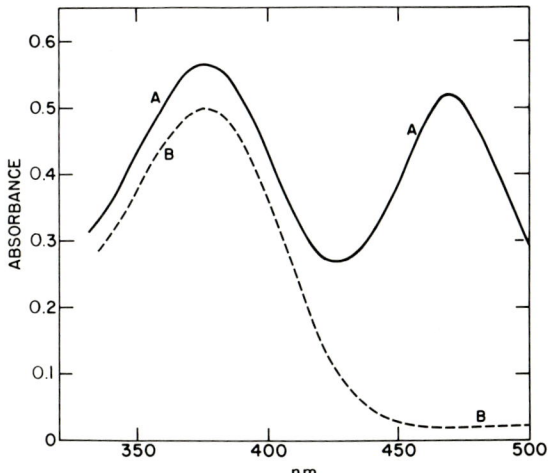

FIG. 5. Curve A: Spectrum of 3'-methoxy-4',7-dihydroxyflavylium chloride (1.04×10^{-2} grams/liter) after standing 1 hour in aqueous solution of pH 3.8. Curve B: Sodium bisulfite (20 ppm) added to solution and spectrum taken within 1 minute. (From Jurd, 1964b. Copyright © 1964 by Institute of Food Technologists.)

fications in which anthocyanins actually exist in pigmented plant cells. Thus, sulfite bleaching of anthocyanins, a process of some commercial importance, was once believed to involve reduction of the pigment or combination of sulfur dioxide with a chalcone formed from the pigment. As indicated in Fig. 5, however, addition of sodium bisulfite to flavylium salts at equilibrium with their chalcones results in the immediate disappearance of the flavylium cation, while the chalcone is unaffected. Sulfite decoloration is due to nucleophilic attack on the flavylium cation by one of the negative ions of sulfurous acid (Jurd, 1964b; Timberlake and Bridle, 1967, 1968). The product formed is probably the chromen-4-sulfonic acid:

As previously indicated, pure anthocyanins are stable only in strongly acid solutions. Because of the instability of their anhydro bases (Fig. 6),

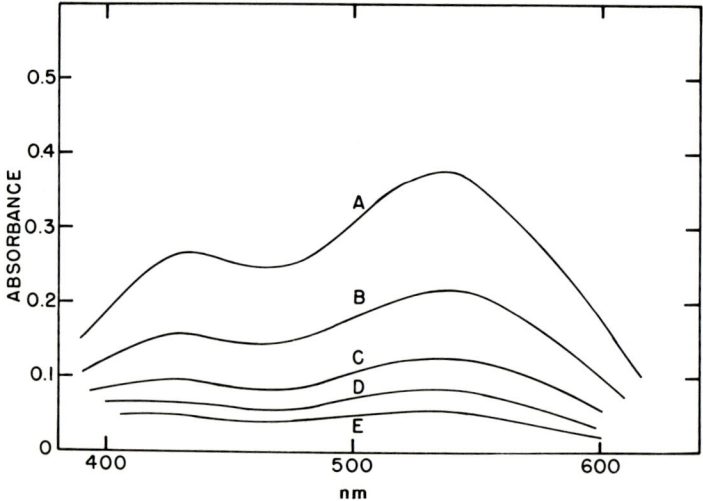

FIG. 6. Visible spectra of cyanidin 3-glucoside (concentration = $3.5 \times 10^{-5} M$) in aqueous buffer, pH 6.10. Spectra measured after: Curve A, 1 minute; B, 10 minutes; C, 20 minutes; D, 30 minutes; and E, 60 minutes. (From Jurd and Asen, 1966, by permission of Pergamon, Oxford and New York.)

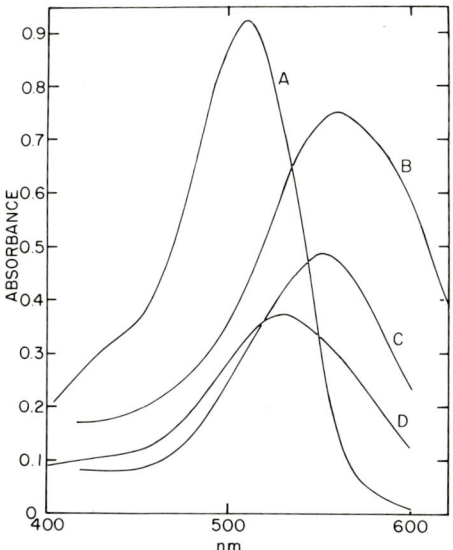

FIG. 7. Effects of pH on visible spectra of cyanidin 3-glucoside (concentration = $3.5 \times 10^{-5} M$) + AlCl: ($8.3 \times 10^{-4} M$). Spectra measured in aqueous buffer solutions after 1 minute. Curve A: pH 1.90, γ_{max} 510; B, pH 5.36, γ_{max} 558; C, pH 3.90; D, pH 3.50.

these compounds rapidly become almost colorless in the approximate pH range from 4 to 6. Since the pH of many highly pigmented flower and fruit cells lie within this pH range, it is clear that anthocyanins actually exist in the natural state in these cells in stabilized forms, whose structures have not yet been unambiguously determined. Much detailed work in the field of anthocyanin-copigment complexes has recently been described by Asen et al. (1970), Bayer (1966), and Hayashi and his co-workers (Yazaki and Hayashi, 1967; Takeda et al., 1966, 1968). Although an adequate review of this subject is beyond the scope of this paper, it should be mentioned that metals are involved in the formation of at least some of these natural pigment complexes. Although pure anhydro bases are quite unstable, these substances yield complexes with metals such as aluminum and iron, which are remarkedly stable in the higher pH range (Jurd and Asen, 1966; Asen et al., 1969). The spectra shown in Fig. 7 indicate the effect of pH on the formation of a colored complex of cyanidin 3-glucoside and aluminum ions. Maximum complex formation occurs at about pH 5.5, and at this pH the cerise complex is stable on long standing in diffuse light.

Peroxide Oxidation of Flavylium Salts

It has been suggested (Sondheimer and Kertesz, 1948, see also 1952) that accelerated decomposition of anthocyanins in the presence of ascorbic acid is due to their oxidation by hydrogen peroxide, formed by ascorbic acid autoxidation. Although it has not been established that peroxides are in fact involved in this particular degradation, there is no doubt that flavylium salts are very susceptible to nucleophilic attack by hydrogen peroxide. In addition to Karrer's oxidation of malvin to malvone (XXVI), more recent studies have shown that, depending on reaction conditions,

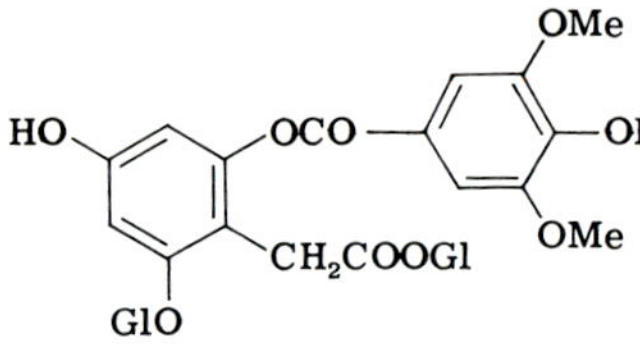

XXVI

3-substituted flavylium salts are oxidized to colorless enolic benzoates of type (XXVII), benzoyl esters (XXVIII) similar to malvone, or 3-substituted 2-phenylbenzofurans (XXIX) (Jurd, 1964a, 1966a). Oxidation of model flavylium salts lacking a 3-substituent gave crystalline ketodiols of type (XXX) (Jurd, 1968a). These ketodiols cyclize with strong acids to yield 3-hydroxyflavylium salts, and upon reduction with sodium borohydride and subsequent acid cyclization they form flavan-3,4-diols.

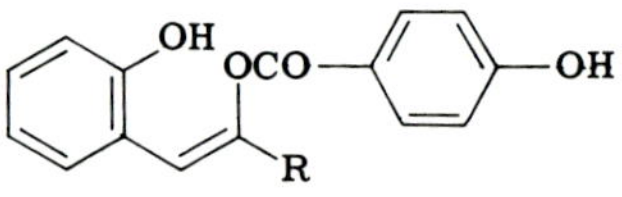

XXVII 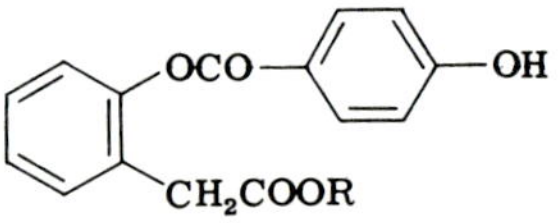   XXVIII

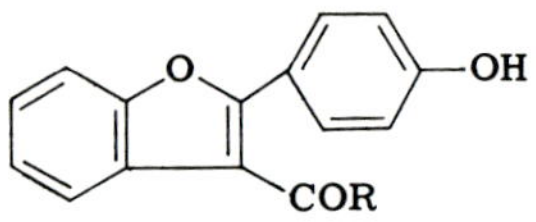

XXIX 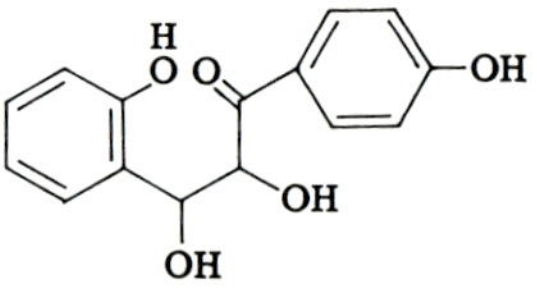   XXX

Chemistry of Flavylium Salts

The oxidation of flavylium salts to enolic benzoates (XXVII), probably involves a Baeyer-Villiger mechanism.

Baeyer-Villiger mechanism

On the basis of this proposal an alternate migration of the B phenyl ring to O^+ could occur in the initial stages. This would be expected to lead to loss of the B ring and formation of a coumarin. Although coumarin formation in the peroxide oxidation of model flavylium salts has not yet been demonstrated, it is of interest that Hrazdina (1970) has recently determined that, upon heating in aqueous solution, malvin (XXXI) and other anthocyanidin 3,5-diglucosides yield the coumarin 3,5-diglucoside (XXXII).

Hrazdina, furthermore, has detected this unusual coumarin glucoside in grape juice and in wines rich in anthocyanidin 3,5-diglucosides.

XXXI   XXXII

It is probable that enolic benzoates (XXVII), are intermediates in the formation of benzoates of type (XXVIII) from flavylium salts, since the former readily isomerize to benzoates of the latter type.

The 3-substituted 2-phenyl benzofurans (XXIX), are formed only by oxidation of flavylium salts in the presence of alcohols. The recognition of the formation of this type of oxidation product led to a very simple synthesis of natural coumarinobenzofurans by oxidation of appropriate 2'-hydroxyflavylium salts (Jurd, 1963b; Jurd, 1965a; Livingston et al., 1964). Coumestrol (XXXIV), for example, can be prepared in 40 percent yields by peroxide oxidation of 7,2',4'-trihydroxy-3-methoxyflavylium chloride (XXXIII).

Flavylium Salt–Polyphenol Condensation Reactions

Flavylium salts condense readily with amino acids (Shriner and Sutton, 1963), β-diketones, and polyphenols to form colorless 4-substituted flav-2-enes of types (XXXV), (XXXVI), and (XXXVII). These reactions are of interest because they provide a mechanistic basis for the proposal

that similar condensations may account for loss of anthocyanin pigment in some plant extracts and also for the formation of some types of natural dimeric and polymeric flavonoids.

The 4-substituted flav-2-enes formed first in the condensation are highly reactive and usually undergo further structural changes, depending on the nature of the 4-substituent, e.g., the phloroglucinol product (XXXVI), cyclizes to a stable flavan derivative (XXXVIII) (Jurd and Waiss, 1965). The flavylium–catechin condensation, on the other hand, involves a complex oxidation–reduction sequence (Jurd, 1967). The 4-substituted flav-

2-ene, XXXVII, initially formed is oxidized by a second molecule of the flavylium salt, the latter thereby being reduced to a simple, colorless flav-2-ene. This reaction sequence is illustrated by the condensation of 8-methoxy-4′-hydroxy-3-methylflavylium chloride (XXXIX), with D-catechin (XL), to give the crystalline pigment (XLI), and colorless flav-2-ene (XLII).

The flavylium–catechin condensation provides experimental support for the suggestion by Whalley (1961) that the biosynthesis of the complex, natural anhydro base, dracorubin (VII), involves an oxidative flavylium–flavan condensation.

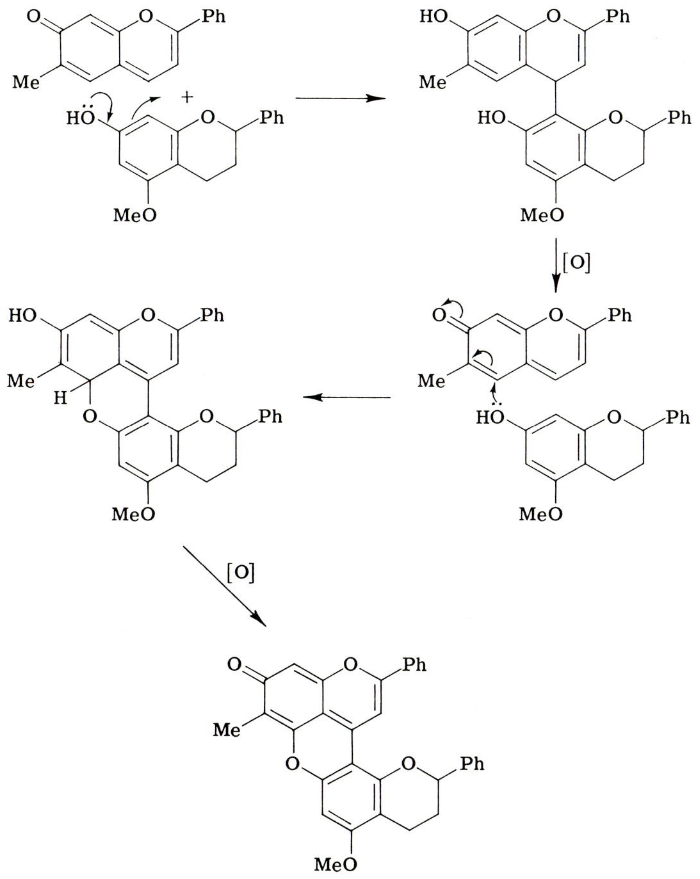

VII

Flavylium–flavan condensation

Although it has not yet been demonstrated experimentally, it would seem that similar anthocyanin–catechin or anthocyanin–leucoanthocyanin condensations could account, at least in part, for the formation of other natural dimers and for the slow incorporation of anthocyanins into red tannin polymers, such as are formed during the aging of red wines (Jurd, 1969b).

Flavylium salts condense very readily with dimedone (XLIII), to yield flavan derivatives (XLIV), and 9-phenacyl-5-ketotetrahydroxanthenes (XLV) (Jurd and Bergot, 1965; Jurd, 1965b, 1966b).

Since ascorbic acid (XLVI) has a β-diketonic structure similar to that of dimedone, it is possible that decomposition of anthocyanins in fruit products in the presence of ascorbic acid may involve the same type of condensation.

In this connection, it is noteworthy that it has been reported that the sodium salt of ascorbic acid reacts with the benzyl carbonium ion, which is analogous to the flavylium cation, to form a C-benzyl product (XLVII) (Buncel et al., 1965).

XLVII

Synthesis and Some Reactions of Flavenes

In connection with flavylium condensation reactions, it has been mentioned above that 4-substituted flav-2-enes are unstable and, therefore, susceptible to structural modification. Flav-2-enes and flav-3-enes, unsubstituted at position 4, are also highly reactive, colorless compounds, which are easily oxidized and reduced under a variety of conditions to form flavylium salts or flavans. With the recognition that these substances may be active intermediates in the synthesis and biosynthesis of anthocyanins and catechins (Clark-Lewis and Skingle, 1967), the chemistry of flavenes has received much attention in the last few years.

Flav-2-enes may be synthesized by reduction of flavylium salts with lithium aluminum hydride (Karrer and Seyhan, 1950; Gramshaw et al., 1958) or with sodium borohydride (Jurd and Waiss, 1965; Jurd, 1967). In addition to monomeric flav-2-enes (XLVIII), the sodium borhydride reduction yields unsymmetrical dimers of type (IL), apparently as a result of nucleophilic attack of the flav-2-ene on unreacted flavylium salt (Reynolds and Van Allan, 1967).

Reynolds and Van Allan, and Jurd and Waiss (1968) have also shown that reduction of flavylium salts by metals such as zinc or magnesium does not give monomeric flav-2-enes. Colorless, crystalline α- and β-bisflav-2-enes of types (L) and (LI) are formed and these isomeric bisflavenes readily undergo thermal interconversion.

XLVIII

IL

L

LI

Flav-3-enes (LII), can be synthesized by sodium borohydride reduction of 2'-hydroxychalcones in isopropanol (Clark-Lewis et al., 1967; Clark-Lewis and Jemison, 1968). Lithium aluminum hydride reduction of methylated flavonols yields separable mixtures of the isomeric flav-2-enes and flav-3-enes (Waiss and Jurd, 1968). Flav-3-enes are oxidized by air in the presence of acids to flavylium salts, and may be catalytically hydrogenated to flavans.

In acid solutions in the presence of air flav-2-enes are only partially oxidized to flavylium salts. In aqueous acid solutions they are chiefly hydrolyzed to dihydrochalcones, eg., (LIII) → (LIV), whereas in non-aqueous media they may disproportionate to the flavylium salt and flavan, e.g., (LIII) → (XXXIX) + (LV).

Flav-2-enes also differ from flav-3-enes in that the former, but not the latter, are oxidized almost quantitatively by quinones to flavylium salts (Jurd, 1966c; Waiss and Jurd, 1968) e.g., (VIII) → (LVI). Flavylium salts

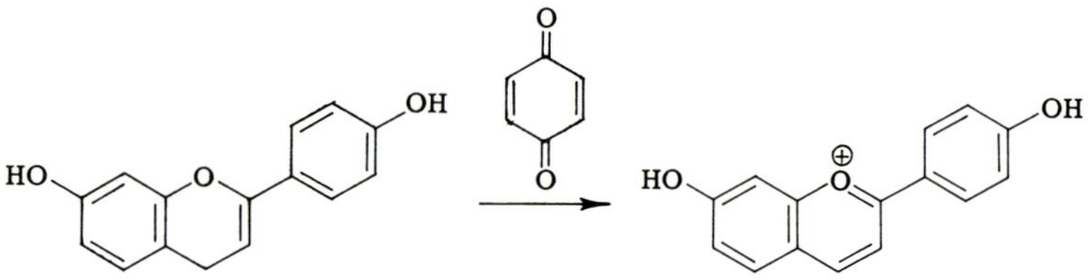

VIII

LVI

may replace benzoquinone in this reaction and intermolecular flavene–flavylium salt oxidation–reduction may occur (Jurd, 1967) e.g., (VIII) + (XXXIX) → (LVI) + (LIII).

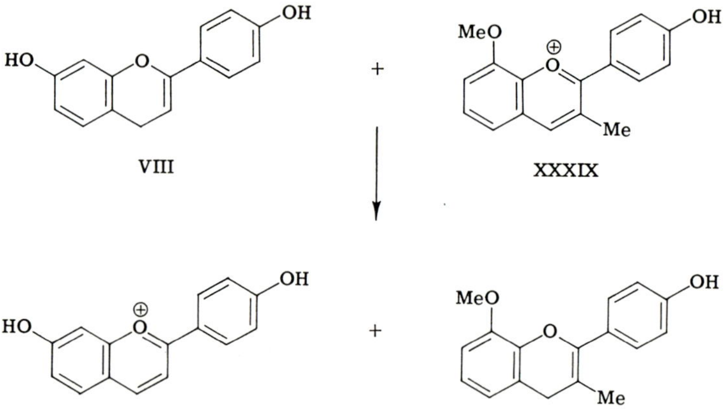

VIII

XXXIX

LVI

LIII

The reduction of 2′-hydroxychalcones by metal hydrides to flav-3-enes, and their ready conversion to flavylium salts, provides a chemical model for a plausible biosynthetic pathway, proposed by Clark-Lewis and Skingle (1967), leading from 2′-hydroxychalcones to anthocyanidins and catechins.

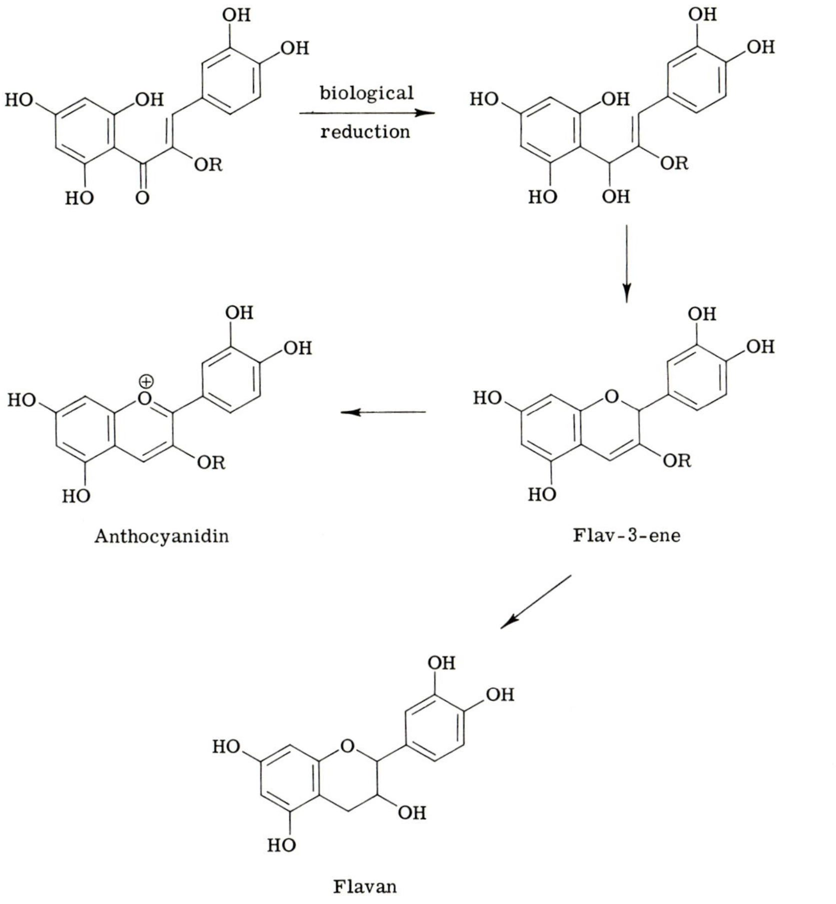

Chemistry of Flavylium Salts

Anthocyanidin

Flav-3-ene

Flavan

In connection with flav-3-ene and anthocyanidin biosynthesis, an alternative mechanism has been suggested (Jurd, 1968b) in which oxidation of the reactive allylic methylene group of natural cinnamylphenols (LVII), leads to flav-3-enes, and these may then disproportionate to flavylium salts and flavans.

The mechanistic feasibility of this suggestion has recently been demonstrated by Cardillo et al. (1969), who found that o-cinnamylphenol (LVIII) is oxidized by dichlorodicyanobenzoquinone to the flav-3-ene (LII) and that this, with acid, disproportionates to a mixture of the flavan and flavylium salt.

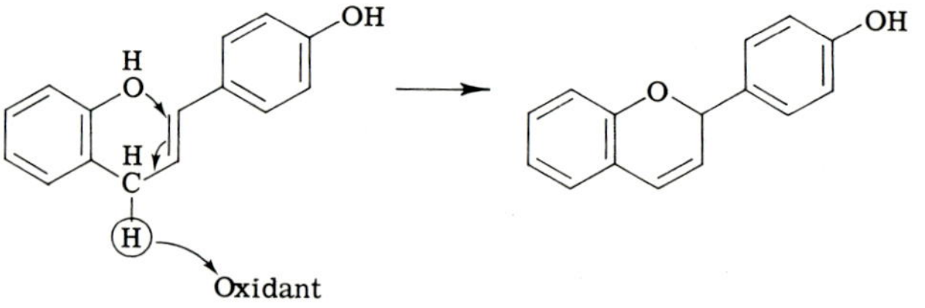

LVII

LVIII DDQ→ LII

Formation of Xanthylium Salts from Proanthrocyanidins

Flavan-3,4-diols and their natural condensation products, e.g., (LIX), yield monomeric anthocyanidins when heated with alcoholic mineral acids (Swain, 1954; Geissman and Dittmar, 1965; Creasy and Swain, 1965). The formation of red anthocyanidin pigments is always accompanied by the formation of other unidentified pigments. Thus, Swain and

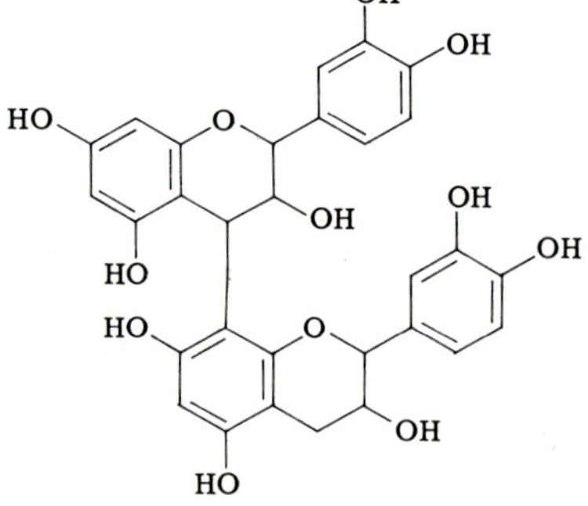

LIX

Hillis (1959) noted that the spectra of anthocyanidins (λ_{max} 550 nm) formed from proanthocyanidins by hot butanol-HCl showed a pronounced shoulder at 450 nm. With hot aqueous or acetic HCl, the absorbance at the λ_{max} of the anthocyanidin decreased and the shoulder at 450 nm developed into a definite peak. Similar ill-defined yellow products with λ_{max} 450 nm (and a minor peak at 365 nm) were formed by acid treatment of apple- (Ito and Joslyn, 1965) and grape-seed (Peri, 1967) proanthocyanidins. The formation of yellow-orange compounds on acid hydrolysis of some tannins has also been observed (Hillis, 1964; Roux, 1957; Roux and Evelyn, 1958).

The nature of the yellow products formed from natural proanthocyanidins is still obscure. However, it has recently been shown that the model proanthocyanidin (LX), when heated in dilute acetic acid, yields a yellow compound(s) which migrates as a discrete species. The spectrum of the

LX   LXI

yellow compound in 1 percent aqueous HCl shows sharp, well-defined maxima at 439 and 267 nm, and minor peaks at 363 and 291 nm, the ratio of the absorbance at λ_{max} 267 nm to the absorbance at λ_{max} 439 nm being 1.09. In all the above details, including the positions of minima, the spectrum of the yellow product is essentially identical with the spectrum of the synthetic tetrahydroxy xanthylium salt (LXI) (Jurd and Somers, 1970). It is possible that the yellow products formed from natural dimeric and polymeric flavan-3,4-diols are also xanthylium salts. The formation of xanthylium salts would suggest that xanthene nuclei might occur in some condensed tannins or in the simpler polyphenols with which they are associated.

REFERENCES

Asen, S., K. H. Norris, and R. N. Stewart. 1969. Absorption spectra and color of aluminum-cyanidin 3-glucoside complexes as influenced by pH. *Phytochemistry* **8**:653.
Asen, S., R. N. Stewart, K. H. Norris, and D. R. Massie. 1970. A stable blue nonmetallic copigment complex of delphanin and C-glycosylflavones in Prof. Blaauw Iris. *Phytochemistry* **9**:619.
Bayer, E. 1966. Complex formation and flower colors. *Angew. Chem. Int. Ed. Engl.* **5**:791.
Bendz, G., O. Märtensson, and L. Terenius. 1962. Moss pigments. 1. The anthocyanins of *Bryum cryophilum* O. Mart. *Acta Chem. Scand.* **16**:1183.
Buncel, E., K. G. A. Jackson, and J. K. N. Jones. 1965. The L-ascorbate ion as an ambient nucleophile. *Chem. Ind. (London)* p. 89.
Cardillo, G., R. Cricchio, and L. Merlini. 1969. Biogenetic-like synthesis of flav-3-enes from o-cinnamylphenols. *Tetrahedron Lett. (London)* 907.
Clark-Lewis, J. W., and R. W. Jemison. 1968. Flavan derivatives. XXIII. Preparation and synthetic applications of flav-3-enes. *Aust. J. Chem.* **21**:2247.
Clark-Lewis, J. W., and D. C. Skingle. 1967. Flavan derivatives. XVIII. Synthesis of hemiketals containing the peltogynol ring system: 3(3')-hydroxyisochromano(4',3':2,3)chromans. Conversion of 2'-hydroxychalcones into flav-3-enes and its biosynthetic implications. *Aust. J. Chem.* **20**:2169.
Clark-Lewis, J. W., R. W. Jemison, D. C. Skingle, and L. R. Williams. 1967. A convenient synthesis of flav-3-enes. *Chem. Ind. (London)* p. 1455.
Creasy, L. L., and T. Swain. 1965. Structure of condensed tannins. *Nature (London)* **208**:151.
Daravingas, G., and R. F. Cain. 1968. Thermal degradation of black raspberry anthocyanin pigments in model systems. *J. Food Sci.* **33**:138.
Decareau, R. V., G. E. Livingston, and C. R. Fellers. 1956. Color changes in strawberry jellies. *Food Technol. (Chicago)* **10**:125.
Geissman, T. A., and H. F. K. Dittmar. 1965. A proanthocyanidin from avocado seed. *Phytochemistry* **4**:359.
Gramshaw, J., A. W. Johnson, and T. J. King. 1958. Synthesis of flavan-2:3-diols-(dihydro-α:2-dihydroxy-chalcones). *J. Chem. Soc.* 4040.
Harborne, J. B. 1965. Anthocyanins in ferns. *Nature (London)* **207**:984.
Harborne, J. B. 1967. "Comparative Biochemistry Flavonoids," pp. 1–30. Academic Press, New York.
Harper, K. A. 1968. Structural changes of flavylium salts. IV. Polargraphic and spectrophotometric examination of pelargonidin chloride. *Aust. J. Chem.* **21**:221.
Harper, K. A., and B. V. Chandler. 1967. Structural changes of flavylium salts. 1. Polarographic and spectrophotometric examination of 7,4'-dihydroxyflavylium perchlorate. *Aust. J. Chem.* **20**:731, 745.
Hillis, W. E. 1964. The formation of polyphenols in trees. *Biochem. J.* **92**:516.
Hrazdina, G. 1970. Personal communication.
Ito, S., and M. A. Joslyn. 1965. Apple leucoanthocyanins. *J. Food Sci.* **30**:44.
Jurd, L. 1962. Rearrangement of 5,8-dihydroxyflavylium salts. *Chem. Ind. (London)* p. 1197.
Jurd, L. 1963a. Anthocyanins and related compounds. 1. Structural transformations of flavylium salts in acidic solutions. *J. Org. Chem.* **28**:987.
Jurd, L. 1963b. The synthesis of coumestrol from a flavylium salt. *Tetrahedron Lett. (London)* p. 1151.

Jurd, L. 1964a. Anthocyanins and related compounds. 111. Oxidation of substituted flavylium salts to 2-phenylbenzofurans. *J. Org. Chem.* **29**:2602.

Jurd, L. 1964b. Reactions involved in sulfite bleaching of anthocyanins. *J. Food Sci.* **29**:16.

Jurd, L. 1965a. Synthesis of 7-hydroxy-5',6'-methylenedioxybenzofurano(3',2':3,4)coumarin (medicagol). *J. Pharm. Sci.* **54**:1221.

Jurd, L. 1965b. Anthocyanidins and related compounds. VIII. Condensation reactions of flavylium salts with 5,5-dimethyl-1,3-cyclohexanedione in acid solutions. *Tetrahedron* **21**:3707.

Jurd, L. 1966a. Anthocyanidins and related compounds. X. Peroxide oxidation products of 3-alkylflavylium salts. *Tetrahedron* **22**:2913.

Jurd, L. 1966b. Anthocyanidins and related compounds. IX. Synthesis of 9-phenacyl-5-ketotetrahydroxanthenes. *J. Org. Chem.* **31**:1639.

Jurd, L. 1966c. Quinone oxidation of flavenes and flavan-3,4-diols. *Chem. Ind. (London)* p. 1683.

Jurd, L. 1967. Anthocyanidins and related compounds. XI. Catechinflavylium salt condensation reactions. *Tetrahedron* **23**:1057.

Jurd, L. 1968a. Anthocyanidins and related compounds. XIII. Hydrogen peroxide oxidation of flavylium salts. *Tetrahedron* **24**:4449.

Jurd, L. 1968b. Biogenetic-like syntheses of benzylstyrenes and neoflavanoids. *Experientia* **24**:858.

Jurd, L. 1969a. Anthocyanins and related compounds. XV. The effects of sunlight on flavylium salt–chalcone equilibrium in acid solutions. *Tetrahedron* **25**:2367.

Jurd, L. 1969b. Review of polyphenol condensation reactions and their possible occurrence in the aging of wines. *Amer. J. Enol. Viticult.* **20**:191.

Jurd, L., and S. Asen. 1966. The formation of metal and copigment complexes of cyanidin 3-glucoside. *Phytochemistry* **5**:1263.

Jurd, L., and B. J. Bergot. 1965. Anthocyanidins and related compounds. VII. Reactions of flavylium salts with 5,5-dimethyl-1,3-cyclohexanedione at pH 5.8. *Tetrahedron* **21**:3697.

Jurd, L., and T. A. Geissman. 1963. Anthocyanins and related compounds 11. Structural transformations of some anhydro bases. *J. Org. Chem.* **28**:2394.

Jurd, L., and T. C. Somers. 1970. The formation of xanthophylium salts from proanthocyanidins. *Phytochemistry* **9**:419.

Jurd, L., and A. C. Waiss, Jr. 1965. Anthocyanins and related compounds. VI. Flavylium salt–phloroglucinol condensation products. *Tetrahedron* **21**:1471.

Jurd, L., and A. C. Waiss, Jr. 1968. Anthocyanidins and related compounds. XIV. Reductive dimerization of flavylium salts by metals. *Tetrahedron* **27**:2801.

Karrer, P., and M. Seyhan. 1950. Uber die Reduktion der Pyryliumsalze mit Lithiumaluminiumhydrid. *Helv. Chim. Acta* **33**:2209.

Livingston, A. L., E. M. Bickoff, R. E. Lundin, and L. Jurd. 1964. Trifoliol, a new coumestan from ladino clover. *Tetrahedron* **20**:1963.

Lukton, A., C. O. Chichester, and G. Mackinney. 1956. The breakdown of strawberry anthocyanin. *Food Technol. (Chicago)* **10**:427.

Markakis, P., G. E. Livingston, and C. R. Fellers. 1957. Quantitative aspects of strawberry pigment degradation. *J. Food Res.* **22**:117.

Meschter, E. E. 1953. Fruit color loss. Effects of carbohydrates and other factors on strawberry products. *J. Agr. Food Chem.* **1**:574.

Nilsson, E. 1967. Moss pigments. Preliminary investigations of the violet pigments in *Sphagnum nemoreum*. *Acta Chem. Scand.* **21**:1942.

Peri, C. 1967. Acid degradation of leucoanthocyanidins. *Amer. J. Enol. Viticult.* **18**:168.
Ponniah, L., and T. R. Seshadri. 1953. Synthesis of isocarajuretin hydrochloride. *Proc. Indian Acad. Sci., Sect. A* **38**:288.
Reynolds, G. A., and J. A. Van Allan. 1967. The reduction of some flavylium salts with sodium borohydride. *J. Org. Chem.* **32**:1897.
Robertson, A., W. B. Whalley, and J. Yates. 1950. The pigments of "Dragon's Blood" resins. III. The constitution of dracorubin. *J. Chem. Soc., London* p. 3117.
Roux, D. G. 1957. Some recent advances in the identification of leuco-anthocyanins and the chemistry of condensed tannins. *Nature (London)* **180**:973.
Roux, D. G., and S. R. Evelyn. 1958. Condensed tannins. 1. A study of complex leucoanthocyanins present in condensed tannins. *Biochem. J.* **69**:530.
Shriner, R. L., and R. Sutton. 1963. Benzopyrylium salts. VII. Reaction of flavylium perchlorate with ethyl esters of α-amino acids. *J. Amer. Chem. Soc.* **85**:3989.
Sondheimer, E., and Z. I. Kertesz. 1948. The anthocyanin of strawberries. *J. Amer. Chem. Soc.* **70**:3476. 288 (1952).
Sondheimer, E., and Z. I. Kertesz. 1952. The kinetics of the oxidation of strawberry anthocyanin by hydrogen peroxide. *Food Res.* **17**:288.
Sperling, W., F. C. Werner, and H. Kuhn. 1966. Vorgänge in Lösungen von Anthocyanidinen bei pH-Änderung und bei Belichtung. *Ber. Bunsenges. Phys. Chem.* **70**:530.
Stafford, H. A. 1965. Flavonoids and related phenolic compounds produced in the first internode of *Sorghum vulgare* Pers. in darkness and in light. *Plant Physiol.* **40**:130.
Starr, M. S., and F. J. Francis. 1968. Oxygen and ascorbic and effect on the relative stability of four anthocyanin pigments in cranberry juice. *Food Technol. (Chicago)* **22**:1293.
Swain, T. 1954. Leucocyanidin. *Chem. Ind. (London)* p. 1144.
Swain, T., and W. E. Hillis. 1959. The phenolic constituents of *Prunus domestica*. 1. The quantitative analysis of phenolic constituents. *J. Sci. Food Agr.* **10**:63.
Takeda, K., S. Mitsui, and K. Hayashi. 1966. Structure of a new flavonoid in the blue complex molecule of commelinin. *Bot. Sci. Jap.* **78**:578.
Takeda, K., N. Saito, and K. Hayashi. 1968. Further studies on the structure of genuine anthocyanins. *Proc. Jap. Acad.* **44**:352.
Timberlake, C. F., and P. Bridle. 1966a. Flavylium salts, anthocyanidins and anthocyanins. 1. Structural transformations in acid solutions. *J. Sci. Food Agr.* **18**:473.
Timberlake, C. F., and P. Bridle. 1966b. Spectral studies of anthocyanin and anthocyanidin equilibria in aqueous solution. *Nature (London)* **212**:158.
Timberlake, C. F., and P. Bridle. 1967. Flavylium salts, anthocyanidins and anthocyanins. 11. Reactions with sulfur dioxide. *J. Sci. Food Agr.* **18**:479.
Timberlake, C. F., and P. Bridle. 1968. Flavylium salts resistant to sulfur dioxide. *Chem. Ind. (London)* p. 1489.
Tinsley, I. J., and A. H. Bockian. 1960. Some effects of sugars on the breakdown of pelargonidin 3-glucoside in model systems at 90°C. *J. Food Res.* **25**:161.
Waiss, A. C., and L. Jurd. 1968. Synthesis of flav-2-enes and flav-3-enes. *Chem. Ind. (London)* p. 743.
Whalley, W. B. 1961. *In* "Chemistry of Natural Phenolic Compounds" (W. D. Ollis, ed.), p. 26. Pergamon, Oxford.
Yazaki, Y., and K. Hayashi. 1967. Analysis of flower colors in *Fuchsia hybrida* in reference to the concept of copigmentation. *Proc. Jap. Acad.* **43**:316.

HISTOCHEMISTRY OF PLANTS IN HEALTH AND DISEASE

D. S. VAN FLEET

University of Georgia, Athens, Georgia

Introduction	165
The Development of a Chemical Barrier	166
Enzymes	167
Phenols, Naphthols, and Anthrols	171
Alkaloids	174
Polyenes and Polyynes	175
Genetics of Endodermal Lipids	176
Enzyme Localization in Health and Disease	182
Analysis of the Localization of Compounds in Health and Disease	187
Conclusions	189
References	190

Introduction

Specialized enzyme systems have evolved in the formation of cell layers of immunity and in the temporary and lesser barriers that afford some degree of resistance to disease. Catalysts and compounds in the localized metabolism of response to pathogens are inherent chemical systems. The inherent and evolved development of localized chemical compounds of defense includes systems of hypersensitivity that afford rapid necrosis of a few cells to sequester invasion and infection. The inherent survival value of chemical compounds of defense is in their localized control of energy loss. A system of reaction to invasion has evolved in those plants with immunity in which there is not a homogeneous distribution of response with large changes in metabolism and excessive energy loss throughout the entire plant.

Immunity and resistance to disease are as complex as random speciation, but immunity to disease may be found in the localization of chemical systems in the barrier layers of epidermis, endodermis, cork, bark, and wound lesions. Immunity and degrees of resistance to disease are controlled by genes that are either functional in immunity or are latent and activated in localized resistance.

The Development of a Chemical Barrier

One of the ancient and relatively invariant plant tissues, the endodermis, is an evolved system of molecular compounds and enzymes that afford either immunity or some degree of resistance to disease. The proendodermis in all roots divides to give rise to the cells of the cortex. At the end of an exact arithmetic number of divisions the proendodermis stops cell division and undergoes internal nuclear division. It becomes a polyploid cell with as high as eight times the base number of chromosomes. All its genes are thus present in multiples.

The endodermis in its polyploidy has the genetic restriction and amplitude of ploidy in its quantitative capacity for function. Through internal polyploidy or endopolyploidy, endodermal cells have more DNA and more genes in replication than other cells of the root. The endodermis appears to be able to produce several compounds that have no exact parallel in quality and quantity in any other tissue, such as antibiotic compounds with conjugated unsaturated bonds. Some of these compounds of unsaturation are oxidized to linoxyn or suberinlike solids (Van Fleet, 1971) and are deposited in the impervious, unperforated, and insoluble Casparian band (Bonnett, 1968) in the radial wall of the endodermis. The Casparian band of material between the radial walls and surrounding each cell is a block to the free diffusion of ions and to the movement of bacteria and fungi into the vascular tissue. The location of this single layer of endodermal cells in all primary roots has the effect of meeting the problems of immunity, resistance, ion selection, and ion exchange.

The compounds formed by this layer of cells follow relationships in phyletic, generic, and species groups; but, overall, endodermal cells throughout the plant kingdom appear to be more closely related in composition than are the plants in which they occur. In addition, the compounds found in the endodermis show similarities between taxonomically unrelated species.

Facts on the enzyme and chemical composition of one layer of cells are hard to come by. However, the normal sequence of enzymes activated and compounds formed in the endodermis appears to have a parallel in the enzymes and compounds localized in zones of wounding and infection.

It seems reasonable to suppose that latent and extant genes evocated in zones of injury and infection may be the same genes that are operating in the formation of the endodermis. Zones of injury frequently have polyploid cells, and some of the same enzymes activated are those that are characteristic of the endodermis. Compounds in zones of infection include molecules with conjugated unsaturated bonds like those produced by the endodermis.

Such compounds possess structures that are not favorable to the normal pathways of metabolism and hence have a role to play in combating the invasion of fungi, bacteria, insects, nematodes, and even mammals. Survival during the course of evolution has been aided by the proliferation of such antibiotic substances in the endodermis, bark, and wound lesions.

This evolution of antibiotics has not been universal; there are no plants that have attained complete immunity to disease or biologically induced injury. The genes of immunity to disease have become localized in their effective function in some few epidermal, endodermal, and lesion tissues, and some degree of genetic resistance to disease has extended in some plants to several cell types within the plant, but there are no genetic archetypes that have given to any family of plants a heritage of full immunity.

Enzymes

Selected systems of enzymes have evolved in the cells comprising the endodermis. Peroxidase and polyphenolase are the most prominent in their intensity and rate of reaction (Van Fleet, 1952, 1961, 1971). Benzidine and related aminophenols used as hydrogen donor indicators for heme enzymes show peroxidase in the protoplast of young cells and bound to the walls of older endodermal cells. The adsorption of the oxidized quinoneimine dyes on heme enzymes (Van Fleet, 1952, 1961, 1962) presents a method for localization studies. The spectrophotometric studies of Hirai (1968) have clearly confirmed the validity of the method. For the determination of peroxidase in the endodermal zone, the microspectroscopic ocular has been used to obtain the characteristic absorption maxima, confirming the results of chemical indicator methods (Van Fleet, 1947, 1961, 1971). The most rapid reaction rates for peroxidase are found in walls and pores lining the endodermis (Fig. 1) and particularly in the walls of the group of four

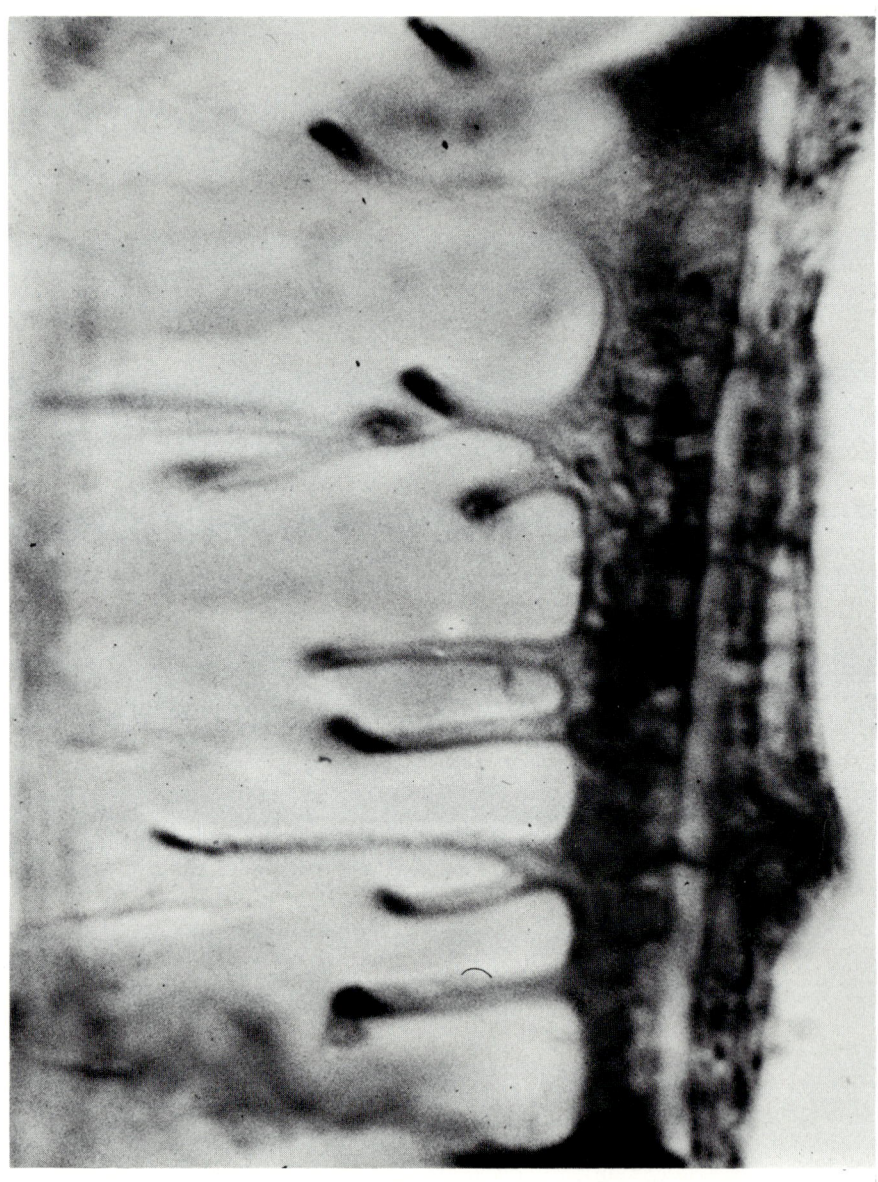

Fig. 1. Unilateral localization of peroxidase in one wall and in plasmodesmatal tubules on one side of an endodermal cell in root of *Smilax glauca*.

endodermal cells making up the endodermal duct canals that produce polyenes, polyynes, and polyeneynes (Van Fleet, 1971). There is a unilateral distribution of peroxidase with the heaviest localization in either the inner or outer tangential walls and less commonly there is a uniform distribution in all walls. Analysis of wall-bound heme has not been completed. The deformation of wall polymers holding the peroxidase requires reagents and solvents that at least partially destroy the enzyme. In general, peroxidase, cytochrome oxidase, and polyphenolases are distributed in the cytoplasm in early stages of development but become localized in a mosaic in cell walls with age (Van Fleet, 1959, 1961, 1971).

Peroxidase is bound in a lipidlike polymer which may be blocked in formation by growing whole plants in an ethylene atmosphere or growing roots in ethylene-saturated water (Van Fleet, 1963). The peroxidase may be coupled with lipids and phenols through aldehyde coupling or, in combination with specific proteins, with tryptophan, phenylalanine, or tyrosine in tubules in developing suberic walls. The formation of polymers of peroxidase with polymer micelles of cell walls may require a specific and as yet unidentified catalyst.

Peroxidase in the endodermis and bundle sheath cells is most pronounced in the roots, and it is equally evident in the endodermis of stems up to the groundline, but is less evident in or absent from the endodermis of aerial stems and leaves. A dark-type peroxidase which is more reactive in diffuse light or in the dark is found in roots. There is an induction period or lag in reaction rate of longer than 4 minutes when dark-type peroxidase is exposed to full light. The dark-type peroxidase produced in roots and on the shaded side of stems is inhibited in reaction rate on exposure to light, and it is either present in low quantity or may not be produced at all in tissue exposed to light, depending on the generic plant type.

Similar light-sensitive isoenzymes of peroxidase are produced in relation to the effect of gravity on the plant (O'Conner, 1971). The effect of gravity on the electric potentials in plant tissues, as shown in the remarkable investigations by Brauner (1969), polarizes the stem or root and modifies potentials across membranes. The upper sides of horizontally placed stems are more negative than the lower. O'Conner (1971) found a similar effect on development of isoenzymes of peroxidase in relation to light and gravitational stimuli.

Polyphenolases and cytochrome oxidase are often less readily detectable in the above-ground or light-exposed endodermis. Polyphenols are oxidized to visible quinones in shaded and darkened stem tissue and in roots grown in the dark. The oxidation of phenols to quinones catalyzed by polyphenolase in the endodermis of some members of the Ericaceae is immediate on wounding the endodermis and may be seen without the assistance of any optical instrument. Polyphenolase is most easily detected in the endodermis under conditions of wounding and unfavorable site locations on north-facing and cold slopes as compared with the same clonal plants grown under favorable conditions. Varietal and histochemical ecotype differences in relation to altitude, latitude, and site location are easily determined in the genus *Smilax* (Van Fleet, 1950, 1961).

The accumulation of phenols in the endodermis has a relationship to β-glucosidase activity, and the concomitant oxidation of phenols to quinones, particularly in the subaerial, wounded, and diseased cortex. A functional β-glucosidase is apparently present in endodermal cells in many plant families while in other plant groups an adventive glucosidase activation in the endodermis results from damage to cortical cells. Several kinds of aglycons of the arbutin, phloretin, and parillin types are localized in the endodermis and are abundant in many ferns and many genera of the Ericaceae. Several direct and indirect methods have been employed in demonstrating the localization of β-glucosidase in the endodermis (Van Fleet, 1952, 1961, 1962). While no one method meets all the criteria of validity, the adductive evidence shows that glucosidases are present in the endodermis and are activated at the slightest injury.

Esterases (lipase, phosphatases, and myrosulfatase) have been demonstrated to be present in the endodermis and in its homolog, the bundle sheath, which surround the vascular bundles of leaves and stems. Activation and function of these enzymes in the endodermis seems to be related to the accumulation of oils and esters which may have significance in defense reactions. The limited references in the literature to esterases of the endodermis is indicative of the difficulty of the research. There is some evidence for nonspecific esterases in the endodermis located in microfibrils paralleling the long axis of the plant over the outside tangential walls of the endodermal cells and forming a continuous network over the surface of the stele.

Phosphorylase, phosphatases, and amylases function in the endodermal boundary between cortex and stele particularly in above-ground stems. Their activity is determined by many factors but most obviously by the rate of carbohydrate supply. Phosphatase–phosphorylase balance in the endodermis of some plants leads to starch accumulation in the day time and transformation to soluble sugar at night. When there are injuries to the cortex, starch does not appear in the endodermis.

Among the significant enzymes of the endodermis are those associated with the production of aromatic and lipid compounds that contain conjugated unsaturated bonds. Perhaps the most important enzymes in the endodermis are those involved in the chain mechanism that leads to the desaturation of lipids. A survey of the enzymes and compounds formed in the endodermis will lead us inevitably back to questions on the inherent localization of enzymes in the endodermis. The following sections on the compounds formed by the endodermis will give the background for a postulate on the genetic basis for the kinds of enzymes and compounds found in the endodermis.

Phenols, Naphthols, and Anthrols

In the young and developing endodermal cells, quinol-quinone compounds are in the reduced state without detectable accumulation of oxidized and polymerized quinones. Oxidized quinones appear in the endodermis on injury or in unfavorable growing conditions. Accumulated phenols, naphthols, and anthrols are all held in a state of reduction in the healthy endodermal cells where the oxidation-reduction potential of surfaces and protoplasts may be too negative to allow phenol oxidation

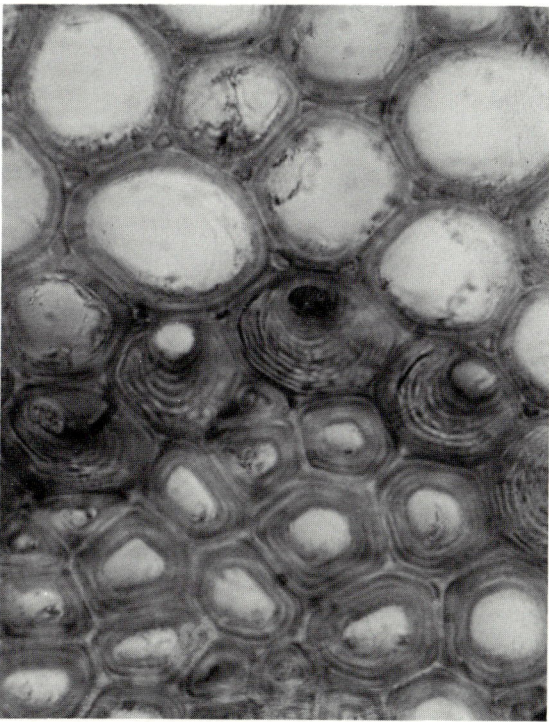

Fig. 2. Phenol-quinone deposits in unilateral layers of endodermal cells between cortex and stele in root of *Smilax glauca*.

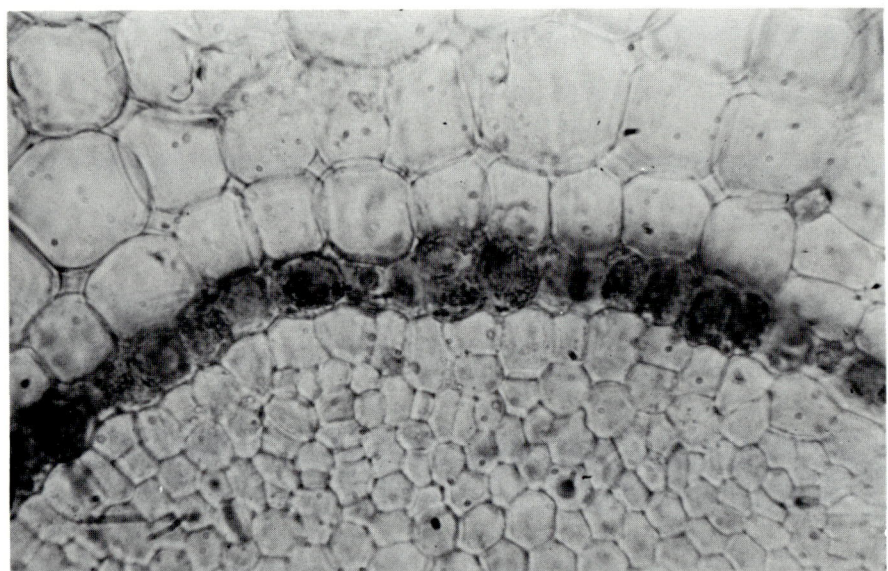

Fig. 3. Localization of hydroxynaphthols in endodermis of root of *Impatiens pallida*. The naphthol compounds were converted to visible colored nitrosonaphthol polymers for photographic purposes. Oxidation of naphthols to naphthoquinones is autocatalytic on wounding and sectioning the cells, but naphthoquinone polymers have a lower optical density and color than the nitrosonaphthols.

by either autoxidation or enzyme catalysis (Figs. 2 and 3). Only on disruption is there evidence of quinones, naphthoquinones, and anthraquinones (Van Fleet, 1955, 1957, 1961). Phenolic compounds, as shown by diazonium coupling and Schiff base reactions with quinones, indicate an increase in phenols with injury and the formation of quinones, just as in infected areas of epidermal and vascular tissue (Oku, 1967).

Microprobe samples of the endodermis in plants of the Juglandaceae, Ericaceae, and Boraginaceae show the characteristic absorption maxima at 250 nm of the ubiquinones. The color forms and broad absorption at 440 nm indicate a marked localization of anthrols–anthraquinones in the endodermis of plants in the Rubiaceae, Rhamnaceae, and Polygonaceae. The localization of these compounds in the endodermis comes at the end of the cell division phase; they may be in either the oxidized or the reduced form (Van Fleet, 1954). Quantities of quinones visible to the eye appear in the endodermis on injury to the cortex. A slow *functional oxidation* of phenols takes place in the endodermis, but there is an accelerated *adventive oxidation* when there are lesions in the adjacent cortical cells. In the healthy aerial stem the redox potential is apparently poised at a point below the level for oxidation of phenolics. Continued oxidation and deposition of quinones takes place, if lesions are formed, in the walls of the endodermis, but the polymerized quinones do not block the synthesis of phenols.

The tautomers of phloretin and other phenolics are heavily localized in the endodermis of roots and the endodermis around every vascular bundle of the leaves and stems (Fig. 4). Whether these phenolics are present as glycosides or aglycons, whether in the oxidized or reduced form, they are heavily localized in one tissue. Whether they are present as glucosides or aglycons is of importance as shown by Goodman et al. (1967) and

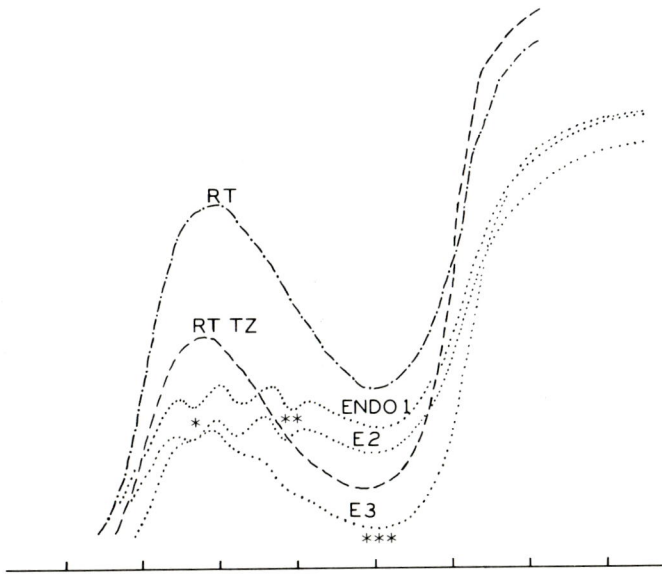

Fig. 4. Transmittance minima of microprobe samples from root of Prunus serotina. RT, from cortical parenchyma; RT TZ, from cortical parenchyma in root–stem transition zone; ENDO 1, from endodermis in primary stage of differentiation; E2, from endodermis in secondary stage of development; E3, from mature endodermis containing oxidized and polymerized phenols. Abscissa is wavelength in nanometers in the ultraviolet. Transmittance minima between 257 (*) and 269 (**) nm in ENDO 1 and E2 are evidence for selective energy absorption bands of the conjugated bonds in phloretinlike phenolics; the same minima are obtained from the endodermis of Onoclea sensibilis and other ferns. The loss of fine structure minima in E3 and the stronger absorption at 280 (***) nm coincides with the visible oxidation of phenols to polymerized and condensed quinones.

others who have found that phloretin has greater antibacterial activity than the glycoside phlorizin.

The increase of phenols in the endodermis under infection was clearly demonstrated in the pioneering research by Best (1944, 1948). He found that scopoletin is present in greatest concentration in the hypodermis and endodermis and increases in these tissues in virus-infected tobacco plants. W. A. Andreae (1948), S. R. Andreae and Andreae (1949), Sequeira (1967), Winkler (1967), and Wender (1970) have given additional information and confirmation of these observations. It has been found that glucoside scopolin is present in larger amounts than the aglycon, but there is a progressive increase in the latter with the spread of disease (Steck, 1967). Inhibition of oxidative phosphorylation and the transport of amino acids through cell membranes by aglycons of various phenols of this type may lead to accumulation of amino acids as shown by Parups (1967). Higher levels of free amino acids in zones of infection and in the endodermis may be a general syndrome of disease.

The oxidized products of aglycons have been shown to inactivate the pectinases released by fungal pathogens (Raa, 1968), and there are many examples of plants in which fungi do not penetrate beyond these compounds in the endodermis. Quinones deposited in the walls of the endodermis would be expected to block the action of esterases of fungi. Hyphae of fungi do not penetrate the catechol tannins in the endodermis of roots of pine (MacDougal and Dufrenoy, 1946). The reversible system, tannins–phenols–quinones, has a broad prescription of probable functions in preventing energy loss, the entry of parasites, the loss of water, and possibly even in systems of wintering-over and frost resistance. However, the complete functional significance of tannin–quinol–quinone molecules in the endodermis has not been established.

Alkaloids

It has been established that some alkaloids are toxic to man and to a lesser degree to insects, mammals, and microorganisms. Colored crystalline precipitates with various reagents give some evidence as to their cellular distribution. In general, alkaloids appear in all cell types of young plant organs but with increasing age become localized in dermal, subepidermal, vascular bundle sheath (endodermal) cells, and in isolated specialized parenchyma cells. Sieve tubes of the phloem are in general free from alkaloids throughout the life of the plant; at least the alkaloid concentration in sieve tubes stays below the level of detection by histochemical methods (Mothes and Romeike, 1958). A general distribution throughout the plant is apparent from general reviews on the subject (James, 1950;

Cromwell, 1955), but accumulation has been observed in the endodermis, pericycle, and bundle sheath, and particularly in wounded tissue. Cells beneath the periderm of roots of two species of *Mahonia* contain berberine, which was found by Greathouse and Watkins (1938) and Greathouse and Rigler (1940) to confer resistance to *Phymatotrichum omnivorum*. The antifungal activity of some alkaloids toward pathogenic fungi has been established but a survey has not been made on the specialized threshold of production of alkaloids in either the endodermis or zones of infection.

POLYENES AND POLYYNES

The endodermis in roots of many species produces molecules with possibly the highest mean unsaturation and caloric value in the organic world. The cells producing these unsaturated lipids are multinucleate and may be 8n or higher as compared with other cells in the plant. Is there a causal relationship between endopolyploidy and the development of conjugated unsaturated bonds in polyenes, polyynes, phenols, and naphthols? One obvious answer is that endopolyploidy affords a redundancy capacity in enzyme and energy production for the formation of highly unsaturated compounds. Similar polyenes are also found in many seeds (where endopolyploidy is at least 3n in the endosperm or aleurone) and in polyploid cells in wound lesions.

Polyploidy, or endopolyploidy, is not the sole causal mechanism in the development of lipids with conjugated unsaturated bonds. The endodermis, located between the cortex and the stele, may be considered as a highly active interface with localized and wall-bound enzymes which function to fill the endodermal ducts with a variety of compounds with conjugated unsaturated bonds.

There are numerous examples of the induction of endopolyploidy on wounding or infection, Kamilova (1959, cited in Gorlenko, 1965) has found polyploid (4n) shoots in tomato plants infected with *Pseudomonas tumefaciens*. Enlargement of nuclei in polyteny and localized polyploidy is one of the first signs of response to wounding and infection. Shahied (1970) has found a functional parallel between the effect of some herbicides and the induction of endopolyploidy by colchicine. He found an increase in enzymes in zones of cortical, endodermal, and pericycle cells which developed extra sets of chromosomes.

The direct browning of tissue on bruising, which brings together phenol and polyphenoloxidase, gives rise to quinones which "tan" the protein and form a protective film over the wound and may block bacteria (Szent-Györgyi, 1968). The distribution of quinones is of obvious importance, but, held as insoluble masses in the vacuole or in loose and random arrangement in the cell, they may not afford protection. Polymerization of phenol quinones with lipids in uniform distribution in cell walls presents a more effective block to the spread of disease.

The nematicidal compounds in the roots of *Tagetes* (Uhlenbroek and Bijloo, 1958, 1959) are produced by the endodermis (Van Fleet, 1971). Samples of 50 μg, taken with microprobes, when placed in 500 μl of hexane in microcuvettes show the characteristic absorption maxima for terthienyl compounds.

The many kinds of polyacetylene compounds in the Compositae, Umbelliferae, and the Araliaceae are produced by the endodermis or by a small group of four endodermal cells that form a duct (Van Fleet, 1971). A preinfectional, antifungal, acetylenic ketoester has been found in shoots of broad beans (Fawcett *et al.*, 1965, 1968). Resistance to halo-blight disease in French beans has been found to be associated with unsaturated lipids in a persuasive analysis made by Deverall (1964).

Localized development of unsaturated lipids on wounding is easily demonstrated and the isomerization and conjugation of unsaturated lipids is apparently common. In the formation of a suberized cicatrix on wounding, the major components appear to be condensed lipids, saturated and unsaturated, and phenolics. The activation of the formation of compounds that undergo oxidative deposition is immediate and the formation of barriers to energy loss is presumably analogous to the formation of the compounds that make up the impervious *Casparian band* around each endodermal cell. A parallel lipid formation and oxidative deposition is found in wound and infection zones. The localized activation of enzymes is as significant as their rate of reaction. Enzymes determine the rate but not the direction of reactions, but when enzymes are localized and bound they are more than catalysts; they control the direction of reaction. The development of a sheathing layer of compounds with conjugated unsaturated bonds may provide a bound system of oxidation–reduction for redox exchange across lipid films.

Genetics of Endodermal Lipids

As mentioned previously, across family and tribal groups the endodermis gives the appearance of having its own genetic constants. Within some families the development of one type of compound extends to all tribal groups. In general, some type of polyunsaturated compound appears to be characteristic of the endodermis in each phyletic group that has been studied. In endodermal cells producing polyacetylenes both generic fixation and phenotypic variation may be found.

An analysis of polyacetylene production by the endodermis was made by the writer to determine the nature of inheritance in the endodermis and to develop a concept of the genetics of the endodermis. *Coreopsis* plants were selected for the analysis because of the extensive and valuable information on this genus from chemical and systematic investigations by the Sörensens (J. S. Sörensen and Sörensen, 1954, 1958a, b; N. A. Sörensen, 1961; N. A. Sörensen *et al.*, 1954), and Jones *et al.* (1954), and because of available natural hybrids and ecotypes of *Coreopsis saxicola* and *Coreopsis grandiflora*. Polyacetylenes samples were taken directly with microprobes in order to determine the nature of the predominant compounds in the endodermis (Van Fleet, 1971). From crosses made

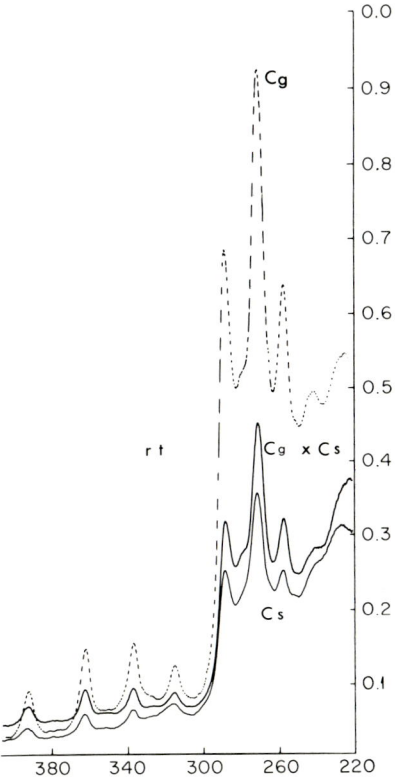

FIG. 5. Ultraviolet absorption in hexane of the tridecadienetetrayne compounds predominant in the endodermis of the root (rt) of: Cg, the *Coreopsis grandiflora* parent; Cs, the *Coreopsis saxicola* parent; Cg × Cs, the hybrid of *Coreopsis grandiflora* × *Coreopsis saxicola*. Ordinate is optical density; abscissa is wavelength in nanometers. From Van Fleet (1971), by permission from the editor of *Advan. Front. Plant Sci.* **26**:109–143.

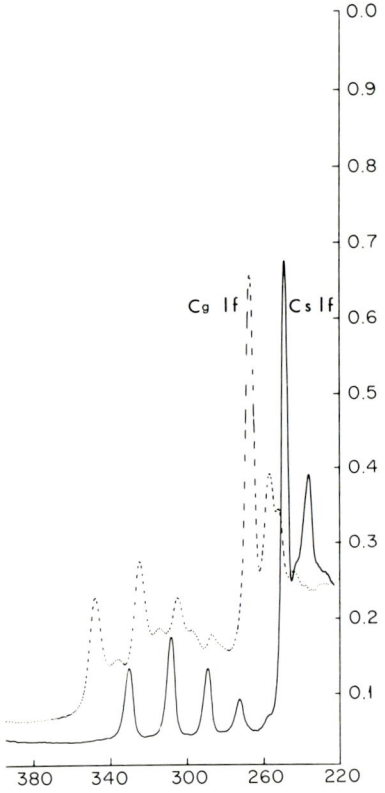

Fig. 6. Ultraviolet absorption in hexane of: Cg 1f, a mixture of phenyldienetriyne and tetraenediyne from the endodermal bundle sheath ducts of the leaf of *Coreopsis grandiflora*; Cs 1f, the predominant phenylheptaenetriyne from the endodermal ducts in the leaf of *Coreopsis saxicola*. From Van Fleet (1971), by permission from the editor of *Advan. Front. Plant Sci.* **26**:109–143.

between *C. saxicola* and *C. grandiflora*, and from field hybrids recognizable as ecotypes, the following data were obtained:

1. The endodermis of the roots of *C. saxicola*, *C. grandiflora*, and their hybrids produces predominantly a tridecadienetetrayne (Fig. 5).

2. The endodermis of stems and leaves of *C. saxicola* and *C. grandiflora* produces predominantly the phenylheptaenetriyne (Fig. 6).

3. The endodermis of stems and leaves of some forms of *C. grandiflora* and in many of the *C. saxicola* × *C. grandiflora* hybrids produces a mixture of tridecatrienetriyne and phenylheptaenediyne.

4. Hybrids and hybrid ecotypes could not be distinguished on a basis of polyacetylenes produced by the endodermis of the roots. They produced predominantly the tridecadienetetrayne, the same as the parents.

5. The following hybrid ecotypes could be distinguished on a basis of polyacetylenes produced by the endodermis of their leaves and stems:
   a. *Echols:* Endodermis of stems and leaves producing predominantly the phenylheptaenediyne.
   b. *Echols G:* Endodermis of stem and leaves producing predominantly the tridecadienetetrayne.
   c. *Eatonton:* Endodermis of leaves producing predominantly the tridecadienetetrayne.
   d. *Eatonton T:* Endodermis of the leaves producing predominantly the tridecatetraenediyne.
   e. *Crawford:* Endodermis of stems and leaves containing complex mixtures of *cis* and *trans* tridecatrienetriyne and the phenylheptaenediyne. Cis isomers showing characteristic double maxima in the fine structure of their ultraviolet absorption spectra were predominant in young stems and gave way to trans isomers with age.
   f. *General:* Endodermis of leaves and stems producing mixtures of phenyltriynes and phenylenetriynes.

In summary, the endodermis of roots of the hybrids produces the same compounds as in the roots of the parents (Fig. 5), whereas the compounds produced by the endodermis of leaves and stems of hybrids are predominantly like those produced by the endodermis in the leaves and stems of the *C. grandiflora* parent (Fig. 7). The endodermis of leaves of hybrid ecotypes produces a mixture of compounds unlike those found in the leaves of either parent. The phenyl compounds are produced more commonly in the aboveground endodermis of parental and hybrid types and the synthesis of the phenyl ring directly from unsaturated aliphatic precursors, as postulated in chemical analysis by Skattebøl and Sörensen (1959), apparently takes place in the aerial endodermis. It is apparent that there are multiples of dominant genes in the ploidy of the endodermis of the root which result in the formation of one type of polyunsaturated compound, with *no* genetic variation apparent. The high level of ploidy in the endodermis of roots and the general invariant development of the tissue gives a picture of genetic fixation. In the aboveground tissue, the greater range of compounds detected presumably reflect a lesser development of endopolyploidy, or greater variation in ploidy permitting the expression of other genes.

Wounding of *C. saxicola* and *C. grandiflora* results in the production of polyacetylene compounds within the wound zone. The compounds formed are similar to the types found in the roots, stems, and leaves of these parents

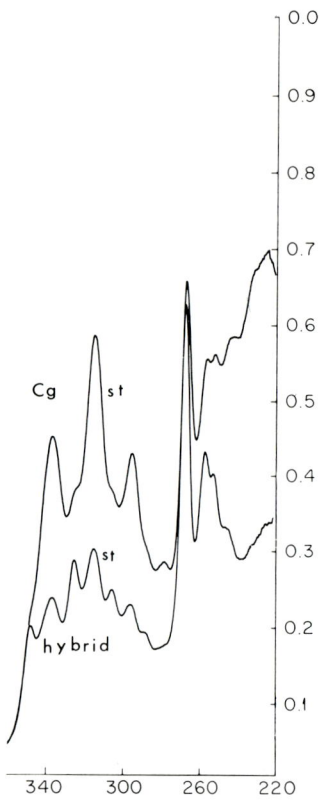

FIG. 7. Ultraviolet absorption in hexane of the predominant phenyleneyne in the stem endodermis from: Cg, *Coreopsis grandiflora*; st hybrid, a hybrid from the cross *Coreopsis grandiflora* × *Coreopsis saxicola*. From Van Fleet (1971), by permission from the editor of *Advan. Front. Plant Sci.* **26**:109–143.

and their hybrids. There are differences in polyacetylenes produced in relation to age of the tissue at the time of wounding. In wound response the balance of passive and active genes in hybrid types appears to be determined by age of the tissue. This analysis of histochemical systems of polyacetylene production has not been completed but similar results have been obtained in hybrids in two other genera.

Conjugated trienes and tetraenes in *Psilotum* and *Tmesipteris* from the belowground and the aerial endodermis, and from wound zones, show no evidence of genetic variation. The differences in the unsaturated lipids in aboveground and belowground tissue indicate a sharp and consistent segregation at the groundline. Characteristic absorption maxima for trienes are found in spectra of the oils from the young endodermis, while tetraenes are found in the mature endodermis (Fig. 8). A mixture of triene and tetraene molecules is produced in the transition zone at the groundline. The predominant polyene from the young endodermis is a hexadecatrienoic acid and occurs as the cis-trans-trans isomer, the trans-trans-trans isomer, or a mixture of these two isomers, but not as the cis-cis-trans isomer. A tetraene is found in the mature endodermis and is either parinaric acid or 2:6-dimethylocta-1:3:5:7-tetraene but the precise absorption maxi-

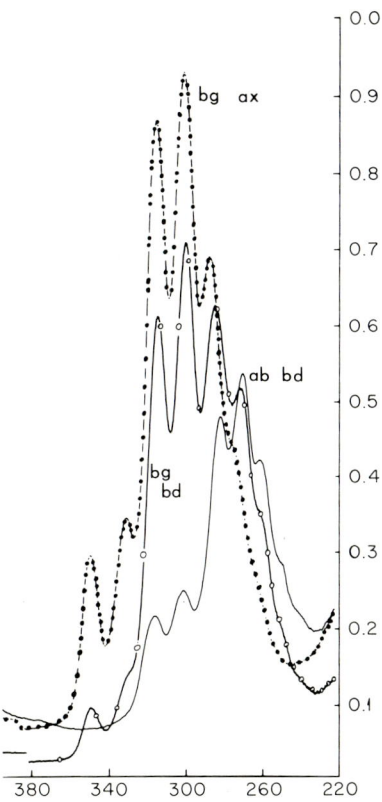

FIG. 8. Ultraviolet absorption in hexane of: bg ax, tetraene from endodermis in belowground axis of *Psilotum nudum*, with main absorption maximum at 300 nm; bg bd, tetraene, from endodermis in belowground, dormant bud of *P. nudum*; ab bd, triene from endodermis in first stage of differentiation in young, aboveground bud. From Van Fleet (1971), by permission from the editor of *Advan. Front. Plant Sci.* **26**:109–143.

mum for the main peak at 300 nm suggests an eicosatetraenoic acid. Tetraenes are produced in greater abundance with increase in age of these endodermal cells. Under etiolation, unfavorable growing conditions, and wounding, trienes are produced by hypodermal and cortical cells. With the great antiquity (380 million years) of these members of the Psilophytales and their extreme polyploidy (244 or more chromosomes) there is genetic fixation. This system of polyenes must have developed early. The fossil record of related forms shows no morphological change in the endodermis during those millions of years. These early land plants must have had a relatively short period of successful selection of histochemical systems.

There is the interrelationship in *Psilotum* with a mycorrhizal fungus in both gametophyte and sporophyte tissue in a balanced commensalism. The outer cortical and epidermal cells of the belowground axis contain the mycorrhizal fungus and the polyene and resin-filled cells of the endodermal zone apparently act as a check to systemic infection. The most recently derived vascular plants, the orchids, show the same type of commensal systems in the cortex and we speculate that selection on a histochemical basis has arrived at the same developmental state both in the beginning and at the end of a long evolutionary spiral.

Enzyme Localization in Health and Disease

One of the most successful steps in evolution has been the localized development of saturated, solid, and insoluble waxes covering the surface of epidermal cells in plants. It is a system that confers some generalized immunity against many pathogens and at the same time functions against water and energy loss as a part of the negative entropy of plants. Horticulturists, plant pathologists, and others, are necessarily engaged in their research in the difficult problem of finding plants with resistance to disease when the natural barriers to disease have been removed. These men must work with plants in which these barriers have been to some degree lost through breeding and selection of desirable characteristics not necessarily linked by inheritance with barriers against pathogens and disease. Many horticultural and agronomic crop plants have dermal coverings that are not as efficient as in wild-type plants. There are few horticultural plants with anything like total immunity to disease.

The enzymes activated or adventively formed as a result of infection are the catalysts of resistance, and it must be emphasized that the localization of these enzymes defines a four-dimensional and heritable pattern of resistance just as all systems of inheritance follow cellular configurations of length, breadth, depth, and time.

Where nematodes penetrate the cortex and stele of the root, enzymes are more active at the sites of infection than in adjacent noninfected tissue as shown in the extensive investigations by Veech and Endo (1969) and Endo and Veech (1969, 1970). Their work emphasizes the increase in enzyme localization within polynuclear cells in the zone of infection. Analysis of the enzymes in the polynuclear syncytia, as a type of polyploidy, is of great importance and is basic to an understanding of other dimensions of the reaction to infection. In the work by Veech and Endo, noticeable increases in enzyme activity were not found outside the polynuclear cells of the typical syncytium in the zone of infection. If a plant has genes for resistance to nematodes what is the significance of ploidy of these genes in the zone of infection? Students of this type of nematode infection have found that resistant plants do not produce syncytia to the same extent as susceptible plants. What does this mean? Any increase in syncytial ploided cells in susceptible plants over the amount in resistant plants would not serve to block infection. Ploidy without any dominant genes of resistance would indeed be a large and empty well, but not to the nematode.

In host–pathogen interactions there is much more than a gene-for-gene mechanism with a gene for resistance in the host and a corresponding gene for virulence in the pathogen. Explanations have been given, on the Jacob-Monod model, of the bridging of inducer and repressor substances by regulator genes of parasite and host (Linskens, 1968). Explanations based on regulator genes are formalistic and conventional, and they are academic in that they apply equally to the function of all systems. Our problem is the analysis of the origin and segregation of chemical compounds of resistance and the origin of the causal factors in their selection. The problem includes an analysis of the localization of enzyme systems.

In the adventive activation of a gene–enzyme system of resistance or in the stimulation of an active gene–enzyme system of resistance there appears to be an involvement of dehydrogenase activity. For example, dehydrogenase activity may not be found directly in the inclusion-bearing cells of tobacco leaves infected with tobacco mosaic virus but adjacent cells may be activated, as shown in the cytochemical studies made by Takahashi and Hirai (1963). Cells adjacent to wound zones show a general increase in several kinds of dehydrogenases in addition to many other kinds of enzymes as demonstrated in the extensive survey made by Surrey (1955). The localization of several kinds of dehydrogenases has been determined by Tschen and Fuchs (1968) with tetrazolium salts in bean leaves after infection by *Uromyces phaseoli*. They found that distribution and activity of the dehydrogenases differed in zones related to infection.

Peroxidase differences in resistant and susceptible plants may be readily demonstrated by physical and chemical methods (Van Fleet, 1947, 1952, 1962). Varieties of tomato plants resistant to Race 1 of *Fusarium oxysporum* f. sp. *lycopersici* have a more active and abundant peroxidase in the phloem and bundle sheath both before and after infection than in the same tissues of more susceptible varieties (Van Fleet, 1947). Rudolph and Stahmann (1964) found an increase in peroxidase in resistant plants whereas it remained unchanged or decreased in susceptible hosts, and they concluded (as others have) that peroxidase activity favors resistance to some pathogens. General conclusions have been reached that peroxidase is stable to the action of toxins produced by most microorganisms (Rubin and Artsikhovskaya, 1963). Localized increase in peroxidase during the advance of morphological symptoms of disease has been observed by other plant pathologists, as in the work of Kljajic *et al.* (1964) on virus-infected strawberry plants and in the analysis of wildfire disease by Lovrekovich *et al.* (1968). A positive correlation between peroxidase activity and degree of resistance in potatoes to late blight was found by Kedar (1959) in an extensive study which showed that major genes for resistance are derived from *Solanum demissum* and minor genes from *S. tuberosum*. Kedar observed that higher peroxidase activity was associated with a higher rate of necrosis as a correlate of resistance and blockage of the spread of the pathogen.

It has been observed that isoenzymes of peroxidase increase in number at the site of lesion formation through an increase in protein synthesis (Farkas and Stahmann, 1966). However, in observations on peroxidase in infected tobacco plants susceptible and resistant to black shank, Veech (1967, 1969) found no detectable differences in the isozymes of peroxidase, although the site of peroxidase localization changed as a result of infection. While Veech sagaciously avoided any general conclusions, it seems reasonable to suggest that differences between resistant and susceptible lines of tobacco must be expressed in some group of compounds or isoenzymes other than peroxidase. In many other species, peroxidase isoenzyme differences are characteristic of wounding and infection. The rapid changes in localized areas must be a result of synthesis or activation of isoenzymes already present; the exchange rate for heme proteins is too low (Shannon, 1968) to account for the rapid development of differences in localized zones.

The denaturation of protein in wound zones increases the kind of protein that has heme-binding capacity. When free heme as hematin is added to wound tissue, the heme absorption bands as seen with the spectroscopic ocular are greatly intensified and heme absorption in wounded phloem is so intense that the tissue becomes pink and false peroxidase reactions are intense (Van Fleet, 1959). Keilin (1960) examined the nature of heme-binding groups in native and denatured hemoglobin and found that denaturation of globin increases its heme-binding capacity at least 6-fold.

Peroxidase indicator reactions are based on the affinity of oxidized amine dye radical intermediates for the kind of protein in heme enzymes (Van Fleet, 1959; Hirai, 1968). Wounded cells have a high affinity for oxidized quinones and absorb the quinones at a preferential rate about 4 times as fast as adjacent unwounded cells, as determined by a recording microdensitometer. Absorption of hemin, hematoporphyrin dihydrochloride, and heme in pyridine, gives the tissue a pink to red color. The development of electrophoretically separated isoenzymes of peroxidase by aminophenol indicators, where reactions take longer than 1 minute for completion, may be a demonstration of bands of proteins that are not a true peroxidases.

It is becoming clearer that, among other things, peroxidase functions as a catalyst to aerobic oxidation, even though there are complicated forms and products of the enzyme (Yamazaki, 1966). Respiratory energy for ion uptake and selection in zones of wounding and infection is associated with heme-containing enzymes. Heme enzymes of the peroxidase type are abundant in phloem companion cells, meristems, wound regeneration tissue, necrotic tissue, the endodermis, and so on. Isoenzymes of peroxidase in localized systems would seem to function in the terminal oxidation balance. Peroxidase is obviously not just a blocker of vectors, pathogens, or toxins.

Different functional forms of peroxidase appear in plants to be associated with different tissues just as in animals. The complex of isoenzymes of peroxidase may all catalyze the same type of reactions, but with differing efficiencies in each location.

The browning of tissue and oxidation of phenols to quinones may be observed in many plants on wounding or following infection. Phenol oxidases have been found by Mace (1964) to be associated with oxidation of dihydroxyphenylalanine to indolequinone in *Fusarium*-infected banana roots. Polyphenoloxidase is three to ten times higher in tomato plants invaded by *Fusarium oxysporum* f. *lycopersici* than in healthy plants. There is a general view that rapid necrosis and hypersensitivity in resistant plants is correlated with higher levels of phenol oxidation. Endo and Veech (1970) have found necrosis and browning associated with plants resistant to nematodes, but they found no localized differences in phenolase. Phenolase detection in this case may have been blocked by high levels of free phenols and quinones in the late stages of necrosis. The absence of phenolase in browning tissue resistant to nematodes in their research might be anticipated. Once the phenol-quinone oxidation reaction is initiated by polyphenolase, the oxidation can proceed by an autocatalytic chain reaction catalyzed by the quinones. Solymosy et al. (1959) found that when leaf tissues of tobacco were damaged with a needle to form lesions similar to those induced by tobacco mosaic virus, the activity of polyphenolases did not show any increase comparable to that in leaves with virus-induced lesions. Solymosy concluded that enhanced polyphenoloxidase activity was a cause, not a result, of cellular breakdown and was the mechanism responsible for inhibiting and localizing the pathogen.

Autoxidation of lecithin, linoleic and linolenic acids, and related unsaturated compounds takes place in all lesions to some degree. Suberic deposits are formed which contain saturated phellonic and phellogenic acids and related fatty acids condensed with phenols. The oxidation of compounds like linolenate to chromophores with highly characteristic fine maxima in their ultraviolet absorption spectra may be found within minutes in some wounded tissues. Oxidative isomerization to conjugated dienes and trienes is common in many plants that have general immunity. Diene ketones and various carbonyl groups formed at this time may account for the false peroxidase reactions in wound lesions. Schiff base and thiobarbituric acid detection of aldehydes (Bernheim et al., 1948) should be made along with rate reactions for true peroxidase.

Successive layers of suberic material may be laid down in lesions, as many have observed, where there are continued advances and penetration of pathogens (De Baun and Nord, 1951). Where suberic deposits are uniform as in cork layers there is rarely penetration by any parasite.

β-Glucosidase is functionally active in the formation of suberic tissues but is usually only adventively activated on wounding or infection in most tissues (Van Fleet, 1962). Alfalfa leaves inoculated with fungi show an increase in β-glucosidase 2 days after inoculation (Olah and Sherwood, 1970). Flavones, isoflavones, anthocyanidins, and other phenolic aglycons are associated with β- glucosidase in many tissues where they accumulate. After glycolysis and oxidation, quinone polymerizations and "tail to tail" polymerizations of gallocatechins then give rise to "phlobatannins" so common in bark.

These phlobatannin polymers probably are one of the highest advances in the attainment of immunity. They are impervious to most pathogens and to all but the most drastic of chemical solvents. The cambium that gives rise to the cork of the bark, and its daughter cells that are beginning to suberize, can be shown to have, after acetone extraction of interfering phenolic substances, β-glucosidase and polyphenoloxidase. Hathway (see review by Endres, 1962) as a part of his extensive analysis of oak bark constituents has presented evidence for the polyphenoloxidase oxidation of pyrogallol groups and a quinone polymerization enzyme in the formation of oak bark.

The protective role of polyphenols and tannins in the general inactivation of fungal extracellular enzymes (the pectic enzymes of cell wall hydrolysis) is now generally recognized (Pridham, 1960), and it must be emphasized that selection by man of palatable plants has obviously resulted in the elimination of genes that control tannin formation. Feenstra's (1960) remarkable investigations of the interactions of genes in the control of phenolic compounds in the seed coat cell layers of French bean is one of the few analyses we have of inheritance and localized formation of polyphenols and tannins in barrier tissues of the seed coat. Aglycons of flavones and flavanol adsorbed in cellulose walls, and the planar structure of the phenolic aglycons which blocks their translocation (Roberts, 1960), results in sharply localized deposition. What is the threshold of action of alleles in controlling the sharply defined differentiation of β-glucosidase in a given tissue?

Analysis of the Localization of Compounds in Health and Disease

Cells may contain only the glucoside, adjacent cells may contain β-glucosidase and the aglycon, and other adjacent cells contain neither the enzyme nor the glucoside. Dominant genes probably determine the development of glucosides and β-glucosidase, but what determines the localization of these compounds in one type of cell? There must be genes that modify location, and these genes are in the line of survival and selection. They are the least known, the most fundamental; they determine the four dimensions of chemical function.

Gene balancing results in patterns of localization that may or may not have survival value. Phlorizin is the principal phenolic in apple leaves and bark, with the highest concentration in the bark of the roots. Arbutin, a phenolic in pears, is abundant in the leaves but present only in small amounts in the roots. Hybridization of apple and pear achieved at the John Innes Horticultural Institution gave a plant in which the normal patterns of distribution of arbutin and phlorizin were superimposed (Williams, 1960). These compounds are not translocated, but are stationary at the locus of formation; arbutin is not translocated from pear across stock-scion junctions in grafts of apple and pear, and no movement of phlorizin has been found in various types of grafts. It would seem, therefore, that there is a breaking of the linkage of genes for formation with those that determine localization. There is the phlorizin–root bark linkage in one parent and arbutin–leaf in the other. The hybrids show a break in localization linkage; but the precise nature of these linkages, whether chromosomal or functional, positive or negative linkages, has yet to be established.

It has been observed that unpigmented plants are generally weak; this suggests that genes controlling pigment production are linked with genes affecting general vigor in growth (Harborne, 1960). Plants lacking anthocyanin as full recessives may lack vigor, and the comparison of growth of unpigmented and pigmented sectors and the comparison of pigmented and unpigmented sibling plants suggests functional but not necessarily chromosomal gene linkages (Van Fleet, 1969).

In the normal functional development and accumulation of phenolics and compounds with conjugated unsaturated bonds in the endodermis we see a linkage of genes for formation and localization. Plants with resistance to disease synthesize and accumulate phenols at a rapid rate as a result of wounding and infection (Clauss, 1961; McClean et al., 1961; Farkas and Kiraly, 1962). Phenolics accumulate in infected zones and oxidize to quinones (Oku, 1964), but any linkage to tissue type seems to be quite general. In the localization of α-methylene butyrolactone in the white skin layers of tulip bulbs resistant to *Fusarium oxysporum*, as found and described by Bergman and Beijersbergen (1968), there would seem to be a localized and functional linkage. The coumarin compounds of the phytoalexins are said to form more rapidly in resistant plants according to the genotype of the host (see review by Metlitskii and Ozeretskovskaya, 1968). We have no information as to which cell types are the more precisely responsive in producing phytoalexins, although the parenchymal tissue lining the pea pod shows a 10-fold increase in phenylalanine ammonia-lyase enzyme in the phenylalanine to pisatin pathway (Hadwiger, 1968). Established linkages might be expected to be found in localized chemical systems of immunity rather than in man-selected garden peas, in which degrees and linkages of resistance may have been lost through selection on a basis of palatability.

Perhaps the most highly evolved of gene linkages in the chemical systems of health and disease in plants are to be found in the mycorrhiza of orchids. Specialized dermal systems hold the fungus in the cortex and, after a period of hyphal storage, a layer of cortical cells in the host sends pseudopodia-like projections from their nuclei into the hyphal mass, which is then digested by the host cells (Burgeff, 1959). This highly complex commensalism is sharply localized; the fungus never becomes systemic and is both nurtured and fed upon by the host. The chemical sequestering systems have yet to be described, and we can only guess at some of the linkages that are developing in these and similar systems that have evolved recently in the Ericaceae and the Orchidaceae.

Conclusions

During plant evolution, survival has been the child of a morganatic marriage of genetic exploration and the localization of barrier compounds. The result of this combination of inheritance and localization of compounds has been the development of a capacity to produce adventively, from latent and extant genes, a system of localized reactions to reduce the effectiveness of pathogens. In the most recently derived and largest of plant families, the orchids, barriers to pathogens have evolved to the point of both blocking and utilizing the mycorrchizal fungus; the host becomes dependent upon the fungus localized in the roots between barrier dermal layers. In the surviving members of the oldest and first of the land plants, *Psilotum* and *Tmesipteris*, similar mechanisms of immunity and reliance on pathogens have been operating for millions of years.

Compounds associated with disease may be slightly different in each family of plants, but in fact there is something common to the molecules found in defense reactions. No one compound confers complete immunity or even resistance to all pathogens (Kuć, 1967), but in conditions of disease the same types of molecules and enzymes may be found in unrelated plant groups. The location of these compounds in barrier tissues and lesion zones is the same in unrelated plants, and it is perhaps not remarkable that the same molecular and enzyme types are found in both ancient and derived barrier systems of antibiosis.

Each histochemical tissue barrier may be thought of as resulting from a system of inheritance which includes latent ancestral and modern extant genes of immunity that may be functional at all times or activated only under untoward and adventive circumstances. Degrees of lesser resistance would seem to be a loss of genes of resistance through the intensive selection by man and other agencies of characteristics unrelated to disease.

The response to wounding and disease is inherent but, as in all mechanisms of inheritance, variations in the molecular compounds and isoenzymes of disease are also inherent. Some enzymes are increased while others are suppressed. Inherent factors process the quantitative enzyme complement and its localization. Tissues that are different in their normal function and morphology are inherently evocated to activate or form enzyme systems and compounds common to many unrelated tissues in unrelated plants.

The enzymes and compounds of disease may be produced in a single cell layer or in zones of cells. The response to infection is not random and is not general in most plants. A three-dimensional zone of response develops in the initial and final barriers of infection (Tomiyana *et al.*, 1967). Localized formation of compounds of resistance in zones seems to have survival value whereas generalized reactions are more common to susceptibility with distributed and logarithmic energy losses.

In all types of wounding and infection extant and ancient genes are conjured up like "the evocation of childhood memories." What seem to be ancestral or latent genes function in a localized way within the extant gene complex in plants having degrees of immunity and resistance. The combined genetic expression of latent and extant genes is the production of antibiotic substances in several cell types or in localized barriers. The initial steps in formation are found in changes in enzyme systems followed by the localized development of antibiotic compounds. Homogenizing and extracting whole plant stems, roots, and leaves to obtain such compounds is fundamental and essential to their systematic identification, but the discovery of the causal factors for their formation must come from analysis of enzyme and compound localization in relation to the somatic system of inheritance.

Genetic analysis in general is based on morphological characters and at the chemical level analysis has been in relation to general biosynthesis. As Bu'Lock (1965) has pointed out, chemical examination has been detailed but unrelated to events of cellular genetics and differentiation. The combined expression of extant and latent genes, gene linkages, and gene modifiers must be examined in localized enzyme activation, polyploidy, polyteny, suberization, the formation of polyenes and polyynes, and many other compounds.

ACKNOWLEDGMENTS

A part of the work reported here was supported by the National Science Foundation (Contract GB1138).

REFERENCES

Andreae, S. R., and W. A. Andreae. 1949. The metabolism of scopoletin by healthy and virus infected potato tubers. *Can. J. Res., Sect. C* **27**:14–22.

Andreae, W. A. 1948. The isolation of a blue fluorescent compound scopoletin from Green Mountain potato tubers, infected with leafroll virus. *Can. J. Res., Sect. C* **26**:31–34.

Bergman, H. H., and J. C. M. Beijersbergen. 1968. A fungitoxic substance extracted from tulips and its possible role as a protectant against disease. *Neth. J. Plant Pathol.* **74**:Suppl. 1, 157–162.

Bernheim, F. M., M. L. C. Bernheim, and K. M. Wilbur. 1948. The reaction between thiobarbituric acid and oxidation products of certain lipides. *J. Biol. Chem.* **174**: 257–264.

Best, R. J. 1944. Studies on a fluorescent substance present in plants. II. Isolation of the substance in a pure state and its identification as 6-methoxy-7-hydroxy-1:2-benzopyrone. *Aust. J. Exp. Biol. Med. Sci.* **22**:251–255.

Best, R. J. 1948. Studies on a fluorescent substance present in plants. The distribution of scopoletin in tobacco plants and some hypotheses on its part in metabolism. *Aust. J. Exp. Biol. Med. Sci.* **26**:223–230.

Bonnett, H. T. 1968. The root endodermis: fine structure and function. *J. Cell Biol.* **37**:199–205.

Brauner, L. 1969. The effect of gravity on the development of electric potentials in plant tissues. *Endeavour* **28**:17–21.

Bu'Lock, J. D. 1965. "The Biosynthesis of Natural Products." McGraw-Hill, New York.

Burgeff, H. 1959. Mycorrhiza of orchids. *In* "The Orchids" (C. Withner, ed.), pp. 361–395. Ronald Press, New York.

Clauss, E. 1961. Die phenolischen Inhaltsstoffe der Samenschalen von *Pisum sativum* L. und ihre Bedeutung für Resistenz gegen die Erreger der Fusskrankheit. *Naturwissenschaften* **48**:106.

Cromwell, B. T. 1955. The alkaloids. *In* "Moderne Methoden der Pflanzenanalyse" (K. Paech and M. V. Tracey, eds.), Vol. 4, pp. 367–516. Springer-Verlag, Berlin and New York.

De Baun, R. M., and F. F. Nord. 1951. The resistance of cork to decay by wood-destroying molds. *Arch. Biochem. Biophys.* **33**:270–276.

Deverall, B. J. 1964. Studies on the physiology of resistance to the halo-blight disease of French beans. *In* "Host-Parasite Relations in Plant Pathology" (Z. Kiraly and G. Urizsy, eds.), pp. 127–130. Publ. of Res. Inst. for Plant Prot., Budapest.

Endo, B. Y., and J. A. Veech. 1969. The histochemical localization of oxidoreductive enzymes of soybeans infected with the root-knot nematode *Meloidogyne incognita acrita*. *Phytopathology* **59**:418–425.

Endo, B. Y., and J. A. Veech. 1970. Morphology and histochemistry of soybean roots infected with *Heterodera glycines*. *Phytopathology* **60**:1493–1498.

Endres, H. 1962. I. Lieferung. II. Gerbstoffe von H. Endres, F. N. Howes, und C. von Regel. *In* "Die Rohstoffe des Pflanzenreichs" (J. von Wiesner, ed.), pp. 75–162. Cramer, Weinheim.

Farkas, G. L., and Z. Kiraly. 1962. Role of phenolic compounds in the physiology of plant diseases and disease resistance. *Phytopathology. Z.* **44**:105–150.

Farkas, G. L., and M. A. Stahmann. 1966. On the nature of changes in peroxidase isoenzymes in bean leaves infected by southern bean mosaic virus. *Phytopathology* **56**:669–677.

Fawcett, C. H., R. L. Spencer, R. L. Wain, Sir E. R. H. Jones, M. LeQuan, C. P. Page, and V. Thaller. 1965. An antifungal acetylenic keto-ester from a plant of the *Papillionaceae* family. *Chem. Commun.* pp. 422–423.

Fawcett, C. H., D. M. Spencer, R. L. Wain, A. G. Fallis, Sir E. R. H. Jones, M. LeQuan, C. B. Page, V. Thaller, D. C. Shubrook, and P. M. Whitman. 1968. Natural acetylenes. Part XXVII. An antifungal acetylenic furanoid ketoester, wyerone, from shoots of the broad bean. *J. Chem. Soc. C*:2455–2462.

Feenstra, W. J. 1960. The genetic control of the formation of phenolic compounds in the seedcoat of *Phaseolus vulgaris* L. *In* "Phenolics in Plants in Health and Disease" (J. B. Pridham, ed.), pp. 127–131. Pergamon, Oxford.

Goodman, R. N., Z. Kiraly, and M. Zaitlin. 1967. "The Biochemistry and Physiology of Infections Plant Disease." Van Nostrand-Reinhold, Princeton, New Jersey.

Gorlenko, M. V. 1965. "Bacterial Diseases of Plants," 2nd Ed. Isr. Program for Sci. Transl., Jerusalem. (Available from U.S. Dep. of Commerce, Springfield, Virginia.)

Greathouse, G. A., and N. E. Rigler. 1940. The chemistry of resistance of plants to *Phymatotrichum* root rot. Influence of alkaloids on growth of fungi. *Phytopathology* **30**:475–485.

Greathouse, G. A., and G. M. Watkins. 1938. Berberine as a factor in the resistance of *Mahonia trifoliata* and *M. swaseyi* to *Phymatotrichum* root rot. *Amer. J. Bot.* **25**:743–748.

Hadwiger, L. A. 1968. Changes in plant metabolism associated with phytoalexin production. *Neth. J. Plant Pathol.* **74**:Suppl. 1, 163–169.

Harborne, J. B. 1960. The genetic variation of anthocyanin pigments in plant tissue. In "Phenolics in Plants in Health and Disease" (J. B. Pridham, ed.), pp. 109–117. Pergamon, Oxford.

Hirai, K. 1968. Specific affinity of oxidized amine dye (radical intermediate) for heme enzymes: study in microscopy and spectrophotometry. *Acta Histochem. Cytochem.* **1**:43–55.

James, W. O. 1950. Alkaloids in the plant. In "The Alkaloids" (R. H. F. Manskee and H. L. Holmes, eds.), Vol. 1, pp. 16–90. Academic Press, New York.

Jones, E. R. H., J. M. Thompson, and M. C. Whiting. 1954. Synthesis of dodeca-1:11-diene-3:5:7:9-tetrayne. *Acta Chem. Scand.* **8**:1944.

Kedar, N. 1959. The peroxidase test as a tool in the selection of varieties resistant to late blight. *Amer. Potato J.* **36**:315–324.

Keilin, J. 1960. Nature of the haem-binding groups in native and denatured haemoglobin and myoglobin. *Nature (London)* **187**:365–371.

Kljajic, R., M. Babovic, and M. Plesničar. 1964. Peroxidase and polyphenoloxidase activity and absorption of P^{32} in the leaves of some varieties of strawberries infected with strawberry crincle virus. In "Host-Parasite Relations in Plant Pathology" (Z. Kiraly and G. Ubrizsy, eds.), pp. 69–72. Publ. of Res. Inst. for Plant Prot., Budapest.

Kuć, J. 1967. Shifts in oxidative metabolism during pathogenesis. In "The Dynamic Role of Molecular Constituents in Plant-Parasite Interaction" (C. J. Mirocha and I. Uritani, eds.), pp. 183–202. Amer. Phytopathol. Soc., Bruce, St. Paul, Minnesota.

Linskens, H. F. 1968. Host-pathogen interaction as a special case of interrelations between organisms. *Neth. J. Plant Pathol.* **74**:Suppl. 1, 1–8.

Lovrekovich, L., H. Lovrekovich, and M. A. Stahmann. 1968. The importance of peroxidase in the wildfire disease. *Phytopathology* **58**:193–198.

McClean, J. G., D. J. LeTourneau, and J. W. Guthrie. 1961. Relation of histochemical tests for phenols to *Verticillium* wilt resistance of potatoes. *Phytopathology* **51**:84–89.

MacDougal, D. T., and J. Dufrenoy. 1946. Criteria of nutritive relations of fungi and seed-plants in mycorrhizae. *Plant Physiol.* **21**:1–10.

Mace, M. E. 1964. Phenol oxidases and their relation to vascular browning in Fusarium-invaded banana roots. *Phytopathology* **54**:840–842.

Metlitskii, L. V., and O. L. Ozeretskovskaya. 1968. "Plant Immunity." Plenum, New York.

Mothes, K., and A. Romeike. 1958. Die Alkaloide. In "Handbuch der Pflanzen Physiologie" (W. Ruhland, ed.), pp. 989–1049. Springer-Verlag, Berlin and New York.

O'Conner, J. 1971. On the influence of light and gravity on the peroxidase enzyme system of *Nicotiana*. Ph.D. Thesis. Univ. of Georgia, Athens, Georgia.

Oku, H. 1964. Host-parasite relation in *Helminthosporium* leaf spot disease of rice plant from the viewpoint of biochemical nature of the pathogen. In "Host-Parasite Relations in Plant Pathology" (Z. Kiraly and G. Ubrizsy, eds.), pp. 183–191. Publ. of Res. Inst. for Plant Prot., Budapest.

Oku, H. 1967. Role of parasite enzymes and toxins in development of characteristic symptoms in plant disease. In "The Dynamic Role of Molecular Constituents in

Plant-Parasite Interaction" (C. J. Mirocha and I. Uritani, eds.), pp. 237–255. Amer. Phytopathol. Soc., Bruce, St. Paul, Minnesota.

Olah, A. F., and R. T. Sherwood. 1970. The relation of glycosidase activity to flavone and isoflavone accumulation in fungal-infected alfalfa. *Amer. J. Bot.* **57**(6):763.

Parups, E. V. 1967. Effect of various plant phenols on protein synthesis in excised plant tissues. *Can. J. Biochem.* **45**:427–433.

Pridham, J. B., ed. 1960. "Phenolics in Plants in Health and Disease," Proc. Plant Phenolics Group Symp. Pergamon, Oxford.

Raa, J. 1968. Polyphenols and natural resistance of apple leaves against *Venturia inaequalis*. *Neth. J. Plant Pathol.* **74**:Suppl. 1, 37–45.

Roberts, E. A. H. 1960. Effect of glycosylation on the enzymic oxidation and translocation of flavonoids. *Nature (London)* **185**:536–537.

Rubin, B. A., and Y. V. Artsikhovskaya. 1963. "Biochemistry and Physiology of Plant Immunity" (Transl. from Russ. by H. Wareing; E. Griffeths, Transl. ed.). Macmillan, New York.

Rudolph, K., and M. A. Stahmann. 1964. Interaction of peroxidase and catalases between *Phaseolus vulgaris* and *Pseudomonas phaseolica*. *Nature (London)* **204**: 474–475.

Sequeira, L. 1967. Accumulation of scopolin and scopoletin in tobacco plants infected by *Peudomonas solanacearum*. *Phytopathology* **57**:830.

Shahied, S. I. 1970. Histochemical and biochemical effects of the herbicide trifluralin on plant roots. Ph.D. Thesis, Univ. of Georgia, Athens, Georgia.

Shannon, L. M. 1968. Plant isoenzymes. *Annu. Rev. Plant Physiol.* **19**:187–210.

Skatteböl, L., and N. A. Sörensen. 1959. Studies related to naturally occurring acetylene compounds. XXVII. The synthesis of a mixture of trideca-1:3:11-triene-5:7:9-triyne and transphenylhepta-1:3-diyn-5-ene. A novel cyclisation reaction. *Acta Chem. Scand.* **13**:2101–2106.

Sörensen, J. S., and N. A. Sörensen. 1954. Studies related to naturally occurring acetylene compounds. XVII. Four new polyacetylenes from garden varieties of *Coreopsis*. *Acta Chem. Scand.* **8**:1741–1756.

Sörensen, J. S., and N. A. Sörensen. 1958a. Studies related to naturally occurring acetylene compounds. XXII. Correctional studies on the constitution of the polyacetylenes of some annual *Coreopsis* species. *Acta Chem. Scand.* **12**:756–764.

Sörensen, J. S., and N. A. Sörensen. 1958b. Studies related to naturally occurring acetylene compounds. XXIII. 1-Phenylhepta-1:3:5-triyne from *Coreopsis grandiflora* Hogg × Sweet. *Acta Chem. Scand.* **12**:765–770.

Sörensen, N. A. 1961. Some naturally occurring acetylenic compounds. *Proc. Chem. Soc., London*, pp. 98–110.

Sörensen, N. A., T. Brunn, and L. Skatteböl. 1954. Studies related to naturally occurring acetylene compounds. XVIII. The synthesis of some phenylacetylenes related to Compositae compounds. *Acta Chem. Scand.* **8**:1757–1762.

Solymosy, F., G. L. Farkas, and Z. Kiraly. 1959. Biochemical mechanism of lesion formation in virus-infected plant tissues. *Nature (London)* **184**:706–707.

Steck, W. 1967. The biosynthetic pathway from caffeic acid to scopolin in tobacco leaves. *Can. J. Biochem.* **45**:1995–2003.

Surrey, K. 1955. Enzyme localization in regeneration meristem in Coleus. Ph.D. Thesis, Univ. of Missouri, Columbia, Missouri.

Szent-Györgyi, A. 1968. Bioelectronics. *Science* **152**:988–990.

Takahashi, T., and T. Hirai. 1963. A cytochemical study of tobacco cells infected with tobacco mosaic virus. *Virology* **19**:431–440.

Tomiyana, K., R. Sakai, T. Sakuma, and N. Ishizaka. 1967. The role of polyphenols in the defense reaction in plants induced by infection. *In* "The Dynamic Role of Molecular Constituents in Plant Parasite Interaction" (C. J. Mirocha and I. Uritani, eds.), pp. 165–182. Amer. Phytopathol. Soc., Bruce, St. Paul, Minnesota.

Tschen, J., and W. H. Fuchs. 1968. Endogene Aktivität der Enzyme in rostinfizierten Bohnenprimär blättern. *Phytopathol. Z.* **63:**187–192.

Uhlenbroek, J. H., and J. D. Bijloo. 1958. Investigations on nematicides. I. Isolation and structure of a nematicidal principle occurring in *Tagetes* roots. *Rec. Trav. Chem. Pays-Bas.* **77:**1004–1009.

Uhlenbroek, J. H., and J. D. Bijloo. 1959. Investigations on nematicides. II. Structure of a second nematicidal principle isolated from *Tagetes* roots. *Rec. Trav. Chim. Pays-Bas.* **78:**382–390.

Van Fleet, D. S. 1947. The distribution of peroxidase in differentiating tissues of vascular plants. *Biodynamica* **6:**125–140.

Van Fleet, D. S. 1950. The cell forms, and their common substance reactions, in the parenchymal vascular boundary. *Bull. Torrey Bot. Club* **77:**340–353.

Van Fleet, D. S. 1952. The histochemical localization of enzymes in vascular plants. *Bot. Rev.* **18:**354–398.

Van Fleet, D. S. 1954. Cell and tissue differentiation in relation to growth (plants). *In* "Dynamics of Growth Processes" (E. J. Boell, ed.), pp. 111–129. Princeton Univ. Press, Princeton, New Jersey.

Van Fleet, D. S. 1955. The significance of the histochemical localization of quinones in the differentiation of plant tissues. *Phytomorphology* **4:**300–310.

Van Fleet, D. S. 1957. Histochemical studies of phenolase and polyphenols in the development of the endodermis in the genus *Smilax*. *Bull. Torrey Bot. Club* **84:**9–28.

Van Fleet, D. S. 1959. An analysis of the histochemical localization of peroxidase related to the differentiation of plant tissues. *Can. J. Bot.* **37:**449–458.

Van Fleet, D. S. 1961. Histochemistry and function of the endodermis. *Bot. Rev.* **27:**165–220.

Van Fleet, D. S. 1962. Histochemical localization of enzymes in plants. *In* "Handbuch der Histochemie" (W. Graumann, and K. Neumann, eds.), Vol. VII/2, Ch. 7. pp. 1–38. Fischer, Jena.

Van Fleet, D. S. 1963. Control of cellular differentiation in plants by distribution of enzymes in a lipid-bound system. *Nature (London)* **200:**889.

Van Fleet, D. S. 1969. An analysis of the histochemistry and function of antohcyanin. *Advan. Front. Plant Sci.* **23:**65–89.

Van Fleet, D. S. 1971. Enzyme localization and the genetics of polyenes and polyacetylenes in the endodermis. *Advan. Front. Plant Sci.* **26:**109–143.

Veech, J. A. 1967. The electrophoresis and histochemical localization of certain respiratory enzymes of tobacco infected by *Phytophthora parasitica* var. *nicotianae*. Ph.D. Thesis, Univ. of Georgia, Athens, Georgia.

Veech, J. A. 1969. Localization of peroxidase in infected tobaccos susceptible and resistant to black shank. *Phytopathology* **59:**566–571.

Veech, J. A., and B. Y. Endo. 1969. The histochemical localization of several enzymes of soybeans infected with the root-knot nematode *Meloidogyne incognita acrita*. *J. Nematol.* **1:**265–276.

Wender, S. H. 1970. Effects of some environmental stress factors on certain phenolic compounds in tobacco. *In* "Recent Advances in Phytochemistry" (C. Steelink and V. C. Runeckles, eds.), Vol. 3, pp. 1–29. Appleton, New York.

Williams, A. H. 1960. The distribution of phenolic compounds in apple and pear trees.

In "Phenolics in Plants in Health and Disease" (J. B. Pridham, ed.), pp. 3–7. Pergamon, Oxford.

Winkler, B. C. 1967. Quantitative analysis of coumarins by thin layer chromatography, related chromatographic studies, and the partial identification of a scopoletin glycoside present in tobacco tissue culture. Ph.D. Thesis, Univ. of Oklahoma, Norman, Oklahoma.

Yamazaki, I. 1966. Function of peroxidase as an oxygen-activating enzyme. *In* "Biological and Chemical Aspects of Oxygenases" (K. Bloch and O. Hayaishi, eds.), pp. 433–449. Maruzen, Tokyo.

SECONDARY PLANT SUBSTANCES AND INSECTS

L. M. SCHOONHOVEN

Department of Entomology, Agricultural University, Wageningen, The Netherlands

Introduction	197
Feeding Stimulants	198
Nutrients and Host Plant Recognition	199
Chemoperception of Plant Constituents	209
Molecular Structure and Stimulating Effectiveness	210
Toxicity of Secondary Plant Substances	213
Conclusion	217
References	218

Introduction

For a long time secondary plant substances have been considered to represent metabolic waste products, which, because of the lack of appropriate excretion organs, were "dumped" inside the plant at certain places, such as vacuoles or cell walls. Although some botanists (Muller, 1967; Mothes, 1969) still support this view, our present knowledge indicates that they have a definite function in the survival of the plant species concerned. The question as to the significance of secondary plant substances cannot be answered satisfactorily by the physiologist, unless he takes into account observations on the relationships between the plant and its habitat with particular reference to the numerous organisms that constantly try to utilize this source of bound energy.

The function of secondary plant substances in the ecology of a plant species was recognized by some entomologists, notably Dethier (1947, 1954) and Fraenkel (1953), and an increasing understanding of the relationships between plants and insects, as they have been developed during evolution, culminated in Fraenkel's (1959) paper on the "Raison d'être of secondary plant substances." Since then much new information has become available, hence it seems appropriate to reconsider the subject and to see where we stand now.

For the purpose of this paper the term "secondary plant substances" will be used for those compounds that are not universally found in higher plants, but are restricted to certain plant taxa, or occur in certain plant taxa at much higher concentrations than in others. They are of no nutritional significance to insects.

Feeding Stimulants

Plant taxa are characterized not only by morphological features, but also by the presence or absence of typical chemicals. Phytophagous insects often restrict themselves to a certain plant species, genus, or family, and, although physical aspects do play some role in the recognition process, chemical stimuli determine to a great extent whether or not a plant is acceptable. Many secondary plant substances appear to function as such chemical signals: some insect species, adapted to a particular plant, may use these cues to recognize their host plants, others may be repelled by the same compounds. The best-known example is sinigrin, which, together with several other mustard oil glucosides, strongly enhance feeding in a number of insect species belonging to various orders, and which in nature are bound to cruciferous plants. This compound, on the contrary, may suppress feeding activity in insects that are not specific Cruciferae-feeders, such as the peach aphid, *Myzus persicae* (Wearing, 1968). Moreover, fission products of these mustard oil glucosides, e.g., allyl isothiocyanate, are also involved in guiding stringent Cruciferae-feeders (and even their parasites) to their food, whereas, for instance, silkworm larvae are repelled by this chemical (Ishikawa and Hirao, 1965).

Not only feeding by the larvae, but also oviposition by the adult female, is governed by these compounds. *Pieris brassicae* females will oviposit on any substrate provided it contains sinigrin, be it either filter paper (David and Gardiner, 1962) or a broad bean plant that was allowed to suck up a sinigrin solution (Ma, 1969). The sinigrin is apparently perceived via tarsal chemoreceptors. Adults of the cabbage root fly, likewise, may be provoked to lay their eggs in the absence of cabbage plants when they are provided with a source of allyl isothiocyanate or the glycoside (Traynier, 1967; Zohren, 1968).

In Table 1 (pp. 200, 201) are listed several secondary plant substances that stimulate feeding in certain insects; and in Table 2 (pp. 202–204) a number of compounds are given, which, if presented in otherwise edible substrates, strongly discourage specific insects from feeding. Several compounds occur in both tables.

It is noteworthy that feeding stimulants evoke maximal responses at a certain optimal concentration. This applies not only to secondary plant substances (Sinha and Krishna, 1970; Nayar and Fraenkel, 1963b), but also to nutritive stimulants, such as glucose (e.g., Hsiao and Fraenkel, 1968a). The same principle also holds for volatile attractants.

Plant odors are often involved in leading specific insect species toward the plant on which they feed and/or oviposit. Table 3 (pp. 206, 207) lists a number of representative examples of odorous attractants. For further information on odorous attractants and repellents the reader is referred to the extensive reviews by Dethier (1947) and Jacobson (1966).

With the aforementioned information it is attractive to postulate the following principle: the insect is guided to its specific food by the typical odor of its host plant. Feeding (or oviposition) is then stimulated by the presence of a substance characteristic for this plant species, provided that strong deterrents are lacking. Generally, however, host plant selection appears to be a more complicated and, consequently, a more subtle process. Several instances of a detailed analysis of the selection procedure have revealed that the insect has the capability to perceive a fairly large number of the vast aggregate of chemical constituents present in its host plant. The alfalfa seed chalcid, *Bruchophagus roddi*, is attracted by the odor of 38 chemicals out of the 95 compounds known to occur in its food plant, whereas 9 compounds are repellent and the remainder are nonstimulating (Kamm and Fronk, 1964). The boll weevil, *Anthonomus grandis*, is also attracted to its food, the cotton bud, by an aroma composed of several components. This mixture contains not only atypical compounds like (−)-α-pinene and (−)-limonene, but also (+)-β-bisabolol, thus far reported only in cotton (Minyard et al., 1969). Likewise, feeding in this insect is stimulated by a mixture of at least 14 compounds, including gossypol, a terpenoid isolated from cotton. This substance, when tested alone, has only a very weak feeding stimulating effect (Hedin et al., 1968).

Nutrients and Host Plant Recognition

In the past the emphasis on secondary plant substances as the only factors by which insects differentiate between plants has been sometimes

TABLE 1

SECONDARY PLANT SUBSTANCES THAT ACT AS FEEDING STIMULANTS IN CERTAIN INSECT SPECIES AS CONCLUDED FROM BEHAVIORAL AND/OR ELECTROPHYSIOLOGICAL EVIDENCE

Plant constituent	Insect species	Insect order[c]	Reference
Sinigrin[a]	Pieris brassicae	Lep.	Verschaffelt (1910)
Sinigrin	P. brassicae[b]	Lep.	Schoonhoven (1967)
Sinigrin	Plutella maculipennis	Lep.	Thorsteinson (1953)
Sinigrin	Brevicoryne brassicae	Hem.	Wensler (1962)
Sinigrin	Athalia proxima	Hym.	Bogawat and Srivastava (1968)
Oxalic acid	Phaedon cochleariae	Col.	Tauton (1965)
Quebrachitol	Gastroidea viridula	Col.	Renner (1970)
Gossypol	Serrodes partita	Lep.	Hewitt et al. (1969)
Cucurbitacin	Anthonomus grandis	Col.	Maxwell et al. (1966)
Cucurbitacin	Diabrotica undecimpunctata	Col.	Chambliss and Jones (1966)
Catalposide	Aulacophora foveicollis	Col.	Sinha and Krishna (1969)
Salicin	Ceratomia catalpae	Lep.	Nayar and Fraenkel (1963a)
Salicin	Plagiodera versicolora	Col.	Matsumoto (1969)
Populin	Laothoe populi[b]	Lep.	Schoonhoven (1970)
Hypericin	L. populi[b]	Lep.	Schoonhoven (1970)
	Chrysomela brunsvicensis[b]	Col.	Rees (1969)

Morin	Bombyx mori	Lep.	Hamamura et al. (1962)
Morin	Hyphantria cunea	Lep.	Ishikawa et al. (1969)
Quercetin	Anthonomus grandis	Col.	Hedin et al. (1968)
Quercitrin	A. grandis	Col.	Hedin et al. (1966)
Quercitrin	Heliothis virescens	Lep.	Guerra and Shaver (1969)
Quercitrin	H. zea	Lep.	Guerra and Shaver (1969)
Isoquercitrin	Bombyx mori	Lep.	Hamamura et al. (1962)
Isoquercitrin	H. virescens	Lep.	Guerra and Shaver (1969)
Isoquercitrin	H. zea	Lep.	Guerra and Shaver (1969)
Isoquercitrin	Anthonomus grandis	Col.	Hedin et al. (1968)
Rutin	H. virescens	Lep.	Guerra and Shaver (1969)
Rutin	H. zea	Lep.	Guerra and Shaver (1969)
Catechin xylopyranoside	Scolytus multistriatus	Col.	Doskotch et al. (1970)
Lupeyl cerotate	S. multistriatus	Col.	Doskotch et al. (1970)
p-Hydroquinone	S. multistriatus	Col.	Norris (1970)
Sparteine	Acyrthosiphon spartii	Hem.	Smith (1957)
Phaseolutanin	Epilachna varivestis	Col.	Nayar and Fraenkel (1963b)
Linamarin	E. varivestis	Col.	Nayar and Fraenkel (1963b)
Lotaustrin	E. varivestis	Col.	Nayar and Fraenkel (1963b)

[a] In most cases, various related mustard oil glucosides provoked similar effects.
[b] Based on electrophysiological evidence only.
[c] Lep., Lepidoptera; Hem, Hemiptera; Hym., Hymenoptera; Col., Coleoptera.

TABLE 2
Secondary Plant Substances That Act As Feeding Deterrents in Certain Insect Species as Concluded from Behavioral and Electrophysiological Evidence

Plant constituent	Insect species	Insect order[b]	Reference
Sinigrin	*Myzus persicae*	Hem.	Wearing (1968)
Glucotropaeolin	*Manduca sexta*[a]	Lep.	Schoonhoven (1970)
Isothiocyanate	*Melanoplus sanguinipes*	Orth.	Pawlowski et al. (1968)
Ammonium nitrate	*Sitona cylindricollis*	Col.	Akeson et al. (1969)
Azadirachtin	*Schistocerca gregaria*	Orth.	Butterworth and Morgan (1968)
Azadirachtin	*Heliothis zea*	Lep.	McMillian et al. (1969)
Azadirachtin	*Spodoptera frugiperda*	Lep.	McMillian et al. (1969)
Meliantriol	*Schistocerca gregaria*	Orth.	Lavie et al. (1967)
Ecdysterone	*Pieris brassicae*[a]	Lep.	Ma (1969)
Ecdysterone	*Bombyx mori*[a]	Lep.	Ma (1970)
Ponasterone	*Pieris brassicae*[a]	Lep.	Ma (1969)
Inokosterone	*Pieris brassicae*[a]	Lep.	Ma (1970)
Scillaren	*Schistocerca gregaria*	Orth.	Chauvin and Mentzer (1951)
Digitonin	*Melanoplus bivittatus*	Orth.	Harley and Thorsteinson (1967)
Diosgenin	*M. bivittatus*	Orth.	Harley and Thorsteinson (1967)
Salicin	*Manduca sexta*[a]	Lep.	Schoonhoven (1969)
Salicin	*Bombyx mori*[a]	Lep.	Ishikawa (1966)
6-Methoxybenzoxazolinone	*Pyrausta nubilalis*	Lep.	Beck (1960)
Phenylbenzothiazole	*P. nubilalis*	Lep.	Beck (1960)
p-Benzoquinone	*Scolytus multistriatus*	Col.	Norris (1970)
Juglone	*S. multistriatus*	Col.	Gilbert et al. (1967)
Coumarin	*Listroderes costirostris*	Col.	Matsumoto (1962)
Veratrine	*Melanoplus bivittatus*	Orth.	Harley and Thorsteinson (1967)
Rutin	*Anthonomus grandis*	Col.	Hedin et al. (1966)

Rutin	*Bombyx mori*[a]	Lep.	Ishikawa (1966)
Quercitrin	*B. mori*	Lep.	Ishikawa (1966)
Morin	*Anthonomus grandis*	Col.	Hedin *et al.* (1966)
Hordenine	*Melanoplus bivittatus*	Orth.	Harley and Thorsteinson (1967)
Caffeine	*Manduca sexta*[a]	Lep.	Schoonhoven (1970)
Morphine	*Pieris brassicae*[a]	Lep.	Ma (1969)
Atropine	*P. brassicae*[a]	Lep.	Ma (1969)
Atropine	*Melanoplus bivittatus*	Orth.	Harley and Thorsteinson (1967)
Quinine	*Bombyx mori*[a]	Lep.	Ishikawa (1966)
Quinine	*Pieris brassicae*[a]	Lep.	Ma (1969)
Berberine	*P. brassicae*[a]	Lep.	Ma (1969)
Berberine	*Bombyx mori*[a]	Lep.	Ishikawa (1966)
Lobeline	*Melanoplus bivittatus*	Orth.	Harley and Thorsteinson (1967)
Pilocarpine	*Bombyx mori*[a]	Lep.	Ishikawa (1966)
Pilocarpine	*Pieris brassicae*[a]	Lep.	Ma (1969)
Solanine	*Empoasca fabae*	Hem.	Dahlman and Hibbs (1967)
Tomatine	*E. fabae*	Hem.	Dahlman and Hibbs (1967)
Tomatine	*Manduca sexta*[a]	Lep.	Schoonhoven (1970)
Tomatine	*Leptinotarsa decemlineata*[a]	Col.	Sturckow (1959)
Demissin	*L. decemlineata*[a]	Col.	Sturckow (1959)
Demissidine	*Empoasca fabae*	Hem.	Dahlman and Hibbs (1967)
Solanidine	*E. fabae*	Hem.	Dahlman and Hibbs (1967)
Tomatidine	*E. fabae*	Hem.	Dahlman and Hibbs (1967)
Leptin	*E. fabae*	Hem.	Dahlman and Hibbs (1967)
Leptin	*Leptinotarsa decemlineata*	Col.	Sturckow and Löw (1961)
Nicandrenone	*L. decemlineata*	Col.	Hsiao and Fraenkel (1968b)
Nicotine	*Bombyx mori*[a]	Lep.	Ishikawa (1966)
Nicotine	*Pieris brassicae*[a]	Lep.	Ma (1969)
Nornicotine	*Melanoplus bivittatus*	Orth.	Harley and Thorsteinson (1967)
Lupinine	*M. bivittatus*	Orth.	Harey and Thorsteinson (1967)

(Continued)

TABLE 2—Continued

SECONDARY PLANT SUBSTANCES THAT ACT AS FEEDING DETERRENTS IN CERTAIN INSECT SPECIES AS CONCLUDED FROM BEHAVIORAL AND ELECTROPHYSIOLOGICAL EVIDENCE

Plant constituent	Insect species	Insect order[b]	Reference
Strychnine	*Bombyx mori*[a]	Lep.	Ishikawa (1966)
Strychnine	*Pieris brassicae*[a]	Lep.	Ma (1969)
Brucine	*P. brassicae*[a]	Lep.	Ma (1969)
Brucine	*Bombyx mori*[a]	Lep.	Ishikawa (1966)
Hyoscine	*B. mori*[a]	Lep.	Ishikawa (1966)
Isoboldine	*Trimeresia miranda*		Wada and Munakata (1968)
Isoboldine	*Prodenia litura*	Lep.	Wada and Munakata (1968)

[a] Based on a combination of behavioral and electrophysiological evidence.
[b] Hem., Hemiptera; Lep., Lepidoptera; Col., Coleoptera; Orth., Orthoptera.

too strong, a consequence of the hypothesis that all plants are in nutritional respects essentially alike (Fraenkel, 1959, 1969). This hypothesis, however, seems to be invalid if nutrients are considered quantitavely. In this respect different species, or even subspecies, may differ enormously. There is increasing evidence that many insects are unable to adapt within a short period to nutrient ratios which differ too much from those encountered in their natural food (House, 1969).

Common nutritive plant constituents often have significant effects on food intake. Thus, sucrose or glucose stimulates feeding in most, if not all, phytophagous insects. In the above-mentioned *Pieris brassicae* larvae or in the aphid *Brevicoryne brassicae*, sugar evokes a much stronger feeding response than sinigrin when added to a tasteless substrate (Ma, 1969; Moon, 1967). Often enough characteristic feeding stimulants, as referred to in Table 1, do not even manifest themselves at all in the absence of a sugar.

Numerous other nutritive compounds may act as feeding stimulants (Thorsteinson, 1960) and the Colorado potato beetle, for instance, shows biting and feeding responses to various amino acids, vitamins, steroids, and lipids, in addition to some sugars (Hsiao and Fraenkel, 1968a). This agrees with the "dual discrimination theory," which Kennedy and Booth (1951) expressed several years ago. They concluded that both secondary plant substances and nutritive factors are involved in host plant recognition.

Relative quantities of several nutrients may vary considerably in different plant species. Sucrose levels in several grasses vary, depending on the species, between 1.1 and 11.9 percent dry weight (Ojima and Isawa, 1969), whereas corn contains only 0.5–2.0 percent (Loomis, 1935). Even different varieties of one species may show a great divergence in this respect: the amounts of sucrose in leaves of 33 apple varieties fluctuates between 0.26 and 2.29 percent of the fresh weight (Johnston *et al.*, 1968). In laboratory tests various insects certainly show a preference for an optimal sugar concentration in their diet, and, although more difficult to establish, this factor most probably affects food selection under natural conditions as well. Thus, increasing sugar levels seem, besides other factors, to increase the susceptibility of a plant species to attack by the Japanese beetle (Metzger *et al.*, 1934).

Also, the variation of sugar quantities between different parts of a particular plant is alleged to account for the aggregation of potato leafhoppers, *Empoasca fabae*, on the midrange leaves of the potato plant, which contain about 25 percent more sucrose than other parts of the plant (Hibbs *et al.*, 1964). Protein levels, likewise, may influence the acceptability of a plant. The polyphagous armyworm, *Prodenia liturata*, shows, in short-lasting

TABLE 3

Volatile Secondary Plant Substances That Attract Certain Insects to Their Host Plants

Plant constituent	Insect species	Insect order[a]	Reference
Allyl isothiocyanate	*Erioischia brassicae*	Dipt.	Traynier (1967)
Allyl isothiocyanate	*Phorbia floralis*	Dipt.	Schnitzler and Müller (1969)
Allyl isothiocyanate	*Listroderes obliquus*	Col.	Sugiyama and Matsumoto (1959)
Allyl isothiocyanate	*Phyllotreta cruciferae*	Col.	Feeny et al. (1970)
Allyl isothiocyanate	*P. striolata*	Col.	Feeny et al. (1970)
Allyl isothiocyanate	*Psylliodes chrysocephala*	Col.	Queinnec (1967)
Allyl isothiocyanate	*Plutella maculipennis*	Lep.	Gupta and Thorsteinson (1960)
Allyl isothiocyanate	*Pieris rapae*	Lep.	Hovanitz et al. (1963)
n-Propylmercaptan	*Hylemya antiqua*	Dipt.	Matsumoto and Thorsteinson (1968)
n-Propyl disulfide	*H. antiqua*	Dipt.	Matsumoto and Thorsteinson (1968)
α-Pinene	*Dendroctonus pseudotsugae*	Col.	Rudinsky (1966)

Limonene	*D. pseudotsugae*	Col.	Rudinsky (1966)
Camphene	*D. pseudotsugae*	Col.	Rudinsky (1966)
α-Pinene	*Gnathotrichus sulcatus*	Col.	Rudinsky (1966)
Camphene	*G. sulcatus*	Col.	Rudinsky (1966)
α-Pinene	*Anthonomus grandis*	Col.	Minyard *et al.* (1969)
Limonene	*A. grandis*	Col.	Minyard *et al.* (1969)
Caryophyllene	*A. grandis*	Col.	Minyard *et al.* (1969)
Bisabolol	*A. grandis*	Col.	Minyard *et al.* (1969)
α-Terpineol	*Blastophagus piniperda*	Col.	Kangas *et al.* (1965)
Methylchavicol	*Papilio ajax*	Lep.	Dethier (1941)
Anethole	*P. ajax*	Lep.	Dethier (1941)
Citral	*Bombyx mori*	Lep.	Hamamura (1965)
Linalool	*B. mori*	Lep.	Hamamura (1965)
Nerol	*B. mori*	Lep.	Ishikawa and Hirao (1965)
Coumarin	*Listroderes costrostris*	Col.	Matsumoto (1962)
p-Methylacetophenone	*Chilo suppressalis*	Lep.	Munakata and Okamoto (1967)

[a] Dipt., Diptera; Col., Coleoptera; Lep., Lepidoptera.

TABLE 4

Phytophagous Insect Species with a Limited Food Range, Which Have Been Reared for One or More Generations on a Meridic Diet, Lacking Host-Specific Chemicals

Insect species	Insect order[a]	Common name	Reference
Anthonomus grandis	Col.	Boll weevil	Vanderzant and Davich (1958)
Leptinotarsa decemlineata	Col.	Colorado potato beetle	Wardojo (1969)
Pieris brassicae	Lep.	Cabbage white	Wardojo (1969)
Manduca sexta	Lep.	Tobacco hornworm	Hoffman et al. (1966)
Pectinophora gossypiella	Lep.	Pink bollworm	Vanderzant and Reiser (1956)
Carpocapsa pomonella	Lep.	Codling moth	Sender (1970)
Chilo suppressalis	Lep.	Rice stem borer	Hirano (1963)
Trichoplusia ni	Lep.	Cabbage looper	Chippendale and Beck (1965)
Papilio xuthus	Lep.	—	Nakayama et al. (1969)
Diatraea saccharalis	Lep.	Sugar cane borer	Bowling (1967)
Acyrthosiphon pisum	Hem.	Pea aphid	Auclair and Cartier (1963)
Rhagoletis pomonella	Dipt.	Apple maggot	Neilson (1969)
Dacus cucurbitae	Dipt.	Melon fruit fly	Pant et al. (1959)
D. oleae	Dipt.	Olive fly	Moore (1962)
Hylemya antiqua	Dipt.	Onion maggot	Friend and Patton (1956)

[a] Col., Coleoptera; Lep., Lepidoptera; Hem., Hemiptera; Dipt., Diptera.

feeding tests, a preference for plant species with high protein contents (Soo Hoo and Fraenkel, 1966). Amino acids are in other cases of crucial importance: the peach aphid, *Myzus persicae*, shows continued feeding activity on an artificial diet provided it contains, among other components, a certain amount of methionine (Harrewijn, 1970).

The observation that several oligophagous and monophagous insects may be cultured on an artificial diet devoid of any host plant-specific feeding stimulant also indicates that feeding activity is not solely governed by secondary plant substances. Examples of restricted feeders which develop on a meridic diet are shown in Table 4. Probably most phytophagous insects can be grown on artificial diets, provided all nutritional and physical requirements are met (Vanderzant, 1966). This conclusion, drawn from a highly artificial situation, is of limited value for the understanding of the role of secondary plant substances under natural conditions and might lead to an unwarranted underestimation.

Chemoperception of Plant Constituents

The capability to recognize the presence of various chemicals in a plant resides primarily in the insect's chemoreceptors. These have been localized on antennae, legs, and various mouth parts. In a few cases a beginning has been made with the analysis of their characteristics by means of electrophysiological methods. The caterpillar of *P. brassicae* may serve again as an illustration (Schoonhoven, 1969). Like all lepidopterous larvae, it has on each maxilla two chemoreceptive hairs. By means of two receptor cells in these hairs, which will be discussed in somewhat greater detail in a later section, it discerns the presence of mustard oil glucosides. Compounds which, on the contrary, strongly counteract feeding activity, appear to stimulate a different receptor cell. They include several types of alkaloids (Ma, 1969). A third cell is stimulated by some anthocyanins, such as pelargonin, malvin, and cyanidin mannoside. As might be expected from the foregoing paragraphs, the animal also possesses specialized sugar-sensitive cells and receptors for amino acids, which, upon stimulation, positively affect the amount of food intake. Many other caterpillar species also have receptors which are specifically stimulated by inositol.

Olfactory receptors have been investigated in only a limited number of insects. Up to now no evidence has been found for the presence of highly specialized receptors sensitive to characteristic host plant odors. Instead the animal seems to discriminate various plant odors by combining the information obtained via a number of "generalists," olfactory cells with fairly wide specificity spectra. Because the receptors differ slightly in their spectra, an odor stimulates only a certain number of them and becomes coded in this way. This system allows for very subtle discrimination. For further details the reader is referred to Schoonhoven (1968).

Molecular Structure and Stimulating Effectiveness

Since we now have information, albeit still scanty, about the specificity of a few contact chemoreceptors, it may be worthwhile to consider the molecular structure of these compounds in relation to their stimulating capacity.

First there is the receptor for mustard oil glucosides in the larvae of *P. brassicae*. No other type of compound has been found to stimulate this cell. The seven glucosides tested have as a common formula: $R-CNO-SO_3^--S-\beta$-glucosyl. R may vary as shown in Fig. 1, in which the compounds are shown in their order of stimulating capacity. Despite the narrow and clear spectrum of this cell, it does not seem possible to correlate the degree of activity provoked by each type of molecule with the rather variable character of the R group.

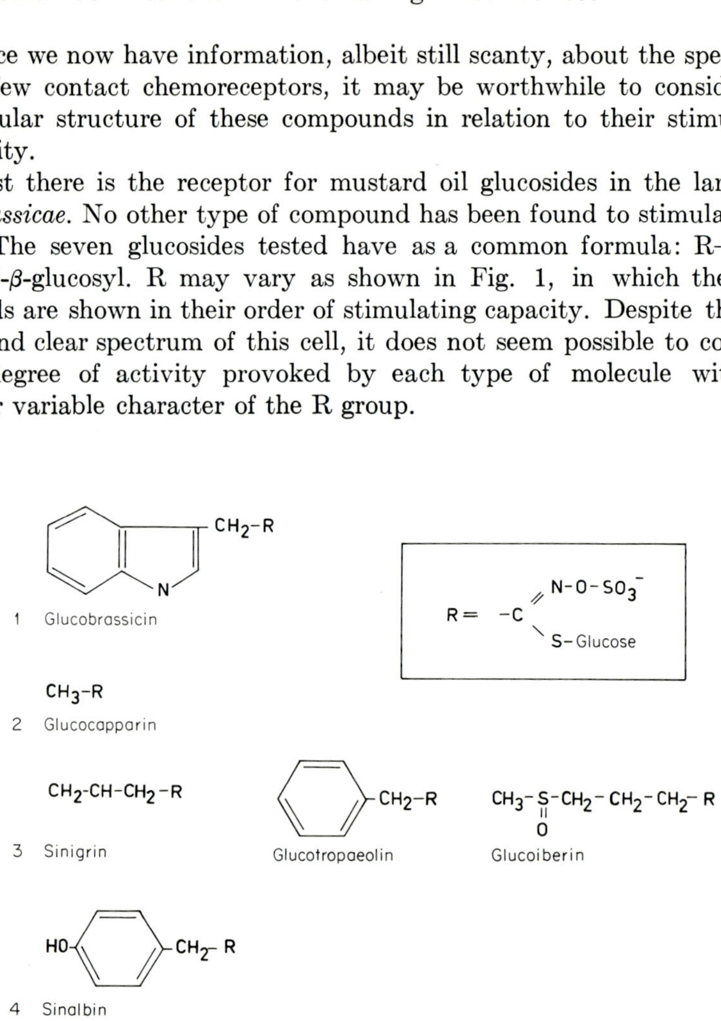

Fig. 1. Chemicals that stimulate the mustard oil glucoside receptor in the lateral maxillary sensillum styloconicum of larvae of *Pieris brassicae*. The numbers indicate the relative effectiveness of the stimuli.

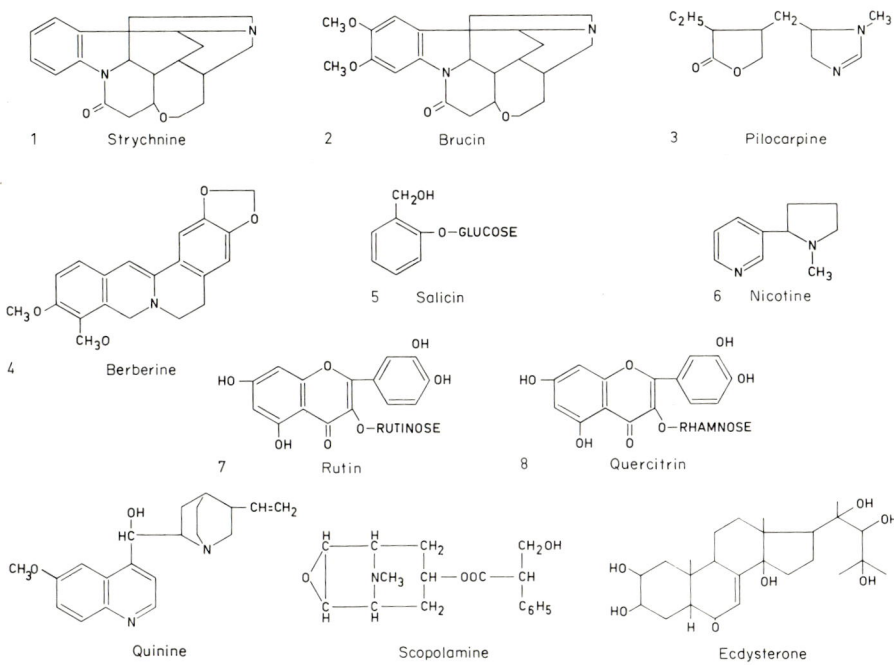

FIG. 2. Chemicals that stimulate a deterrent receptor cell in *Bombyx mori* larvae. The numbers indicate the relative effectiveness of the stimuli. After data from Ishikawa (1966) and Ma (1970), by permission.

The same insect possesses a second receptor type which responds only to glucotropaeolin and sinalbin. Thus, in addition to the typical–CNO–SO_3^-–S–β-glucosyl group, the molecule should contain a benzene ring. This marks this cell as highly specific. The insect's behavior with regard to the various mustard oil glucosides may be explained by the combined nervous input from both receptor types (Schoonhoven, 1967).

In the silkworm, *Bombyx mori*, a number of distasteful substances may stimulate a particular very sensitive receptor cell. In contrast to the above-mentioned example, it is impossible to indicate which part of the stimulating molecule is crucial for an interaction with the receptor. Indeed, the variation may be very great, as is shown in Fig. 2. Still more amazingly, it appears that some closely related compounds are inert to this receptor. Thus, quercetin-3-rutinose (rutin) and quercetin-3-rhamnose (quercitrin) are stimulating, whereas quercetin-3-glucose (isoquercitrin) is inactive (Ishikawa, 1966). Apparently, the sugar moiety exerts a decisive influence. Morin, an isomer of quercetin, is also neutral to this

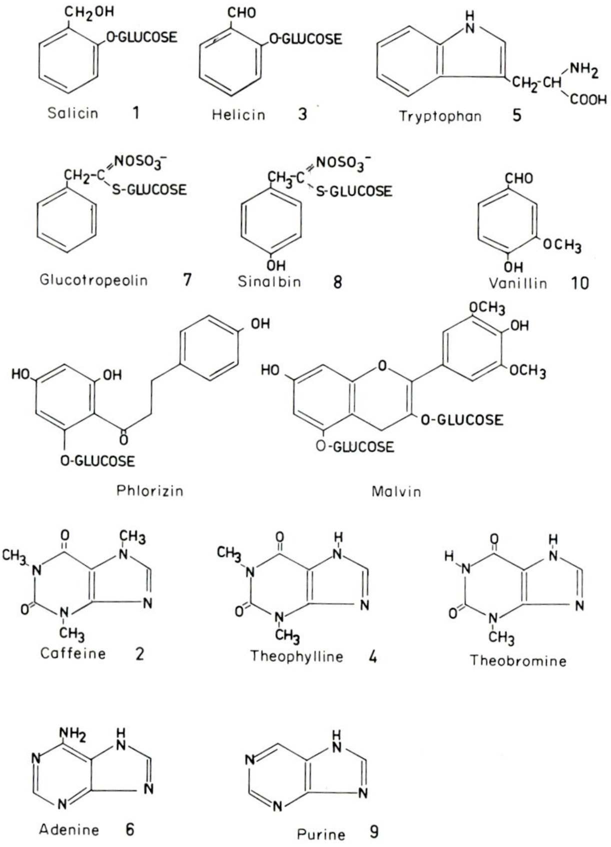

FIG. 3. Chemicals that stimulate a contact chemoreceptory cell in the lateral maxillary sensillum styloconicum of *Manduca sexta* larvae. The numbers indicate the relative effectiveness of the stimuli.

deterrent receptor. This compound has been isolated from the wood of some Moraceae, and isoquercitrin occurs in mulberry leaves, the food of this caterpillar. Such capricious but apparently subtle sensitivity spectra are by no means uncommon in the chemical sense. We may recall the observation that L-alanine is a feeding stimulant for the European corn borer, whereas β-alanine is deterrent (Beck, 1960). A comparable example is found in man: 3-nitro-*o*-toluidine is tasteless, 5-nitro-*o*-toluidine is sweet, and 2-nitro-*p*-toluidine is slightly bitter. In these cases nothing is known at the receptor level.

Larvae of *P. brassicae* are equipped with a receptor type, which in many respects closely resembles the one found in *B. mori*. It also functions as a deterrent receptor (Ma, 1969).

The larva of *Manduca sexta* has in one of the maxillary sensilla a cell that shows a more intelligible sensitivity spectrum: all effective compounds (with the possible exception of solanine) contain either an aromatic ring or a purine skeleton. The stimulants are given in Fig. 3 with the order of their relative effectiveness indicated. It appears especially in the purine series that increasing the number of side chains on the basic skeleton leads to an increased stimulating effect, reaching its maximum in caffeine. The threshold of electrophysiological response for caffeine and salicin is about 10^{-6} to 10^{-7} M. The two types of chemicals stimulating this receptor suggests the presence of two different types of receptor sites. This idea is corroborated by the finding that a mixture of salicin and caffeine does not give a response intermediate between the reactions of each compound alone at double the concentration, but a somewhat lower response than the weakest stimulus (caffeine) would have induced alone. Thus, there appears to be some form of inhibition. When, on the other hand, caffeine and theophylline are mixed, the predicted intermediate reaction is observed. Comparable inhibitory reactions have been recorded before from the sugar receptor of a blowfly (Omand and Dethier, 1969). The function of the receptor is uncertain; salicin and some of the other compounds, when tested in behavioral experiments, showed some feeding inhibitory activity, although only at concentrations of about 10^{-3} M or higher.

From the foregoing examples we may conclude that in some cases the insect's chemoreceptor is tuned to certain distinct classes of chemicals, whereas in other cases (e.g., the deterrent receptor of the silkworm) no logical grouping of the stimulating compounds seems as yet to be possible.

Toxicity of Secondary Plant Substances

Frequently secondary plant substances not only are repugnant to various insects, but also show toxic effects after ingestion. Although often difficult to establish (since their uptake is hampered by their unpalatability) a large variety of potent poisons are found among, for instance, the alkaloids. However, all poisons are not necessarily distasteful and a plant may indubitably contain toxins, which are tasteless to certain insect species. In some, probably exceptional, cases such substances may even stimulate feeding. The latter has been observed in larvae of the diamondback moth, which are in this way deceived by gluconasturtiin and glyconapin, which, like all mustard oil glucosides, are strong feeding stimulants (Nayar and Thorsteinson, 1963). Some compounds, such as saponin, tomatine, and

TABLE 5
Secondary Plant Substances That Are Poisonous or Inhibit Growth in Insects

Plant constituent	Insect species	Insect order[a]	Reference
Gluconasturtiin	*Plutella maculipennis*	Lep.	Nayer and Thorsteinson (1963)
Gluconappin	*P. maculipennis*	Lep.	Nayar and Thorsteinson (1963)
Gossypol	*Spodoptera exigua*	Lep.	Bottger and Patana (1966)
Gossypol	*Heliothis zea*	Lep.	Bottger and Patana (1966)
Gossypol	*Trichoplusia ni*	Lep.	Bottger and Patana (1966)
Gossypol	*Estigmene acrea*	Lep.	Bottger and Patana (1966)
Gossypol	*Heliothis virescens*	Lep.	Lukefahr and Houghtaling (1969)
Gossypol	*Pectinophora gossypiella*	Lep.	Shaver and Lukefahr (1969)
Azadirachtin	*Heliothis zea*	Lep.	McMillian et al. (1969)
Azadirachtin	*Spodoptera frugiperda*	Lep.	McMillian et al. (1969)
Digitonin	*Melanoplus bivittatus*	Orth.	Harley and Thorsteinson (1967)
Diosgenin	*M. bivittatus*	Orth.	Harley and Thorsteinson (1967)
Saponin	*M. bivittatus*	Orth.	Harley and Thorsteinson (1967)
Picrotoxin	*Leptinotarsa decemlineata*	Col.	Buhr et al. (1958)
Benzothiazole	*Pyrausta nubilalis*	Lep.	Beck (1960)
6-Methoxybenzoxazoline	*P. nubilalis*	Lep.	Beck (1960)
Tannin	*Operophtera brumata*	Lep.	Feeny (1968)
Myristicin	*Epilachna varivestis*	Col.	Lichtenstein and Casida (1963)
Myristicin	*Acyrthosiphon pisum*	Hem.	Lichtenstein and Casida (1963)
Veratrine	*Leptinotarsa decemlineata*	Col.	Buhr et al. (1958)
Rutin	*Heliothis zea*	Lep.	Shaver and Lukefahr (1969)
Rutin	*H. virescens*	Lep.	Shaver and Lukefahr (1969)
Rutin	*Pectinophora gossypiella*	Lep.	Shaver and Lukefahr (1969)
Quercitrin	*P. gossypiella*	Lep.	Shaver and Lukefahr (1969)

Secondary Plant Substances and Insects

Compound	Insect	Order	Reference
Quercitrin	*Heliothis zea*	Lep.	Shaver and Lukefahr (1969)
Quercitrin	*H. virescens*	Lep.	Shaver and Lukefahr (1969)
Isoquercitrin	*H. virescens*	Lep.	Shaver and Lukefahr (1969)
Isoquercitrin	*H. zea*	Lep.	Shaver and Lukefahr (1969)
Isoquercitrin	*Pectinophora gossypiella*	Lep.	Shaver and Lukefahr (1969)
Quercetin	*P. gossypiella*	Lep.	Shaver and Lukefahr (1969)
Quercetin	*Heliothis virescens*	Lep.	Shaver and Lukefahr (1969)
Quercetin	*H. zea*	Lep.	Shaver and Lukefahr (1969)
Morin	*H. zea*	Lep.	Shaver and Lukefahr (1969)
Morin	*H. virescens*	Lep.	Shaver and Lukefahr (1969)
Morin	*Pectinophora gossypiella*	Lep.	Shaver and Lukefahr (1969)
Delphinin	*Leptinotarsa decemlineata*	Col.	Buhr *et al.* (1958)
Colchicin	*L. decemlineata*	Col.	Buhr *et al.* (1958)
Physostigmine	*L. decemlineata*	Col.	Buhr *et al.* (1958)
Solanine	*Melanoplus bivittatus*	Orth.	Harley and Thorsteinson (1967)
Tomatine	*M. bivittatus*	Orth.	Harley and Thorsteinson (1967)
Nicotine	*Leptinotarsa decemlineata*	Col.	Buhr *et al.* (1958)
Nornicotine	*Melanoplus bivittatus*	Orth.	Harley and Thorsteinson (1967)
Lupinine	*M. bivittatus*	Orth.	Harley and Thorsteinson (1967)
Lupinine	*Acyrthosiphon pisum*	Hem.	Krymańska (1967)
Lupanine	*A. pisum*	Hem.	Krymańska (1967)
13-Hydroxylupanine	*A. pisum*	Hem.	Krymańska (1967)
17-Hydroxylupanine	*A. pisum*	Hem.	Krymańska (1967)
Sparteine	*A. pisum*	Hem.	Krymańska (1967)
Indole-3-acetonitrile	*Pyrausta nubilalis*	Lep.	Smissman *et al.* (1961)
Cocculolidine	*Nephotettix cincticeps*	Hem.	Wada and Munakata (1968)
Cocculolidine	*Callosobruchus chinensis*	Col.	Wada and Munakata (1968)
Aconitin	*Leptinotarsa decemlineata*	Col.	Buhr *et al.* (1958)
Quassin	*L. decemlineata*	Col.	Buhr *et al.* (1958)

c Lep, Lepidoptera; Orth., Orthoptera; Col., Coleoptera; Hem., Hemiptera.

solanine, are tasteless but deleterious to the grasshopper *Melanoplus bivittatus* (Harley and Thorsteinson, 1967). It must be concluded that in its natural environment this insect recognizes plants with such constituents by other cues.

Toxic compounds are widespread over the plant kingdom. McIndoo (1945), 25 years ago listed 1180 plant species that contain some insect poison, and classical insecticides, such as nicotine, pyrethrin, and rotenone, are all of botanical origin. A selection of some recently published insect growth inhibitors is given in Table 5. The fact, however, that in nature hardly any plant species can be found without some insect species thriving on it, indicates that in each particular case the insect has found a way to circumvent possible noxious effects. As yet very little is known about the physiological background of the detoxification mechanisms. Studies on the tobacco hornworm showed that nicotine does not reach the central nervous system in this insect, because the enveloping membrane is almost impermeable to nicotine (Yang and Guthrie, 1969). Moreover, most of the nicotine in the food is not absorbed by the gut wall at all, but is excreted with the feces. In some other insects nicotine is metabolically detoxified (Hodgson et al., 1965). In larvae of *Tortrix viridana*, feeding on oak, special cells in the gut wall absorb tannin, which is subsequently stored in a bound form (Hollande, 1923).

In some cases the insects are not only adapted to the presence of certain toxins in their food, but even exploit them by accumulating them in specialized tissues, which results in the insect becoming distasteful to several predators, e.g., birds. Thus, the grasshopper *Poekilocerus bufonius* accumulates some cardenolides, like calactin and calotropin, in a poison gland (von Euw et al., 1967). The monarch butterfly also obtains several cardenolides from its larval food plants, *Asclepias* sp., and stores them either unchanged or with only minor metabolic transformation. Here, the amount of poison per insect may come to as much as 1.8 times the lethal dose for a cat (Reichstein et al., 1968). The cinnabar moth retains several of the poisonous alkaloids present in the plants on which it feeds, *Senecio vulgaris* and *S. jacobaeae* (Aplin et al., 1968). The beetle *Aromia moschata*, which is found on some Salicaceae, has specialized glands from which it ejects, in case of attack, a salicylaldehyde-containing secretion. The active principle herein is most probably derived from salicin or populin, constituents of its food (Hollande, 1909). The accumulation of hypericin in another beetle, *Chrysomela brunsvicensis*, possibly has a similar protective value (Rees, 1969). This very interesting phenomenon of poison storage renders an estimation of the relevance of secondary plant substances within a certain biocenose still more complicated.

Conclusion

Many secondary plant substances cannot be considered as metabolic end products, which gradually accumulate in the plant, but rather seem to represent a group of chemicals that in some way are still directly connected with physiological processes in the plant. Several cases of unexpected high turnover rates and intermediate functions in essential metabolic processes have been reported recently (e.g., Loomis, 1967; Zenk, 1967). Sometimes the concentration of a specific substance increases during the growing season, but in other instances it may decrease. Oak leaves are relatively free of tannin when young, but during the summer the tannin concentration rises gradually, reaching a maximum in the autumn at about 5 percent dry weight of the total leaf. Development of the winter moth, *Operophtera brumata*, and probably of other insects as well, is seriously hampered by this substance (Feeny, 1968). The amount of solanine in potato leaves, on the contrary, is reduced during the growing season to less than 5 percent of its original value (Dijkstra and Reestman, 1943; Passeschnitschenko, 1957). The reduction of salicin concentrations during the summer months in poplar leaves may also exceed 90 percent (Pearl and Darling, 1968). Likewise, the concentration of glucobrassicin in cabbage leaves decreases gradually (Schraudorf, cited in Kutáček, 1964).

Clearly, the exact role of many secondary plant substances in the total physiology of the individual plant still waits to be revealed.

With regard to the function of these compounds in the ecological relations of the plant species, the picture seems to be less obscure (see, however, Hegnauer, 1969). In the foregoing sections it has been shown that many secondary plant substances may have a deterrent and/or toxic effect on an insect. In a number of cases, however, these substances stimulate feeding in a particular insect species. Other organisms in the environment, such as vertebrates, nematodes, and pathogenic fungi, are likewise affected. Gossypol and 6-methoxybenzoxazolinone, known to inhibit growth in some insects, also suppress development in some fungi (Bell, 1967; Smissman *et al.*, 1961), and many more instances of comparable inhibitory effects on fungi have been reported (e.g., Rubin and Artsikhovskaya, 1963). On the other hand, microorganisms may "recognize" their host plants by the chemicals which are released. Alkyl sulfides, emitted by onions and attractive to the onion maggot (Matsumoto and Thorsteinson, 1968), stimulate germination of sclerotia in the parasitic fungus *Sclerotium cepivorum* (Coley-Smith and King, 1969). Furthermore, typical volatiles or leachates from the leaves of plants and exudates from the roots may affect the growth and development of surrounding conspecific or other plants. For examples of such allelopathic interactions, see, e.g., Muller (1967) and Tukey (1969).

For a long time the classification of plants has been done almost exclusively according to morphological characteristics, as could be expected in view of our well developed visual sense. No less interesting, however, is the chemical dimension, which in plants certainly shows great perspectives. There is little doubt that during evolution a chemical diversification has taken place under the pressure of surrounding organisms. Fraenkel (1959) and Dethier (1953) have stressed the role which insects played in this chemical evolution. The development of a typical chemical resulted generally in a protection against attack by insects and other organisms. Some insect species, however, became resistant and developed a predilection for the substance. This could lead to an intimate relationship between a plant species and its commensals, from which, in general, both partners benefit. It seems very likely that such processes occurred repeatedly, resulting in a stepwise coevolution, during which angiosperms and insects ultimately attained an enormous degree of diversification (Ehrlich and Raven, 1964).

In conclusion it may be stated that an insect selects only some compounds out of a multitude of chemicals found in a plant species to identify suitable host plants. The number of "key chemicals" involved may vary between perhaps only a few (in certain, though not all, cases of monophagy) and a fairly large number. Compounds of general occurrence (including nutrients) may serve as "key chemicals," as well as secondary plant substances.

Acknowledgments

The author is indebted to Dr. M. Rudrum and Dr. J. de Wilde for helpful comments on the typescript.

References

Akeson, W. R., F. A. Haskins, and H. J. Gorz. 1969. Sweetclover-weevil feeding deterrent B: isolation and identification. *Science* **163**:293–294.

Aplin, R. T., M. H. Benn, and M. Rothschild. 1968. Poisonous alkaloids in the body tissues of the cinnabar moth (*Callimorpha jacobaeae* L.). *Nature (London)* **219**: 747–748.

Auclair, J. L., and J. J. Cartier. 1963. Pea aphid: rearing on a chemically defined diet. *Science* **142**:1068–1069.

Beck, S. D. 1960. The European corn borer, *Pyrausta nubilalis* (Hübn.), and its principal host plant. VII. Larval feeding behavior and host plant resistance. *Ann. Entomol. Soc. Amer.* **53**:206–212.

Bell, A. A. 1967. Formation of gossypol in infected or chemically irritated tissues of *Gossypium* species. *Phytopathology* **57**:759–764.

Bogawat, J. K., and B. K. Srivastava. 1968. Discovery of sinigrin as a phagostimulant by *Athalia proxima* Klug. (Hymenoptera: Tenthredinidae). *Indian J. Entomol.* **30**:89.

Bottger, G. T., and R. Patana. 1966. Growth, development and survival of certain Lepidoptera fed gossypol in diet. *J. Econ. Entomol.* **59**:1166–1168.

Bowling, C. C. 1967. Rearing of two lepidopterous pests of rice on a common artificial diet. *Ann. Entomol. Soc. Amer.* **60**:1215–1216.

Buhr, H., R. Toball, and K. Schreiber. 1958. Die Wirkung von einigen pflanzlichen Sonderstoffen, insbesondere von Alkaloiden, auf die Entwicklung der Larven des Kartoffelkäfers (*Leptinotarsa decemlineata* Say). *Entomol. Exp. Appl.* **1**:209–224.

Butterworth, J. H., and E. D. Morgan. 1968. Isolation of a substance that suppresses feeding in locusts. *Chem. Commun.* pp. 23–24.

Chambliss, O. L., and C. M. Jones. 1966. Cucurbitacins: specific insect attractants in Cucurbitaceae. *Science* **153**:1392–1393.

Chauvin, R., and C. Mentzer. 1951. Contribution à l'étude des substances naturelles et de synthèse repulsives pour les acridiens. *Bull. Off. Nat. Anti-Acrid., Paris* **1**:5–14.

Chippendale, G. M., and S. D. Beck. 1965. A method for rearing the cabbage looper, *Trichoplusia ni*, on a meridic diet. *J. Econ. Entomol.* **58**:377–380.

Coley-Smith, J. R., and J. E. King. 1969. The production by species of *Allium* of alkyl sulphides and their effect on germination of sclerotia of *Sclerotium cepivorum* Berk. *Ann. Appl. Biol.* **64**:289–301.

Dahlman, D. L., and E. T. Hibbs. 1967. Responses of *Empoasca fabae* (Cicadellidae: Homoptera) to tomatine, solanine, leptine I, tomatidine, solanidine and demissidine. *Ann. Entomol. Soc. Amer.* **60**:732–740.

David, W. A. L., and B. O. C. Gardiner. 1962. Oviposition and the hatching of the eggs of *Pieris brassicae* (L.) in a laboratory culture. *Bull. Entomol. Res.* **53**:91–109.

Dethier, V. G. 1941. Chemical factors determining the choice of food plants by *Papilio* larvae. *Amer. Natur.* **75**:61–73.

Dethier, V. G. 1947. "Chemical Insect Attractants and Repellents." The Blakiston Company, Philadelphia, Toronto.

Dethier, V. G. 1953. Host plant perception in phytophagous insects. *Trans. 9th Int. Congr. Entomol. Amsterdam*, 1951 **2**:81–88.

Dethier, V. G. 1954. Evolution of feeding preferences in phytophagous insects. *Evolution* **8**:33–54.

Dijkstra, N. D., and A. J. Reestman. 1943. Het gebruik van aardappelloof als veevoeder. *Landbouwk. Tijdschr.* **55**:191–210.

Doskotch, R. W., S. K. Chatterji, and J. W. Peacock. 1970. Elm bark derived feeding stimulants for the smaller European elm bark beetle. *Science* **167**:380–382.

Ehrlich, P. R., and P. H. Raven. 1964. Butterflies and plants: a study in coevolution. *Evolution* **18**:586–608.

Feeny, P. P. 1968. Effect of oak leaf tannins on larval growth of the winter moth *Operophtera brumata*. *J. Insect Physiol.* **14**:805–818.

Feeny, P. P., K. L. Paauwe, and N. J. Demong. 1970. Flea beetles and mustard oils: host plant specificity of *Phyllotreta cruciferae* and *P. striolata* adults (Coleoptera: Chrysomelidae). *Ann. Entomol. Soc. Amer.* **63**:832–841.

Fraenkel, G. S. 1953. The nutritional value of green plants for insects. *Trans. 9th Int. Congr. Entomol., Amsterdam*, 1951 **2**:90–100.

Fraenkel, G. S. 1959. The raison d'être of secondary plant substances. *Science* **129**:1466–1470.

Fraenkel, G. S. 1969. Evaluation of our thoughts on secondary plant substances. *Entomol. Exp. Appl.* **12**:473–486.

Friend, W. G., and R. L. Patton. 1956. Studies on vitamin requirements of larvae of the onion maggot, *Hylemya antiqua* (Mg.), under aseptic conditions. *Can. J. Zool.* **34**:152–162.

Gilbert, B. L., J. E. Baker, and D. M. Norris. 1967. Juglone (5-hydroxy-1,4-naphthoquinone) from *Carya ovata*, a deterrent to feeding by *Scolytus multistriatus*. *J. Insect Physiol.* **13**:1453–1459.

Guerra, A. A., and T. N. Shaver. 1969. Feeding stimulants from plants for larvae of the tobacco budworm and bollworm. *J. Econ. Entomol.* **62**:98–100.

Gupta, P. D., and A. J. Thorsteinson. 1960. Food plant relationships of the diamondback moth (*Plutella maculipennis* (Curt.)). II. Sensory regulation of oviposition of the adult female. *Entomol. Exp. Appl.* **3**:305–314.

Hamamura, Y. 1965. On the feeding mechanism and artificial food of silkworm, *Bombyx mori*. *Mem. Konan Univ., Sci. Ser.* **8**:Art. 38, 17–22.

Hamamura, Y., K. Hayashiya, K. Naito, K. Matsuura, and J. Nishida. 1962. Food selection by silkworm larvae. *Nature (London)* **194**:754–755.

Harley, K. L. S., and A. J. Thorsteinson. 1967. The influence of plant chemicals on the feeding behaviour, development and survival of the two-striped grasshopper, *Melanoplus bivittatus* (Say), Acrididae: Orthoptera. *Can. J. Zool.* **45**:305–319.

Harrewijn, P., and J. P. W. Noordink. 1971. Taste perception of *Myzus persicae* in relation to food uptake and developmental processes. *Entomol. Exp. Appl.* **14**:413–419.

Hedin, P. A., A. C. Thompson, and J. P. Minyard. 1966. Constituents of the cotton bud. III. Factors that stimulate feeding by the boll weevil. *J. Econ. Entomol.* **59**:181–185.

Hedin, P. A., L. R. Miles, A. C. Thompson, and J. P. Minyard. 1968. Constituents of a cotton bud. X. Formulation of a boll weevil feeding stimulant mixture. *J. Agr. Food Chem.* **16**:505–513.

Hegnauer, R. 1969. Probleme der Chemotaxonomie. Über chemische Variation bei Pflanzen und deren Ursachen und biologische Bedeutung. *Farm. Aikak.* **7–8**:151–161.

Hewitt, P. H., V. B. Whitehead, and J. S. Read. 1969. Quebrachitol: a phagostimulant for the larvae of the moth, *Serrodes partita*. *J. Insect Physiol.* **15**:1929–1934.

Hibbs, E. T., D. L. Dahlman, and R. L. Rice. 1964. Potato foliage sugar concentration in relation to infestation by the potato-leafhopper, *Empoasca fabae* (Homoptera: Cicadellidae). *Ann. Entomol. Soc. Amer.* **57**:517–521.

Hirano, C. 1963. Effect of dietary unsaturated fatty acids on the growth of larvae of *Chilo suppressalis* (Walker) (Lepidoptera: Pyralidae). *Nippon Oyo Dobutsu Konchu Gakkai-Shi* **7**:59–62.

Hodgson, E., L. S. Self, and F. E. Guthrie. 1965. Adaptation of insects to toxic plants. *Proc. 12th Int. Congr. Entomol., London, 1964* p. 210.

Hoffman, J. D., F. R. Lawson, and Y. Yamamoto. 1966. Tobacco hornworms. In "Insect Colonization and Mass Production" (C. N. Smith, ed.), pp. 479–486. Academic Press, New York.

Hollande, A. C. 1909. Sur la fonction d'excrétion chez les insectes salicicoles et en particulier sur l'existence des dérivés salicylés. *Ann. Univ. Grenoble* **21**:459–513.

Hollande, A. C. 1923. L'origine et la nature des cristaux des cellules caliciformes de l'intestin moyen de la chenille tordeuse du chêne (*Tortrix viridana* L.); fonction d'arrêt de ces cellules. *Arch. Anat. Microsc.* **19**:349–369.

House, H. L. 1969. Effects of different proportions of nutrients on insects. *Entomol. Exp. Appl.* **12**:651–669.

Hovanitz, W., V. C. S. Chang, and G. Honch. 1963. The effectiveness of different isothiocyanates on attracting larvae of *P. rapae*. *J. Res. Lepidoptera* **1**:249–260.

Hsiao, T. H., and G. Fraenkel. 1968a. The influence of nutrient chemicals on the feeding behavior of the Colorado potato beetle, *Leptinotarsa decemlineata* (Coleoptera: Chrysomelidae). *Ann. Entomol. Soc. Amer.* **61**:44–54.

Hsiao, T. H., and G. Fraenkel. 1968b. The role of secondary plant substances in the food specificity of the Colorado potato beetle. *Ann. Entomol. Soc. Amer.* **61**:485–493.

Ishikawa, S. 1966. Electrical response and function of a bitter substance receptor associated with the maxillary sensilla of the larva of the silkworm, *Bombyx mori* L. *J. Cell. Physiol.* **67**:1–12.

Ishikawa, S., and T. Hirao. 1965. Studies on olfactory sensation in the larvae of the silkworm, *Bombyx mori*. III. Attractants and repellents of hatched larvae. *Bull. Sericult. Exp. Sta., Tokyo* **20**:21–36.

Ishikawa, S., T. Hirao, and N. Arai. 1969. Chemosensory basis of hostplant selection in the silkworm. *Entomol. Exp. Appl.* **12**:544–554.

Jacobson, M. 1966. Chemical insect attractants and repellents. *Annu. Rev. Entomol.* **11**:403–422.

Johnston, F. B., L. P. Spangelo, R. Watkins, and M. M. Hammill. 1968. Relationships of sugar, acid and phenolic constituents of apple leaves and fruits. *Can. J. Plant Sci.* **48**:473–477.

Kamm, J. A., and W. D. Fronk. 1964. Olfactory response of the alfalfa-seed chalcid, *Bruchophagus roddi* Guss., to chemicals found in alfalfa. *Wyo. Agr. Exp. Sta., Bull.* **413**:36 pp.

Kangas, E., V. Perttunen, H. Oksanen, and M. Rinne. 1965. Orientation of *Blastophagus piniperda* L. (Col., Scolytidae) to its breeding material. Attractant effect of α-terpineol isolated from pine rind. *Ann. Entomol. Fenn.* **31**:61–73.

Kennedy, J. S., and C. O. Booth. 1951. Host alternation in *Aphis fabae* Scop. I. Feeding preferences and fecundity in relation to the age and kind of leaves. *Ann. Appl. Biol.* **38**:25–64.

Krymańska, J. 1967. Rola alkaloidów w odporności niektórych odmian lubinu na mszycę grochową (*Acyrthosiphon pisum* Harris.). *Biul. Inst. Ochr. Rosl.* **36**:237–247.

Kutáček, M. 1964. On the distribution of glucobrassicin, the precursor of ascorbigen, indolyl acetonitrile and thiocyanate ions in plants of the Brassicaceae family (In Russ.). *Fiziol. Rast.* **11**:867–872.

Lavie, D., M. K. Jain, and S. R. Shpan-Galbrielith. 1967. A locust phagorepellent from two *Melia* species. *Chem. Commun.* pp. 910–911.

Lichtenstein, E. P., and J. E. Casida. 1963. Naturally occurring insecticides. Myristicin, an insecticide and synergist occurring naturally in the edible parts of parsnips. *J. Agr. Food Chem.* **11**:410–415.

Loomis, W. D. 1967. Biosynthesis and metabolism of monoterpenes. *In* "Terpenoids in Plants" (J. B. Pridham, ed.), pp. 59–82. Academic Press, New York.

Loomis, W. E. 1935. The translocation of carbohydrates in maize. *Iowa State Coll. J. Sci.* **9**:295–306.

Lukefahr, M. J., and J. E. Houghtaling. 1969. Resistance of cotton strains with high gossypol content to *Heliothis* spp. *J. Econ. Entomol.* **62**:588–591.

Ma, W. C. 1969. Some properties of gustation in the larva of *Pieris brassicae*. *Entomol. Exp. Appl.* **12**:584–590.

Ma, W. C. 1970. Unpublished observations.

McIndoo, N. E. 1945. Plants of possible insecticidal value. *U.S. Dep. Agr., Bur. Entomol. Plant Quarantine* **ET661**:1–286.

McMillian, W. W., M. C. Bowman, J. K. Starks, and B. R. Wiseman. 1969. Extract of Chinaberry leaf as a feeding deterrent and growth retardant for larvae of the corn earworm and fall armyworm. *J. Econ. Entomol.* **62:**708–710.

Matsumoto, Y. 1962. A dual effect of coumarin, olfactory attraction and feeding inhibition, on the vegetable weevil adult, in relation to the uneatability of sweet clover leaves. Studies on host plant discrimination of the leaf-feeding insects. VI. *Nippon Oyo Dobutsu Konchu Gakkai-Shi* **6:**141–149.

Matsumoto, Y. 1969. Some plant chemicals influencing the insect behaviors. *Proc. 11th Int. Congr. Bot., Seattle* p. 143. (Abstr.)

Matsumoto, Y., and A. J. Thorsteinson. 1968. Olfactory response of larvae of the onion maggot *Hylemya antiqua* Meigen (Diptera: Anthomyiidae) to organic sulfur compounds. *Appl. Entomol. Zool.* **3:**107–111.

Maxwell, F. G., H. N. Lafever, and J. N. Jenkins. 1966. Influence of the glandless genes in cotton on feeding, oviposition and development of the boll weevil in the laboratory. *J. Econ. Entomol.* **59:**585–588.

Metzger, P. A., P. A. Van der Meulen, and C. W. Mell. 1934. The relation of the sugar content and odor of clarified extracts of plants to their susceptibility to attack by the Japanese beetle. *J. Agr. Res.* **49:**1001–1008.

Minyard, J. P., D. D. Hardee, R. C. Gueldner, A. C. Thompson, G. Wiygul, and P. A. Hedin. 1969. Constituents of the cotton bud. Compounds attractive to the boll weevil. *J. Agr. Food Chem.* **17:**1093–1097.

Moon, M. S. 1967. Phagostimulation of a monophagous aphid. *Oikos* **18:**96–101.

Moore, I. 1962. Further investigations on the artificial breeding of the olive fly—*Dacus oleae* Gmel.—under aseptic conditions. *Entomophaga* **7:**53–57.

Mothes, K. 1969. Die Alkaloide im Stoffwechsel der Pflanze. *Experientia* **25:**225–239.

Muller, C. H. 1967. Die Bedeutung der Allelopathie für die Zusammensetzung der Vegetation. *Z. Pflanzenkr. Pflanzenpathol. Pflanzenschutz* **74:**333–346.

Munakata, M., and D. Okamoto. 1967. Varietal resistance to rice stem borers in Japan. *In* "The Major Insect Pests of the Rice Plant," Proc. Symp. IRRI pp. 419–430. John Hopkins Press, Baltimore, Maryland.

Nakayama, I., S. Nagasawa, and H. Shimizu. 1969. Rearing of larvae of *Papilio xuthus* L. (Lepidoptera, Papilionidae) on artificial diets without leaf powder. *Nippon Oyo Dobutsu Konchu Gakkai-Shi* **13:**86–89.

Nayar, J. K., and G. Fraenkel. 1963a. The chemical basis of host selection in the Catalpa sphinx, *Ceratomia catalpae* (Lepidoptera, Sphingidae). *Ann. Entomol. Soc. Amer.* **56:**119–122.

Nayar, J. K., and G. Fraenkel. 1963b. The chemical basis of host selection in the Mexican been beetle, *Epilachna varivestis* (Coleoptera, Coccinellidae). *Ann. Entomol. Soc. Amer.* **56:**174–178.

Nayar, J. K., and A. J. Thorsteinson. 1963. Further investigations into the chemical basis of insect-host relationships in an oligophagous insect, *Plutella maculipennis* (Curtis). *Can. J. Zool.* **41:**923–929.

Neilson, W. T. A. 1969. Rearing larvae of the apple maggot on an artificial medium. *Ann. Entomol. Soc. Amer.* **62:**1028–1031.

Norris, D. M. 1970. Quinol stimulation and quinone deterrency of gustation by *Scolytus multistriatus* (Coleoptera: Scolytidae). *Ann. Entomol. Soc. Amer.* **63:**476–478.

Ojima, K., and T. Isawa. 1969. The variation of carbohydrates in various species of grasses and legumes. *Can. J. Bot.* **46:**1507–1511.

Omand, E., and V. G. Dethier. 1969. An electrophysiological analysis of the action of carbohydrates on the sugar receptor of the blowfly. *Proc. Nat. Acad. Sci. U.S.* **62**:136–143.

Pant, N. C., S. Ghai, and S. S. Chawla. 1959. A simple unfortified rice medium for mass-rearing of the melon fruit-fly, *Dacus cucurbitae* Coquillett under laboratory conditions. *Curr. Sci.* **28**:288–289.

Passeschnitschenko, W. A. 1957. The solanin and chaconin contents of the potato in the course of the growing season (in Russ.). *Biokhimiya* **22**:981–983.

Pawlowski, S. H., P. W. Riegert, and J. Krymanski. 1968. Use of grasshoppers in bioassay of thioglucosides in rape seed (*Brassica napus*). *Nature (London)* **220**:174–175.

Pearl, I. A., and S. F. Darling. 1968. Studies on the leaves of the family Salicaceae. XI. The hot water extractives of the leaves of *Populus balsamifera. Phytochemistry* **7**:1845–1849.

Queinnec, Y. 1967. Étude des facteurs psychophysiologiques permettant la découverte de la plante-hôte par les larves néonates de l'Altise d'hiver du Colza (*Psylliodes chrysocephala* L.) *Ann. Epiphyt.* **18**:27–74.

Rees, C. J. C. 1969. Chemoreceptor specificity associated with choice of feeding site by the beetle, *Chrysolina brunsvicensis* on its foodplant, *Hypericum hirsutum. Entomol. Exp. Appl.* **12**:565–583.

Reichstein, T., J. von Euw, J. A. Parsons, and M. Rothschild. 1968. Heart poisons in the monarch butterfly. *Science* **161**:861–866.

Renner, K. 1970. Die Zucht von *Gastroidea viridula* Deg. (Col., Chrysomelidae) auf Blättern und Blattpulversubstraten von *Rumex obtusifolius* L. *Z. Angew. Entomol.* **65**:131–145.

Rubin, B. A., and Y. V. Artsikhovskaya. 1963. "Biochemistry and Physiology of Plant Immunity." Pergamon, Oxford.

Rudsinsky, J. A. 1966. Scolytid beetles associated with Douglas fir: response to terpenes. *Science* **152**:218–219.

Schnitzler, W. H., and H. P. Müller. 1969. Über die Lockwirkung eines Senföls (Allylisothiocyanat) auf die grosze Kohlfliege, *Phorbia floralis* Fallén. *Z. Angew. Entomol.* **63**:1–8.

Schoonhoven, L. M. 1967. Chemoreception of mustard oil glucosides in larvae of *Pieris brassicae. Proc. Kon. Ned. Akad. Wetensch. Ser. C* **70**:556–568.

Schoonhoven, L. M. 1968. Chemosensory basis of host plant selection. *Annu. Rev. Entomol.* **13**:115–136.

Schoonhoven, L. M. 1969. Gustation and foodplant selection in some lepidopterous larvae. *Entomol. Exp. Appl.* **12**:555–564.

Schoonhoven, L. M. 1970. Unpublished observations.

Sender, C. 1970. Élevage du Carpocapse des pommes sur un nouveau milieu artificiel non spécifique. *Ann. Zool. Ecol. Anim.* **2**:93–95.

Shaver, T. N., and M. J. Lukefahr. 1969. Effect of flavonoid pigments and gossypol on growth and development of the bollworm, tobacco budworm and pink bollworm. *J. Econ. Entomol.* **62**:643–646.

Sinha, A. K., and S. S. Krishna. 1969. Feeding of *Aulacophora foveicollis* on cucurbitacin. *J. Econ. Entomol.* **62**:512–513.

Sinha, A. K., and S. S. Krishna. 1970. Further studies on the feeding behavior of *Aulacophora foveicollis* on cucurbitacin. *J. Econ. Entomol.* **63**:333–334.

Smissman, E. E., S. D. Beck, and M. R. Boots. 1961. Growth inhibition of insects and a fungus by indole-3-acetonitrile. *Science* **133**:462.

Smith, B. D. 1957. A study of the factors affecting the populations of aphids of *Sarothamnus scoparius.* Ph.D. Thesis, Univ. of London, London.

Soo Hoo, C. F., and G. Fraenkel. 1966. The selection of food plants in a polyphagous insect, *Prodenia eridania* (Cramer). *J. Insect Physiol.* **12**:693–709.

Stürckow, B. 1959. Über den Geschmacksinn und den Tastsinn von *Leptinotarsa decemlineata* Say. *Z. Vergl. Physiol.* **42**:255–302.

Stürckow, B., and I. Löw. 1961. Die Wirkung einiger *Solanum*-alkaloidglykoside auf den Kartoffelkäfer *Leptinotarsa decemlineata* Say. *Entomol. Exp. Appl.* **4**:133–142.

Sugiyama, S., and Y. Matsumoto. 1959. Olfactory responses of the vegetable weevil larvae to various mustard oils. Studies on the host plant discrimination of the leaf-feeding insects. II. *Nogaku Kenkyu* **46**:150–157.

Tauton, M. T. 1965. Agar and chemostimulant concentrations and their effect on intake of synthetic food by larvae of the mustard beetle, *Phaedon cochleariae* Fab. *Entomol. Exp. Appl.* **8**:74–82.

Thorsteinson, A. J. 1953. The chemotactic responses that determine host specificity in an oligophagous insect. (*Plutella maculipennis* (Curt.) Lepidoptera). *Can. J. Zool.* **31**:52–72.

Thorsteinson, A. J. 1960. Host selection in phytophagous insects. *Annu. Rev. Entomol.* **5**:193–218.

Traynier, R. M. M. 1967. Stimulation of oviposition by the cabbage root fly, *Erioischia brassicae. Entomol. Exp. Appl.* **10**:401–412.

Tukey, H. B. 1969. Implications of allelopathy in agricultural plant science. *Bot. Rev.* **35**:1–16.

Vanderzant, E. S. 1966. Defined diets for phytophagous insects. *In* "Insect Colonization and Mass Production" (C. N. Smith, ed.), pp. 273–303. Academic Press, New York.

Vanderzant, E. S., and T. B. Davich. 1958. Laboratory rearing of a boll weevil: a satisfactory larval diet and oviposition studies. *J. Econ. Entomol.* **51**:288–291.

Vanderzant, E. S., and R. Reiser. 1956. Studies of the nutrition of the pink bollworm using purified casein media. *J. Econ. Entomol.* **49**:454–458.

von Euw, J., L. Fishelson, J. A. Parsons, T. Reichstein, and M. Rothschild. 1967. Cardenolides (heart poisons) in a grasshopper feeding on milkweeds. *Nature (London)* **214**:35–39.

Verschaffelt, E. 1910. The cause determining the selection of food in some herbivorous insects. *Proc. Kon. Ned. Akad. Wetensch.* **13**:536–542.

Wada, K., and K. Munakata. 1968. Naturally occurring insect control chemicals. Isoboldine, a feeding inhibitor, and cocculolidine, an insecticide in the leaves of *Cocculus trilobus* DC. *J. Agr. Food Chem.* **16**:471–474.

Wardojo, S. 1969. Artificial diet without crude plant material for two oligophagous leaf feeders. *Entomol. Exp. Appl.* **12**:698–702.

Wearing, C. H. 1968. Responses of aphids to pressure applied to liquid diet behind parafilm membrane. Longevity and larviposition of *Myzus persicae* (Sulz.) and *Brevicoryne brassicae* (L.) (Homoptera: Aphididae) feeding on sucrose and sinigrin solutions. *N. Z. J. Sci.* **11**:105–121.

Wensler, R. J. D. 1962. Mode of host selection by an aphid. *Nature (London)* **195**:830–831.

Yang, R. S. H., and F. E. Guthrie. 1969. Physiological responses of insects to nicotine. *Ann. Entomol. Soc. Amer.* **62**:141–146.

Zenk, M. H. 1967. Biochemie und Physiologie sekundärer Pflanzenstoffe. *Ber. Deut. Bot. Ges.* **80**:573–591.

Zohren, E. 1968. Laboruntersuchungen zu Massenanzucht, Lebensweise, Eiablage und Eiablageverhalten der Kohlfliege, *Chortophila brassica* Bouché (Diptera, Anthomyiidae). *Z. Angew. Entomol.* **62**:139–188.

HERBICIDE METABOLISM IN PLANTS*

D. S. FREAR, H. R. SWANSON, and F. S. TANAKA

Plant Science Research Division, Agricultural Research Service,
United States Department of Agriculture,
Metabolism and Radiation Research Laboratory, Fargo, North Dakota

Introduction	225
N-Demethylation of N-Methylphenylurea Herbicides	226
Glutathione Conjugation of 2-Chloro-s-triazine Herbicides	239
Conclusions	244
Appendix	244
References	245

Introduction

During the last 20 years, studies on the metabolism of herbicides in higher plants have been primarily directed toward the isolation and identification of metabolites and the characterization of metabolic pathways (Kearney and Kaufman, 1969; Casida and Lykken, 1969; Menzie, 1969). Reports of these *in vivo* metabolism studies have shown that herbicides undergo a variety of biotransformations in higher plants and have established oxidation, reduction, hydrolysis, and conjugation as the principal metabolic reactions.

The identification of major metabolic pathways and the development of methods for the isolation and identification of many metabolites provide the necessary basic information and techniques for the phytochemist to

---

* The use of trade names is for the purpose of identification and does not constitute endorsement by the U. S. Department of Agriculture.

consider herbicide metabolism at the molecular level with *in vitro* studies of key plant enzyme systems.

Unfortunately, the use of *in vitro* enzyme systems as a supplement to *in vivo* herbicide metabolism studies in higher plants has received little attention by most investigators. The authors feel, however, that the advantages of the *in vitro* approach to herbicide metabolism studies merit additional consideration. Some of the advantages follow: (1) Specific metabolic reaction steps can be isolated and studied under controlled and defined experimental conditions. (2) Simplified reaction systems reduce the possibility of secondary metabolic products, facilitate the isolation and identification of reaction products and enhance the possibility of isolating and identifying unstable metabolic intermediates. (3) Shortened reaction times and assay procedures provide a rapid means for studying the effect of many substrates and inhibitors on key metabolic reaction steps. (4) Activities of enzyme systems in different plant species and plant tissues can be compared, and the subcellular localization of specific metabolic reaction steps can be determined. (5) The influence of environmental factors, such as microbial activity, evaporation, and photochemical reactions, can be controlled or avoided.

Enzyme systems that catalyze the initial steps in the metabolism of two major classes of selective herbicides are described in this report. A brief discussion of studies with these enzyme systems illustrates some of the advantages, limitations, and problems associated with the *in vitro* approach to investigations of herbicide metabolism in plants.

N-Demethylation of *N*-Methylphenylurea Herbicides

Many reports in the literature have established *N*-demethylation as the initial metabolic reaction in the detoxication of *N*-methylphenylurea herbicides (Fig. 1) and have resulted in the development of methods for the separation and identification of *N*-demethylated reaction products (Menzie, 1969). Recent reports of *in vivo* studies have also indicated that the selectivity of *N*-methylphenylurea herbicides is the result of significant differences in the ability of resistant and susceptible plants to detoxify these potent photosynthetic inhibitors (Geissbühler *et al.*, 1963; Smith and Sheets, 1967; Swanson and Swanson, 1968a,b; Onley *et al.*, 1968; Rodgers and Funderburk, 1968; Nashed *et al.*, 1970). Recovery rates from the inhibition of photosynthesis by these herbicides have been directly correlated with rates of *N*-demethylation to less phytotoxic monomethyl and demethylated metabolites (van Oorschot, 1965; Swanson and Swanson, 1968a; Neptune and Funderburk, 1969).

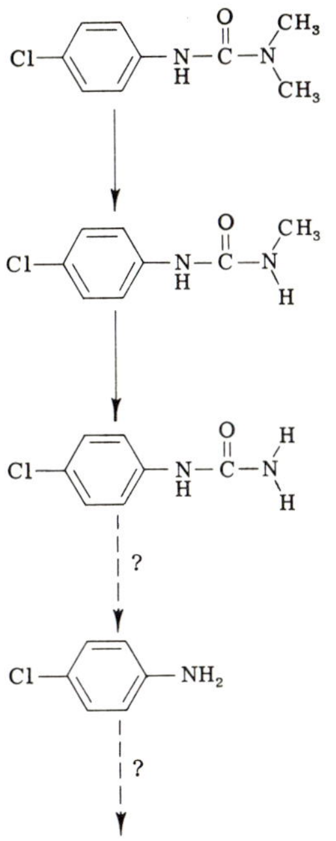

FIG. 1. Pathway for the metabolism of monuron in tolerant plants.

Numerous reports on the metabolism of foreign compounds by animal microsomal enzyme systems (Shuster, 1964; Mason et al., 1965) and limited reports on the metabolism of endogenous compounds by plant microsomal preparations (Nair and Vining, 1965; Galliard and Stumpf, 1966; Zenk, 1967; Russell et al., 1968; Knapp et al., 1969; Stohs, 1969; Murphy and West, 1969) strongly suggest that many of the observed biotransformations of herbicides and other pesticides in plants result from the direct involvement of active microsomal enzyme systems. Direct evidence to support this supposition was recently obtained with the isolation of a microsomal mixed function oxidase system from cotton that catalyzes the N-demethylation of N-methylphenylurea herbicides (Frear, 1968; Frear et al., 1969).

The reaction catalyzed by the cotton N-demethylase system is illustrated in Fig. 2 with monuron (XXII)* as the substrate. The reaction is catalyzed by a mixed function oxidase which requires molecular oxygen and reduced pyridine nucleotides as cofactors. The reaction products have been identified as monomethylmonuron (XXI) and formaldehyde (Frear et al., 1969).

* Numerals refer to compounds listed in the Appendix to this article and in the tables.

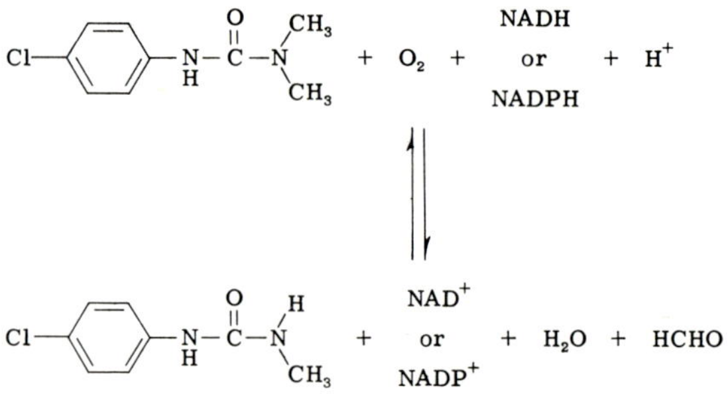

$$\text{Cl}-\underset{}{\underset{}{\bigcirc}}-\underset{H}{N}-\overset{O}{\underset{}{C}}-\underset{CH_3}{\overset{CH_3}{N}} + O_2 + \begin{array}{c}\text{NADH}\\ \text{or}\\ \text{NADPH}\end{array} + H^+$$

$$\updownarrow$$

$$\text{Cl}-\underset{}{\underset{}{\bigcirc}}-\underset{H}{N}-\overset{O}{\underset{}{C}}-\underset{CH_3}{\overset{H}{N}} + \begin{array}{c}\text{NAD}^+\\ \text{or}\\ \text{NADP}^+\end{array} + H_2O + HCHO$$

Fig. 2. Oxidative N-demethylation of monuron by microsomal mixed-function oxidase from cotton.

One molecule of formaldehyde is formed for each molecule of substrate N-demethylated.

Distribution studies of N-methylphenylurea N-demethylase activity in plant tissues have demonstrated some of the advantages as well as the limitations of *in vitro* herbicide metabolism studies. Differential and density gradient centrifugation studies have shown that cotton N-demethylase activity is localized in the microsomal fraction and is primarily associated with the "smooth" fraction of the endoplasmic reticulum (Frear *et al.*, 1969). Similar enzyme distributions have been reported for microsomal N-demethylase systems in animals (Gram and Fouts, 1968).

Differential centrifugation studies on the distribution of urea herbicide N-demethylase activity in cotton leaf homogenates have shown that enzyme activity is either very low or absent in crude cell-free homogenates and in supernatant fractions after the removal of chloroplasts and mitochondria (Frear, 1968). A 2- to 5-fold increase in N-demethylase activity was obtained with washed microsomal preparations and the dialysis or gel filtration of crude microsomal preparations also resulted in greatly increased enzyme activity. These studies all suggest the presence of strong, soluble, endogenous N-demethylase inhibitors in cotton microsomal preparations.

The activity of microsomal N-demethylase in leaf tissues of several plant species is apparently correlated with the relative tolerance of these species to N-methylphenylurea herbicides (Frear *et al.*, 1969). Exceptions to this hypothesis were noted in early studies with microsomal preparations from okra and carrot leaves. No N-demethylase activity was detected in crude microsomal preparations from these two tolerant plant tissues.

Additional studies, however, with infiltrated carrot leaf sections and with Sephadex G–25 treated microsomal preparations from carrot and okra leaves have shown that these tissues do contain active N-demethylase systems. These studies demonstrate that *in vitro* enzyme activity measurements can, and often do, vary considerably between tissues and plant species because of the presence of endogenous inhibitors. All *in vitro* studies should be supported with appropriate *in vivo* studies to establish the level of enzyme activity in the tissues under investigation, and each plant species and tissue should be considered a unique enzyme-isolation problem.

Another interesting effect of endogenous N-demethylase inhibitors has been recently reported (Rusness and Frear, 1970). In addition to N-demethylase activity, microsomal preparations from etiolated cotton hypocotyls also incorporate glucose from uridine diphosphoglucose (UDPG) into water-soluble and insoluble β-1,4-glucan(s) and into water-insoluble β-1,3-noncellulosic linkages. The soluble, low molecular weight endogenous inhibitors separated from cotton and okra microsomal preparations by washing followed by dialysis or gel filtration inhibit water-soluble β-1,4-glucan synthesis and appear to be noncompetitive inhibitors of N-demethylase activity. A marked increase in the ratio of water-insoluble β-1,3:β-1,4-glucosyl linkages is also observed when the separated inhibitor fraction is added back to washed and Sephadex G-25 treated cotton microsome preparations. The concentration of this loosely bound microsomal inhibitor fraction increases with time as excised plant tissues are exposed to air. The possibility of an allosteric response in plant microsomal systems resulting from tissue injury has been suggested (Rusness and Frear, 1970).

Inhibition of cotton N-demethylase activity by carbon monoxide, detergents, sulfhydryl reagents, chelating agents; and electron acceptors the presence of a b_5 type cytochrome and an active NADPH–cytochrome c reductase, all provide indirect evidence for similar microsomal electron transport systems in plants and animals (Frear *et al.*, 1969). Carbon monoxide difference spectra of microsomal N-demethylase preparations from etiolated cotton hypocotyls have recently provided direct evidence to support the presence of P 420 and P 450 type cytochromes (Fig. 3). A similar carbon monoxide difference spectrum has recently been reported for a plant microsomal mixed-function oxidase system that oxidizes kaurene (Murphy and West, 1969). Treatment of cotton microsomal preparations with steapsin apparently results in a significant conversion of P 450 to the P 420 form. A similar conversion was reported for animal microsomal oxidase systems (Omura and Sato, 1964). Calculations based on the molar extinction coefficients reported for cytochrome P 420 in animals (Omura and Sato, 1964) show that P 420 concentrations in cotton N-demethylase preparations vary from 0.17 to 0.60 nmole per milligram of microsomal protein. This range of P 420 concentrations is considerably below those reported for animal systems (Omura and Sato, 1964).

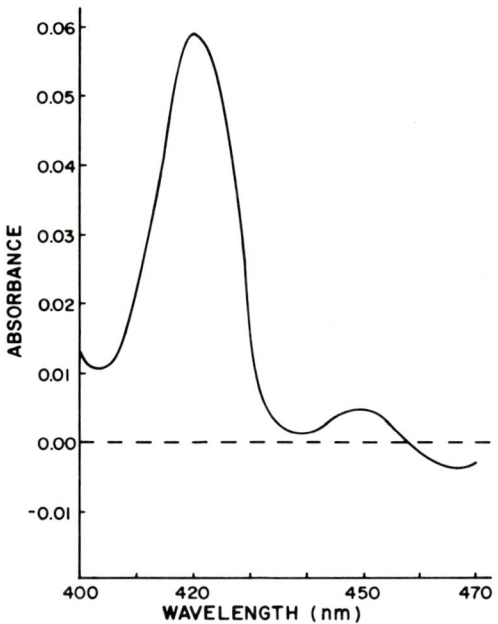

FIG. 3. Carbon monoxide difference spectrum of dithionite reduced, cotton microsomes after treatment with steapsin.

In addition to many similarities, several differences have also been observed between cotton N-demethylase and related animal microsomal oxidase systems (Frear et al., 1969). From the standpoint of enzyme isolation and characterization studies, the most troublesome of these differences are the instability of the N-demethylase system at temperatures above 0–4°C (Frear et al., 1969), the presence of endogenous inhibitors, and the reduced enzyme activities apparently associated with significantly lower levels of electron transport cytochromes.

Substrate specificity studies with ^{14}C-labeled substrates have shown that diuron (XII), fluometuron (XV), monuron (XXII), monomethylmonuron (XXI), fenuron (XIV), monomethylfenuron (XIX), and monomethylfluometuron (XX) are N-demethylated by the cotton N-demethylase system. The specific activities and apparent K_m values for the first three of these substrates have been reported (Frear et al., 1969). A closely related N-methylphenylurea herbicide, linuron (XVII), was not N-demethylated by the cotton N-demethylase system. Studies with infiltrated cotton leaf disks have shown that linuron is not N-demethylated and that linuron inhibition of photosynthesis is not recovered by metabolic detoxication.

Studies with intact carrot plants (Kuratle et al., 1969) and with infiltrated carrot leaf sections have shown that both dimethyl- and methoxymethylphenylurea herbicides are actively N-demethylated by this plant species. Sephadex G-25 microsomal preparations from carrot leaves N-demethylate both dimethyl- and methoxymethylphenylurea substrates. The explanation for this apparently subtle difference in the specificity of N-demethylase systems isolated from the carrot and cotton leaves is unknown.

Other herbicides that have been shown to undergo *in vivo* N-demethylation, N-dealkylation, and sulfoxidation reactions were tested as possible substrates for the cotton N-demethylase system. These compounds showed no activity as substrates, and included diphenamid (XI) (Lemin, 1966; Golab et al., 1967), atrazine (II) (Shimabukuro et al., 1966; Shimabukuro, 1967a,b, 1968), and prometryne (XXVI) (Ebert and Müller, 1968). These limited tests for substrate specificity indicate that the cotton N-demethylase system may be quite specific for dimethyl- and monomethylphenylurea substrates.

Assays for plant enzymes involved in the metabolism of herbicides generally require the use of radioisotope-labeled substrates because of limited substrate solubility and the necessity for the quantitative determination of reaction products at the microgram level. Such a limitation often restricts the number of available compounds for direct substrate specificity studies. This limitation can often be circumvented to some extent by inhibition studies. Strong competitive inhibition by a compound may indicate that the compound also functions as a substrate for the specific enzyme system under study. Therefore, *in vitro* inhibitor studies provide information concerning enzyme mechanisms and possible enzyme substrates. Those compounds which are shown to be effective competitive inhibitors can then be considered for radioisotopic synthesis and direct testing as possible substrates.

The *in vivo* inhibition of N-methylphenylurea herbicide demethylation in cotton leaf disks by several insecticidal carbamates has been reported (Swanson and Swanson, 1968b). The same insecticidal carbamates were also found to be effective inhibitors of the isolated microsomal N-demethylase from cotton (Frear et al., 1969). Kinetic studies with one of these inhibitors, carbaryl (VI), showed that it was a strong competitive inhibitor of fluometuron N-demethylation with an apparent K_i of 1.5×10^{-6} M (Frear et al., 1969). Direct tests to establish whether this insecticidal carbamate inhibitor also functions as a substrate have not been conducted. Such a test should be of interest because *in vivo* studies with cotton and bean plants have reported that carbaryl is oxidized to 1-naphthyl hydroxymethylcarbamate (Kuhr and Casida, 1967).

Several investigators have shown that bidrin (V) and azodrin (III) are

N-demethylated in intact cotton plants (Bull and Lindquist, 1964; Menzer and Casida, 1965; Lindquist and Bull, 1967). However, inhibition studies with these and other organophosphate insecticides including phorate (XXV), parathion (XXIV), and paraoxon (XXIII) failed to demonstrate inhibition of cotton N-demethylation activity at the 0.1 mM level.

Inhibition of cotton N-demethylase activity by several semicarbazide analogs of diuron are shown in Table 1. These compounds have been investigated as herbicides and inhibitors of the Hill reaction (Wilcox and Moreland, 1969). Studies by these investigators have indicated that an alkyl substitution of the 2-nitrogen and a lipophilic side chain were necessary for herbicidal activity, but not for the inhibition of the Hill reaction. The results shown in Table 1 indicate that 4-(3,4-dichlorophenyl)-2-methylsemicarbazide (XXXI) and 4-(3,4-dichlorophenyl)-1,1-dimethylsemicarbazide (XXXII) are the most effective inhibitors of cotton N-demethylase, and 4-(3,4-dichlorophenyl)-1,1,2-trimethylsemicarbazide

TABLE 1
INHIBITION OF FLUOMETURON N-DEMETHYLATION BY SEMICARBAZIDE ANALOGS OF DIURON

Inhibitor[a]	Conc. (mM)	Inhibition (%)	Herbicide activity	Hill reaction inhibition
R—NH—C(=O)—N(CH$_3$)—NH$_2$ XXXI	0.005	85	No	Yes
R—NH—C(=O)—NH—N(CH$_3$)$_2$ XXXII	0.005	53	No	Yes
R—NH—C(=O)—N(CH$_3$)—N(CH$_3$)$_2$ XXXIII	0.005	12	Yes	Yes

[a] R = (3,4-dichlorophenyl).

(XXXIII) is the least effective inhibitor. The latter compound may be a weak inhibitor and also a poor substrate because of the steric and solubility factors associated with the lipophilic side chain. If the more lipophilic trimethyl-substituted semicarbazide compound is a poor substrate for N-demethylase activity, it may be a more effective herbicide. Rapid detoxication by N-demethylation may be less likely, and more of the parent compound may reach the target site for phytotoxicity, the chloroplast. Substitution of a polar amine group on the 2-nitrogen or the placing of the N,N-dimethyl group at a greater distance from the carbonyl group does not seem to affect greatly the binding of these inhibitors at or near the active site of the cotton N-demethylase.

The results of studies on the inhibition of fluometuron N-demethylation by linuron, diuron, monomethyldiuron, and their thio analogs are shown in Table 2. All these compounds were shown to function as strong inhibitors of fluometuron N-demethylation. Linuron, diuron, and their thio analogs are competive inhibitors with apparent K_i values between 1.8×10^{-7} M and 4.8×10^{-7} M. However, as mentioned earlier, linuron does not appear to function as a substrate for cotton N-demethylase. Except for the monomethyl compounds, the thio analogs were stronger inhibitors than the oxygen analogs. Sulfur is generally considered to be a better nucleophile than oxygen because of the greater ease by which it is polarized and desolvated. The relationship of these properties of sulfur and oxygen to the effectiveness of substituted phenylurea analogs as inhibitors has not been determined.

The inhibition of cotton N-demethylase activity by diuron, monuron, and their N-demethylated products are shown in Table 3. These data indicate that progressive N-demethylation results in increasing product inhibition with each N-demethylation step. This suggests that subsequent metabolism of these initial metabolites or the removal of these metabolites from the microsomal N-demethylation site is necessary to prevent product inhibition of the detoxication pathway. The increase in the polarity of N-demethylated metabolites may result in a more favorable partitioning of these compounds into the cytoplasm and away from the lipophilic microsomal membrane system. Further metabolism of substituted phenylurea metabolites to anilines and other polar products have been reported (Onley et al., 1968; Nashed et al., 1970; Nashed and Ilnicki, 1970). The subsequent hydroxylation or hydrolysis of substituted phenylurea metabolites by the cotton N-demethylase system has not been determined.

The effect of other substituted urea compounds on the N-demethylation of fluometuron (XV) are shown in Table 4 (p. 236). These studies indicate that substitution either on the carbonyl oxygen or on the 1-nitrogen results in compounds with little inhibitory activity. This may indicate a require-

TABLE 2
Inhibition of Fluometuron N-Demethylation by Linuron, Diuron, Monomethyldiuron, and Their Thio Analogs

Inhibitor[a]	Concentration (mM)	Inhibition (%)
R—NH—C(=S)—N(OCH₃)(CH₃)	0.010 0.005	93 87
R—NH—C(=O)—N(OCH₃)(CH₃) XVII	0.010 0.005	81 70
R—NH—C(=S)—N(CH₃)(CH₃)	0.010 0.005	88 75
R—NH—C(=O)—N(CH₃)(CH₃) XII	0.010 0.005	48 22
R—NH—C(=S)—N(CH₃)(H)	0.010 0.005	70 40
R—NH—C(=O)—N(CH₃)(H) XVIII	0.010 0.005	74 41

[a] R = (3,4-dichlorophenyl).

TABLE 3
INHIBITION OF FLUOMETURON N-DEMETHYLATION BY DIURON, MONURON, AND THEIR N-DEMETHYLATED PRODUCTS

Inhibitor[a]	Concentration (mM)	Inhibition (%)
R_1—NH—C(=O)—NH$_2$	0.01	87
R_1—NH—C(=O)—N(CH$_3$)(H) XVIII	0.01	74
R_1—NH—C(=O)—N(CH$_3$)(CH$_3$) XII	0.01	48
R_2—NH—C(=O)—NH$_2$	0.01	33
R_2—NH—C(=O)—N(CH$_3$)(H) XXI	0.01	24
R_2—NH—C(=O)—N(CH$_3$)(CH$_3$) XXII	0.01	0

[a] R_1 = (3,4-dichlorophenyl); R_2 = (4-chlorophenyl).

ment for both of these functional groups to be unsubstituted in effective inhibitors and substrates of cotton N-demethylase. Chloroxuron (IX) and 1,1-dimethyl-3-[3-N-tert-butylcarbamoyloxy)phenyl]urea (XXXIV) may be poor inhibitors and possibly poor substrates because of steric interference

TABLE 4
INHIBITION OF FLUOMETURON N-DEMETHYLATION BY SUBSTITUTED N-METHYLPHENYLUREA COMPOUNDS

Inhibitor	Concentration (mM)	Inhibition (%)
XXXV (4-Br, 3-Cl phenyl–NH–C(=O)–N(OCH₃)(CH₃))	0.01	75
(4-Br phenyl–NH–C(=O)–N(OCH₃)(CH₃))	0.01	60
(3,4-diCl phenyl–N=C(OCH₃)–N(CH₃)₂)	0.01	16
IX (4-Cl phenyl–O–phenyl–NH–C(=O)–N(CH₃)₂)	0.01	3
(3,4-diCl phenyl–N=C(oxazolidine)–N–CH₃)	0.01	2
(4-Cl phenyl–N(CH₃)–C(=O)–NH(CH₃))	0.01	0
XXXIV	0.01	0

TABLE 5
INHIBITION OF FLUOMETURON N-DEMETHYLATION BY ANILIDE COMPOUNDS

Inhibitor	Concentration (mM)	Inhibition (%)
3,4-dichlorophenyl-NH-CO-CH$_2$-CH$_3$ (XXVIII)	0.01	94
3,4-dichlorophenyl-NH-CO-C(CH$_3$)=CH$_2$ (X)	0.01	54
phenyl-NH-CO-(carboxin structure) (VII)	0.01	15

either by a bulky side chain or by a substituted ring structure. The substitution of groups which inductively withdraw electrons in the phenyl ring may enhance inhibition. This is indicated in Tables 2 and 4 by the increased inhibition of linuron compared to 3-(3-chloro-4-bromophenyl)-1-methoxy-1-methylurea (XXXV).

Several substituted anilide pesticides were also tested as possible cotton N-demethylase inhibitors because of their structural similarities with N-methylphenylurea herbicides (Table 5). The first two compounds, propanil (XXVIII) and dicryl (X), are selective herbicides, while the third compound is a systemic fungicide, carboxin (VII). Studies such as these illustrate the advantages of key *in vitro* enzyme systems for the rapid screening of a variety of compounds as possible modifiers of *in vivo* herbicide metabolism and detoxication in higher plants.

Studies with a number of other pesticides, which are structurally related to N-methylphenylurea herbicides, have shown that these compounds are either ineffective or, at best, weak inhibitors of fluometuron N-demethylation at the 0.1 mM level. The pesticides tested in these studies included temik (XXX), chlorpropham (VIII), swep (XXIX), methyl 4-chloro-

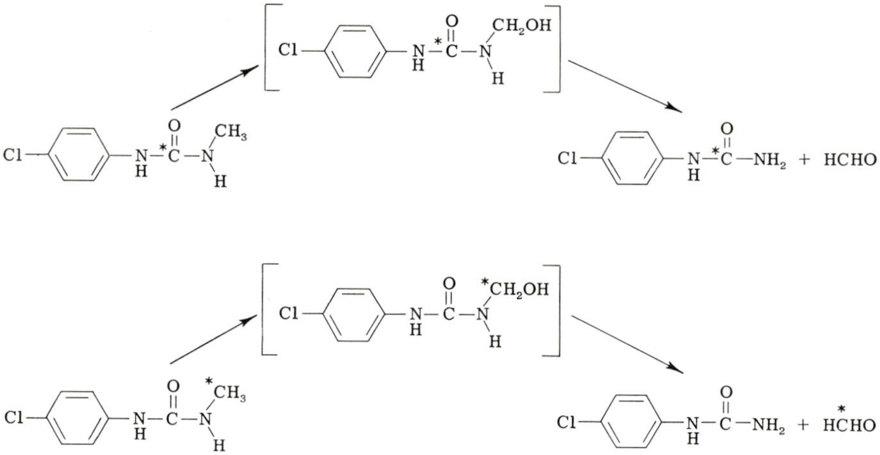

FIG. 4. Identification of N-hydroxymethyl intermediate in the oxidation of monomethylmonuron by cotton N-demethylase.

carbanilate, and diphenamid (XI). The lack of inhibition by diphenamid supports the previously mentioned studies which showed that this compound was not a substrate for the cotton microsomal N-demethylase system.

One of the advantages of *in vitro* enzyme studies is the ability to eliminate secondary reactions and to isolate unstable intermediate reaction products. This advantage was clearly illustrated in the identification of a previously undetected intermediate in the oxidative N-demethylation of monuron. Both dimethyl- and monomethylphenylurea herbicides are N-demethylated by cotton N-demethylase. The N-demethylation of dimethylphenylurea substrates essentially stops with the formation of the monomethylphenylurea products because of the limited amount of monomethylphenylurea available as a substrate for further oxidation. When monomethylphenylurea compounds are used as substrates, however, the reaction proceeds with the oxidation of the remaining methyl group to formaldehyde and the corresponding phenylurea. The N-demethylation of substituted monomethylphenylurea substrates, however, results in the formation of a significant quantity of an unstable intermediate which is slightly more polar than the end product, phenylurea. The formation of this intermediate requires molecular oxygen and reduced pyridine nucleotides. Studies with both carbonyl-^{14}C- and methyl-^{14}C-labeled monomethylmonuron have demonstrated that this oxidized intermediate product contains the N-methyl label and is readily converted to formaldehyde and 4-chlorophenylurea (Fig. 4). In these studies, ^{14}C-labeled formaldehyde was identified as the formaldomethone derivative (Frear et al., 1969), and ^{14}C-labeled 4-chlorophenylurea was identified by dilution with known nonradioactive 4-chlorophenylurea and recrystallization to a constant specific activity. Attempts to further characterize this unstable intermediate are in progress, but the indirect evidence cited above indicates that this intermediate oxidation product is 3-(4-chlorophenyl)-1-hydroxymethylurea.

Several unidentified polar metabolites have been noted in in vivo studies with N-methyphenylurea herbicides. Attempts to isolate and characterize these metabolites have been unsuccessful. Acid hydrolysis and thin-layer chromatography of these unknown metabolic products result in the isolation of additional monomethyl and phenylurea compounds. The identification of an N-hydroxymethyl intermediate with the in vitro cotton N-demethylase system and the reports on the identification of N-hydroxymethyl glycoside metabolites of other N-methyl pesticides in higher plants (Menzie, 1969; Casida and Lykken, 1969; Kearney and Kaufman, 1969) strongly suggest that the unknown polar metabolites of N-methylphenylurea herbicides may be O-glycosides of N-hydroxymethyl intermediates. These O-glycosides may be readily degraded during isolation or by acid hydrolysis to their respective monomethyl and phenylurea compounds.

The authors hope that the identification of a N-hydroxymethyl intermediate together with additional cotton N-demethylase characterization studies will result in a better understanding of the mechanism of N-demethylation by mixed-function oxidase systems and will provide basic information on the significance and characteristics of microsomal enzymes as major sites of pesticide metabolism in higher plants.

Glutathione Conjugation of 2-Chloro-s-triazine Herbicides

Three metabolic and detoxication pathways have been established for 2-chloro-s-triazine herbicides in higher plants (Lamoureux et al., 1970a; Shimabukuro and Swanson, 1969). The principal pathway, in highly tolerant plants, involves the rapid initial conjugation of the herbicide with reduced glutathione (Lamoureux et al., 1970a; Shimabukuro et al., 1970a). A similar pathway for 2-chloro-s-triazine herbicide metabolism has recently been reported in the rat (Hutson et al., 1970).

The soluble enzyme system responsible for this conjugation pathway in plants, a glutathione S-transferase, has been isolated and partially characterized (Frear and Swanson, 1970). The reaction catalyzed by this enzyme system is shown in Fig. 5.

Differential centrifugation, ammonium sulfate precipitation, gel filtration, ion-exchange cellulose chromatography, and preparative acrylamide gel

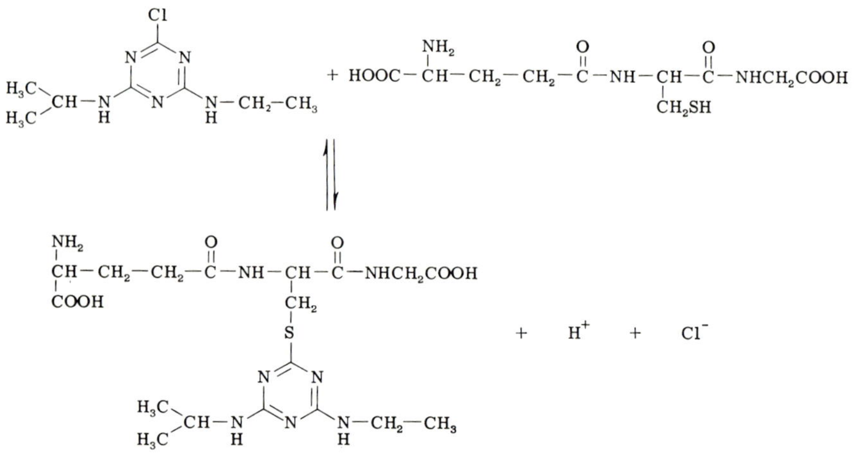

FIG. 5. Glutathione S-transferase catalyzed formation of S-(4-ethylamino-6-isopropylamino-s-triazinyl-2)-glutathione.

electrophoresis have resulted in a 73-fold purification of the enzyme from corn leaves (Table 6). A similar purification, through the Sephadex-200 step, has also been achieved for the glutathione S-transferase system from homogenates of sugar cane leaves. The molecular weight of the purified corn enzyme has been estimated at 40,000 by gel filtration on Sephadex G-200 (Fig. 6). The enzyme fraction eluted from the preparative polyacrylamide gel column after electrophoresis contains at least three protein components. Further purification of this fraction is in progress.

TABLE 6
Purification of Glutathione S-Transferase from Corn Leaves

Fraction	Volume (ml)	Total protein (mg)	Specific activity units[a]	Total units	Yield (%)	Purification (-fold)
15,000 g supernatant	19,000	53,200	0.30	15,990	100	1
40–60 $(NH_4)_2SO_4$ precipitate	300	10,640	1.50	15,960	99	5
Sephadex G-25	50	7,600	2.39	18,160	114	8
Sephadex G-100	10	1,400	7.50	10,500	66	25
DEAE-cellulose	10	588	9.57	5,627	35	32
Sephadex G-200	6	365	13.03	4,756	30	43
Preparative acrylamide electrophoresis	5	90	22.00	1,980	12	73

[a] 1 unit = mμmoles S-(4-ethylamino-6-isopropylamino-2-s-triazinyl-2)-glutathione formed per milligram of protein in 30 minutes.

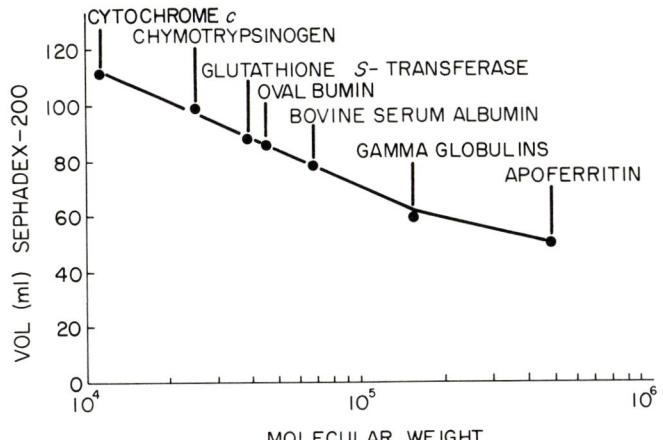

Fig. 6. Molecular weight estimation of partially purified glutathione S-transferase from cotton.

The distribution of glutathione S-transferase activity in crude, cell-free extracts of several plant species and tissues has been determined. Enzyme activity is primarily associated with the foliar tissues of highly resistant species (Frear and Swanson, 1970). Recovery of atrazine-inhibited photosynthesis in infiltrated corn leaf disks is associated with rapid atrazine-glutathione conjugation (Shimabukuro et al., 1970a). The localization of enzyme activity in leaf tissues is also supported by in vivo studies which demonstrate significant differences in the relative activities of the different metabolic pathways in foliar and root-treated plants (Shimabukuro et al., 1970a,b). Striking differences in enzyme activity have also been reported in the foliar tissues of resistant and susceptible inbred lines of corn (Shimabukuro et al., 1970b). These studies have provided direct evidence for herbicide selectivity based on the genetic control of a key enzyme in the major detoxication pathway for substituted 2-chloro-s-triazines in tolerant plants.

Attempts to isolate an active glutathione S-transferase from barley leaf homogenates have been unsuccessful. However, studies with infiltrated barley leaf tissues have demonstrated a slow, but significant rate of glutathione conjugation with both atrazine (II) and propachlor (XXVII) (Lamoureux, 1970; Lamoureux et al., 1971). The inability to demonstrate in vitro glutathione S-transferase activity in barley homogenates may indicate the presence of potent endogenous inhibitors or a low enzyme titer in this plant species.

Substrate specificity studies have indicated that the partially purified glutathione *S*-transferase from corn is specific for reduced glutathione (Frear and Swanson, 1970). A similar specificity for reduced glutathione was reported for an animal glutathione *S*-transferase system (Booth *et al.*, 1961). The specificity of the enzyme for several substituted *s*-triazines has also been reported (Frear and Swanson, 1970). The enzyme appears to be specific for substituted 2-chloro-*s*-triazines. Little enzyme activity was observed either when the chlorine in the 2-position was substituted with methoxy, methylmercapto, or hydroxy groups, or when an alkyl side chain was absent. The inability of substituted 2-hydroxy-*s*-triazines to function as effective glutathione *S*-transferase substrates is supported by *in vivo* studies in corn and sorghum (Shimabukuro, 1968). The reactivities of 2-chloro-*s*-triazine herbicide substrates vary with the substitution of alkyl side chains in the 4 and 6 positions (Frear and Swanson, 1970). Similar differences in reactivity between the various substituted 2-chloro-*s*-triazine herbicides have also been demonstrated *in vivo* and have been shown to vary in leaf tissues from several tolerant plants (Lamoureux, 1970). The significance of these differences in activity is undetermined, but may indicate subtle differences in specificity between glutathione *S*-transferase systems isolated from different leaf tissues.

Several chloro-substituted arylalkyl compounds including benzyl chloride and the herbicides propachlor (XXVII) and barban (IV) react rapidly with glutathione either in the presence or in the absence of enzyme. In studies with barban and benzyl chloride, however, the reaction rates of boiled controls were significantly less than with the glutathione *S*-transferase from corn. Recent *in vivo* studies have shown that propachlor and probably barban form transitory glutathione conjugates in excised corn, sorghum, sugar cane, and barley leaves (Lamoureux *et al.*, 1971). Benzyl chloride has been reported as a substrate for animal glutathione *S*-transferase systems (Boyland and Chasseaud, 1969a).

Inhibition studies have shown that dithiol compounds, such as 2,3-dimercaptopropanol and 1,4-dithiothreitol, inhibit glutathione *S*-transferase while monothiols, such as cysteine and 2-mercaptoethanol, do not inhibit at the 1.0 mM level (Frear and Swanson, 1970). These studies have also shown that 2,3-dimercaptopropanol is a more effective inhibitor than 1,4-dithiothreitol and may indicate a rather critical distance requirement for a bifunctional binding site on the enzyme.

Several alkylating agents which react with thiol, amino, and imidazole groups have also been shown to be effective inhibitors of glutathione *S*-transferase activity at the 1 mM level. These inhibitors include L-1-tosylamido-2-phenylethyl chloromethyl ketone, diphenylcarbamyl chloride, 2,4-dinitrofluorobenzene, 1-chloro-3-tosylamido-7-amino-L-2-heptanone· HCl, and phenylmethanesulfonyl fluoride. Studies with L-1-tosylamido-2-

phenylethyl chloromethyl ketone (Schoellmann and Shaw, 1963), diphenylcarbamyl chloride (Erlanger et al., 1966), and 1-chloro-3-tosylamido-7-amino-L-2-heptanone·HCl (Shaw et al., 1965) have shown that these agents are selective inhibitors for chymotrypsin and trypsin. These selective inhibitors appear to react with a histidine group at the active site of these enzymes.

A number of substituted triazine compounds have been tested as enzyme inhibitors. Herbicidal s-triazines substituted in the 2-position with methoxy or methylmercapto groups are strong inhibitors of glutathione S-transferase, but do not function as substrates. Kinetic studies with prometryne (XXVI) have shown that this herbicide is a competitive inhibitor of atrazine conjugation with an apparent K_i of 2.8×10^{-5} M. The fungicide dyrene (XIII) and the herbicide ipazine (XVI) were also effective enzyme inhibitors, but they have not been tested as possible substrates. Studies with N-dealkylated 2-chloro-s-triazines and substituted 2-hydroxy s-triazines have shown that these metabolites do not function effectively as either inhibitors or substrates. Experiments with 4-amino-6-t-butyl-3-(methylthio)-as-triazin-5-(4H)-one, a representative compound in a new class of triazine herbicides, have shown that this compound was not an effective inhibitor of corn glutathione S-transferase activity.

The inhibition of enzyme activity by the carbanilate and anilide herbicides, barban (IV), propachlor (XXVII), and alachlor (I) as well as by benzyl chloride supports *in vitro* and *in vivo* studies which indicate that these compounds may also function as substrates for glutathione S-transferase from corn.

Similarities between the corn glutathione S-transferase system and animal glutathione S-transferases have been indicated by inhibition studies with several substituted chloro and nitro aryl and arylalkyl compounds. These compounds have been demonstrated as substrates for animal glutathione S-transferases and include 2-amino-5-nitrophenol, 1,2-dichloro-4-nitrobenzene, sulfobromophthalein, and benzyl chloride (Booth et al., 1961; Wit and Leenwangh, 1969; Boyland and Chasseaud, 1969a,b; Cohen et al., 1964; Grover and Sims, 1964). Sulfobromophthalein was found to be a strong competitive inhibitor of atrazine conjugation with an apparent K_i of 2.7×10^{-5} M.

Recent studies with animal glutathione S-transferase systems suggest that there may be several glutathione S-transferase systems with differing substrate specificities and properties (Boyland and Chasseaud, 1969a,b). The presence of glutathione S-aryltransferase, S-alkyltransferase, S-arylalkyltransferase, S-epoxidetransferase, and S-alkenetransferase systems have been reported in animals (Boyland and Chasseaud, 1969a). Inhibition and substrate specificity studies with the partially purified glutathione

S-transferase from corn indicate that this system catalyzes the conjugation of reduced glutathione with both aryl and arylalkyl substrates. A determination of whether or not this broad specificity of the *in vitro* plant system is due to more than one glutathione S-transferase must await further purification and characterization of the enzyme.

Conclusions

It is hoped that the preceding discussion has drawn attention to some of the advantages, limitations, problems, and challenges of *in vitro* enzyme studies as a valuable supplement and extension of *in vivo* herbicide metabolism studies in higher plants.

Recent studies on the isolation and partial characterization of a microsomal mixed-function oxidase and a soluble glutathione S-transferase have been cited as illustrations of this molecular approach to the investigation of herbicide metabolism in plants. The reactions catalyzed by these enzyme systems illustrate two of the major pathways of herbicide metabolism: oxidation and conjugation. The differences in the subcellular location and in the complexity of these enzyme systems demonstrate the diversity of problems and techniques associated with *in vitro* metabolism studies.

The use of herbicides and other pesticides as substrates for the isolation and characterization of plant enzyme systems provides the phytochemist with a powerful research tool for basic studies in an exciting, challenging, and emerging area of plant biochemistry—the metabolism of xenobiotics.

Appendix

COMMON AND CHEMICAL NAMES OF PESTICIDES MENTIONED IN TEXT, FIGURES, AND TABLES

Common name	Chemical name
Alachlor (I)	2-Chloro-2′,6′-diethyl-N-(methoxymethyl)-acetanilide
Atrazine (II)	2-Chloro-4-ethylamino-6-isopropylamino-s-triazine
Azodrin (III)	3-(Dimethoxyphosphinyloxy)-N-methyl-*cis*-crotonamide
Barban (IV)	4-Chloro-2-butynyl-3′-chlorocarbanilate
Bidrin (V)	3-(Dimethoxyphosphinyloxy)-N,N-dimethyl-*cis*-crotonamide
Carbaryl (VI)	1-Naphthyl methylcarbamate
Carboxin (VII)	5,6-Dihydro-2-methyl-1,4-oxathiin-3-carboxyanilide
Chlorpropham (VIII)	Isopropyl 3-chlorocarbanilate

Common and Chemical Names of Pesticides Mentioned in Text, Figures, and Tables (Continued)

Common name	Chemical name
Chloroxuron (IX)	3-[4-(4-Chlorophenoxy)phenyl]-1,1-dimethylurea
Dicryl (X)	3',4'-Dichloro-2-methylacrylanilide
Diphenamid (XI)	N,N-Dimethyl-2,2-diphenylacetamide
Diuron (XII)	3-(3,4-Dichlorophenyl)-1,1-dimethylurea
Dyrene (XIII)	2,4-Dichloro-6-(2-chloroanilino)-s-triazine
Fenuron (XIV)	1,1-Dimethyl-3-phenylurea
Fluometuron (XV)	3-(3-Trifluoromethylphenyl)-1,1-dimethylurea
Ipazine (XVI)	2-Chloro-4-diethylamino-6-isopropylamino-s-triazine
Linuron (XVII)	3-(3,4-Dichlorophenyl)-1-methoxy-1-methylurea
Monomethyldiuron (XVIII)	3-(3,4-Dichlorophenyl)-1-methylurea
Monomethylfenuron (XIX)	3-Methyl-3-phenylurea
Monomethylfluometuron (XX)	3-(3-Trifluoromethylphenyl)-1-methylurea
Monomethylmonuron (XXI)	3-(4-Chlorophenyl)-1-methylurea
Monuron (XXII)	3-(4-Chlorophenyl)-1,1-dimethylurea
Paraoxon (XXIII)	O,O-Diethyl-O-4-nitrophenyl phosphate
Parathion (XXIV)	O,O-Diethyl-O-4-nitrophenyl phosphorothioate
Phorate (XXV)	O,O-Diethyl-S-(ethylthio) methyl phosphorodithioate
Prometryne (XXVI)	2,4-Bis(isopropylamino)-6-methylmercapto-s-triazine
Propachlor (XXVII)	2-Chloro-N-isopropylacetanilide
Propanil (XXVIII)	3,4-Dichloropropionanilide
Swep (XXIX)	Methyl-3,4-dichlorocarbanilate
Temik (XXX)	2-Methyl-3(methylthio) propionaldehyde-O-(methylcarbonyl) oxime

Acknowledgments

The authors wish to thank their colleagues in Pesticide Investigations—Metabolism in Plants, especially Drs. R. H. Shimabukuro and G. L. Lamoureux, for their advice, encouragement, and helpful discussions.

References

Booth, J., E. Boyland, and P. Sims. 1961. *Biochem. J.* **79**:516.
Boyland, E., and L. F. Chasseaud. 1969a. *Biochem. J.* **115**:985.
Boyland, E., and L. F. Chasseaud. 1969b. *Advan. Enzymol. Relat. Areas Mol. Biol.* **32**:173.
Bull, D. L., and D. A. Lindquist. 1964. *J. Agr. Food Chem.* **12**:310.
Casida, J. E., and L. Lykken. 1969. *Annu. Rev. Plant Physiol.* **20**:618.
Cohen, A. J., J. N. Smith, and H. Turbert. 1964. *Biochem. J.* **90**:457.
Ebert, E., and P. W. Müller. 1968. *Experientia* **24**:1.
Erlanger, B. F., A. G. Cooper, and W. Cohen. 1966. *Biochemistry* **5**:190.
Frear, D. S. 1968. *Science* **162**:674.
Frear, D. S., and H. R. Swanson. 1970. *Phytochemistry* **9**:2123.
Frear, D. S., H. R. Swanson, and F. S. Tanaka. 1969. *Phytochemistry* **8**:2157.
Galliard, T., and P. K. Stumpf. 1966. *J. Biol. Chem.* **241**:5806.
Geissbühler, H., C. Haselbach, H. Aebi, and L. Ebner. 1963. *Weed Res.* **3**:277.

Golab, T., R. J. Herberg, S. J. Parka, and J. B. Tepe. 1967. *J. Agr. Food Chem.* **15**:638.
Gram, T. E., and J. R. Fouts. 1968. *In* "Enzymic Oxidations of Toxicants." (E. Hodgson, ed.), p. 47. North Carolina State Univ. Press, Raleigh, North Carolina.
Grover, P. L., and P. Sims. 1964. *Biochem. J.* **90**:603.
Hutson, D. H., E. C. Hoadley, M. H. Griffiths, and C. Donninger. 1970. *J. Agr. Food Chem.* **18**:507.
Kearney, P. C., and D. E. Kaufman. 1969. "Degradation of Herbicides." Dekker, New York.
Knapp, F. F., R. T. Aexel, and H. J. Nicholas. 1969. *Plant Physiol.* **44**:442.
Kuhr, R. J., and J. E. Casida. 1967. *J. Agr. Food Chem.* **15**:814.
Kuratle, H., E. M. Rohn, and C. W. Woodmansee. 1969. *Weed Sci.* **17**:216.
Lamoureux, G. L. 1970. Unpublished observations.
Lamoureux, G. L., R. H. Shimabukuro, H. R. Swanson, and D. S. Frear. 1970a. *J. Agr. Food Chem.* **18**:81.
Lamoureux, G. L., L. E. Stafford, and F. S. Tanaka. 1971. *J. Agr. Food Chem.* **19**, 346.
Lemin, A. J. 1966. *J. Agr. Food Chem.* **14**:109.
Lindquist, D. A., and D. L. Bull. 1967. *J. Agr. Food Chem.* **15**:267.
Mason, H. S., J. C. North, and M. Vannesti. 1965. *Fed. Proc. Fed. Amer. Soc. Exp. Biol.* **24**:1172.
Menzer, R. E., and J. E. Casida. 1965. *J. Agr. Food Chem.* **13**:102.
Menzie, C. N. 1969. Metabolism of Pesticides. *U.S. Fish Wildl. Serv., Spec. Sci. Rep.: Wildl.* **127**.
Murphy, P. J., and C. A. West. 1969. *Arch. Biochem. Biophys.* **133**:395.
Nair, P. M., and L. C. Vinning. 1965. *Phytochemistry* **4**:161.
Nashed, R. B., and R. D. Ilnicki. 1970. *Weed Sci.* **18**:25.
Nashed, R. B., J. E. Katz, and R. D. Ilnicki. 1970. *Weed Sci.* **18**:122.
Neptune, M. D., and H. H. Funderburk, Jr. 1969. *Proc. S. Weed Sci. Soc.* p. 369.
Omura, T., and R. Sato. 1964. *J. Biol. Chem.* **239**:2379.
Onley, J. H., G. Yip, and M. H. Aldridge. 1968. *J. Agr. Food Chem.* **16**:426.
Rodgers, R. L., and H. H. Funderburk, Jr. 1968. *J. Agr. Food Chem.* **16**:434.
Rusness, D. G., and D. S. Frear. 1970. *Proc. N. Dak. Acad. Sci.* **24**:27. (Abstr.)
Russell, D. W., E. E. Conn, A. Stutter, and H. Grisebach. 1968. *Biochim. Biophys. Acta* **170**:210.
Schoellmann, G., and E. Shaw. 1963. *Biochemistry* **2**:252.
Shaw, E., M. Mares-Guia, and W. Cohen. 1965. *Biochemistry* **4**:2219.
Shimabukuro, R. H. 1967a. *J. Agr. Food Chem.* **15**:557.
Shimabukuro, R. H. 1967b. *Plant Physiol.* **42**:1269.
Shimabukuro, R. H. 1968. *Plant Physiol.* **43**:1925.
Shimabukuro, R. H., and H. R. Swanson. 1969. *J. Agr. Food Chem.* **17**:199.
Shimabukuro, R. H., R. E. Kadunce, and D. S. Frear. 1966. *J. Agr. Food Chem.* **14**:392.
Shimabukuro, R. H., H. R. Swanson, and W. C. Walsh. 1970a. *Plant Physiol.* **46**:103.
Shimabukuro, R. H., D. S. Frear, H. R. Swanson, and W. C. Walsh. 1970b. *Plant Physiol.* **47**:10.
Shuster, L. 1964. *Annu. Rev. Biochem.* **33**:571.
Smith, J. W., and T. J. Sheets. 1967. *J. Agr. Food Chem.* **15**:577.
Stohs, S. J. 1969. *Phytochemistry* **8**:1215.
Swanson, C. R., and H. R. Swanson. 1968a. *Weed Sci.* **16**:137.
Swanson, C. R., and H. R. Swanson. 1968b. *Weed Sci.* **16**:481.
van Oorschot, J. L. P. 1965. *Weed Res.* **5**:84.
Wilcox, M., and D. E. Moreland. 1969. *Nature (London)* **222**:878.
Wit, J. G., and P. Leenwangh. 1969. *Biochim. Biophys. Acta* **177**:329.
Zenk, M. H. 1967. *Z. Pflanzenphysiol.* **57**:477.

THE CHEMISTRY OF TEA AND TEA MANUFACTURING

GARY W. SANDERSON
Thomas J. Lipton, Inc.
Englewood Cliffs, New Jersey

Introduction	247
Tea Manufacturing Process	249
Biochemistry of the Tea Flush	251
Chemical Composition	251
Polyphenolic Compounds	252
Enzymes	255
Amino Acids	258
Carbohydrates	259
Organic Acids	260
Chlorophylls and Carotenes	262
Caffeine and Other Purines	263
Minerals	264
Chemistry of Tea Manufacturing	265
Development of Black Tea Characteristics	265
Biochemistry of the Withering Process	266
Fermentation: Tea Catechin Oxidation	268
Fermentation: Tea Aroma and Its Formation	288
Mechanism of Tea Fermentation	302
Tea Cream	303
Conclusions	304
References	306

Introduction

Tea is one of the most widely consumed beverages in the world. Its obscure origins lie in Chinese mythology (Scott, 1964) perhaps 30 centuries before the year 1 A.D. From these legendary beginnings in the tropical forests of the Southern Chinese region, the tea plant has come to be widely cultivated in tropical and near tropical regions of the world. This story is an exciting one and has been described in several different accounts (Gray, 1903; Scott, 1964; Ukers, 1935, 1936). Today there are in the world over 3 million acres of land devoted to tea cultivation with an annual production (excluding China) of about 2.3 billion pounds of tea (International Tea Committee, 1968; U.S. Department of Agriculture, 1969). Tea production is centered in tropical and near tropical areas of the world where rainfall is fairly uniform and exceeds about 50 inches per year, where day length exceeds about 11 hours during most of the year, and where the mean ambient temperatures exceed about 22°C.

As with any food or beverage, the organoleptic properties of tea are due to its peculiar chemical composition. It is the tender, rapidly growing shoot tips of the tea plant, *Camellia sinensis*, (L.) O. Kunze, that are used as the starting material for tea manufacture (Eden, 1965; Harler, 1963; Hainsworth, 1969; Millin and Rustidge, 1967), and it is the biochemical and morphological properties of these tissues that underlie the chemistry of the tea manufacturing process. There are many types and grades of tea available in the markets of the world, but these may be broadly classified as (a) green tea, (b) oolong tea, or (c) black tea. The manufacturing process is somewhat different for each of these basic types of tea but, in chemical terms, the main difference among them stems from differences in the degree of fermentation, or enzymatic oxidation, which they undergo during the tea manufacturing process: green tea undergoes little or no tea fermentation, oolong tea is partially fermented, and black tea is produced as the result of a "complete" tea fermentation. Therefore, the character of the finished tea product depends on the chemical composition of fresh tea flush insofar as it ultimately determines which chemical substances can be formed as a result of the tea manufacturing process. Horticultural factors—such as clone(s) of tea plants cultivated; pruning, plucking, and fertilization practices; and climate—determine the potential of the flush to make any particular tea product, and the manufacturing process determines the extent to which this potential is realized (Sanderson and Kanapathipillai, 1964; Sanderson, 1965c).

It is the intent of this paper to review those aspects of tea chemistry that appear to have a bearing on determining the organoleptic properties of the finished tea product. Further, most of the discussion will be concerned with the chemistry of black tea. The information on this subject up to about 1961 was admirably reviewed by Roberts (1962). Stahl (1962) has compiled much scattered information on tea chemistry, and Millin and Rustidge (1967) have prepared a valuable concise review of tea chemistry. Quite recently Bokuchava and Skobeleva (1969) have published a review of tea chemistry which is especially valuable because of its detailed description of Russian work in the field. The heretofore neglected field of tea aroma chemistry has been reviewed by Gianturco and co-workers (Bondarovich et al., 1967) and by Yamanishi (1968), both of whom have made important contributions to the subject. Finally, Forsyth (1964) has made a valuable contribution in which he discusses the common features in the chemistry of tea, cocoa, and tobacco processing.

Tea Manufacturing Process

The tea manufacturing process is a biochemical one which is controlled to a certain extent by mechanical manipulation of the tea flush. The traditional process for the manufacture of black tea may be broken down into stages called: (a) plucking or harvesting, (b) withering, (c) leaf maceration or rolling, (d) fermentation, (e) firing, and (f) sorting and packaging (Eden, 1965; Hainsworth, 1969; Harler, 1963; Keegel, 1958). The tea manufacturing process is normally carried out in a factory located at, or very near, the site where the tea flush is grown. The entire tea manufacturing process, from plucking through drying, is usually completed in 6 to 24 hours, depending on climatic conditions and the type of manufacturing equipment employed. Green tea is made by a process similar to that used for black tea manufacture except that the fermentation stage is drastically altered by steaming the tea flush prior to the maceration stage, causing the enzymes in the flush to be inactivated (Hainsworth, 1969).

It is only the immature vegetative portions of the tea plant, namely, the rapidly growing shoot tips down to about the second or third unfolded leaf (Fig. 1), which are suitable for use in the manufacture of black tea. This harvestable portion of the tea plant is collectively called tea flush. Plucking, or harvesting, the tea flush fixes the chemical potential of the tea leaf material to make a tea product (Sanderson and Kanapathipillai, 1964) and initiates the withering stage. The tea flush is transported to a tea factory on, or near, the tea plantation where it is spread out partially to desiccate. This process usually takes 4–18 hours, during which time the moisture content of the tea flush is reduced from about 75 percent to about 65–55 percent. The withering process conditions the tissues of the tea flush both physically and chemically for subsequent stages of manufacture (Sanderson, 1964d, 1968). The withered flush is macerated in machines designed to break up the flush, causing the cell contents of the tissues involved to become mixed. This action causes the tea flavanols

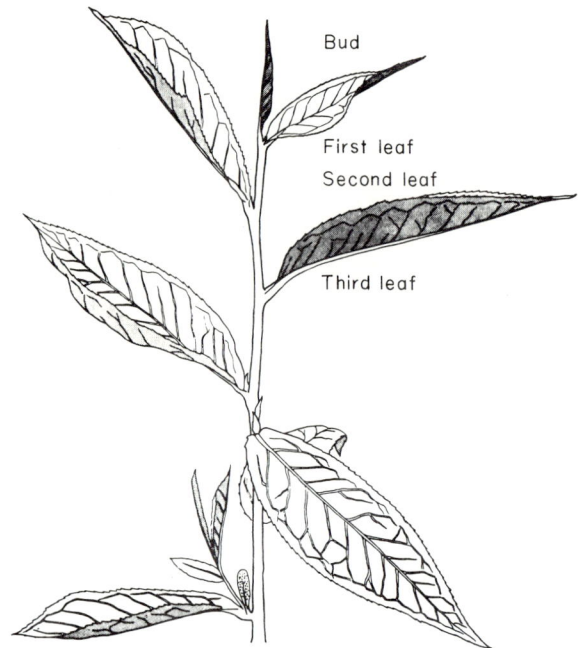

Fig. 1. Stem and leaves of a tea plant. The stem is broken between the second and third leaf to harvest the young shoot (flush) for processing. From Hainsworth (1969). (Reprinted with permission of John Wiley & Sons, Inc., New York.)

and the tea catechol oxidase to come into contact with one another, and the tea fermentation process begins.

The tea fermentation process is allowed to proceed in the macerated flush for about 1–4 hours at ambient temperatures that are usually about 18–30°C. Rather dramatic chemical changes take place during tea fermentation which are in large part responsible for the development of the color and the flavor that are characteristic of the finished tea product. The fermentation stage is terminated by firing (drying) the fermented tea flush. Firing reduces the moisture content of the tea to about 2–5 percent moisture during a drying period of 18–22 minutes at temperatures which increase from about 65°C to about 95°C during the drying cycle. The firing treatment virtually finishes the conversion of the fresh tea flush to a black tea product leaving only subtle aging processes to take place during the 3–15 months that elapse between the time the black tea is made and the time it is consumed.

Sorting and packing of the fired tea product is purely mechanical although it does tend to separate the particles of tea into different grades of tea with differing qualities.

Russian biochemists have developed in the past two decades a "thermal process" to be substituted for the tea fermentation process outlined above. The thermal process is described by Bokuchava and Skobeleva (1969), and it is claimed that this process modifies the chemical changes that take place during tea fermentation in a way which results in the improvement of certain qualities of the finished product.

Biochemistry of the Tea Flush

CHEMICAL COMPOSITION

A general analysis of tea flush compiled by Millin and Rustidge (1967) is shown in Table 1. Other general analyses of tea have been reported by Kursanov (1956) and Vuataz et al. (1959). There are some differences in these analyses which are surely due to differences in the analytical tech-

TABLE 1
COMPOSITION OF FRESH TEA FLUSH—ASSAM VARIETY[a]

Components	Amount found in flush (% dry wt.)
Substances soluble in hot water	
Flavanols	17–30
Flavonols and flavonol glycosides	3–4
Leucoanthocyanins	2–3
Polyphenolic acids and depsides	∼5
Total polyphenols	∼30
Caffeine	3–4
Amino acids	∼4
Simple carbohydrates	∼4
Organic acids	∼0.5
Substances partially soluble in hot water	
Polysaccharides	∼13
Proteins	∼15
Ash	∼5
Substances insoluble in water	
Cellulose	∼7
Lignin	∼6
Lipids	∼3
Pigments	∼0.5
Volatile substances	0.01–0.02

[a] Adapted from Millin and Rustidge (1967).

niques used and to differences in the composition of the materials analyzed. Environmental factors (Ramaswamy, 1963; Sanderson and Kanapathipillai, 1964; Dzhemukhadze and Nakhmedov, 1964; Bhatia and Ullah, 1968) and the clone of tea plants cultivated (Sanderson, 1964c; Sanderson and Kanapathipillai, 1964; Bhatia and Ullah, 1968) have been shown to affect the chemical composition of the tea flush with concomitant effects on the chemistry, and the organoleptic properties, of the final tea product. However, the chemical analysis shown in Table 1 is representative, and it does serve as a basis for the following discussion of tea chemistry.

POLYPHENOLIC COMPOUNDS

Examination of the chemical composition of fresh tea flush (Table 1) reveals that about 30 percent of the dry weight of these tissues is made up of polyphenolic compounds. Characterization of these polyphenols has been carried out by a number of workers. The historical aspects of this work have been reviewed by Roberts (1962), and a compilation of the compounds which have been identified is shown in Table 2. The major components of this fraction are the flavanols (−)-epicatechin (I), (−)-epicatechin-3-gallate (II), (−)-epigallocatechin (III), (−)-epigallocatechin-

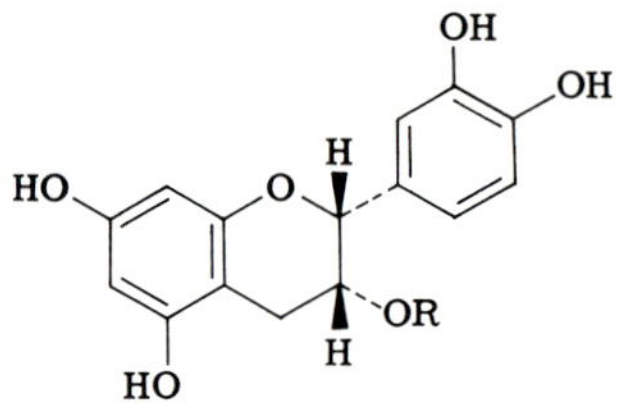

I (−)-Epicatechin; R = H
II (−)-Epicatechin gallate; R = 3, 4, 5-trihydroxybenzoyl

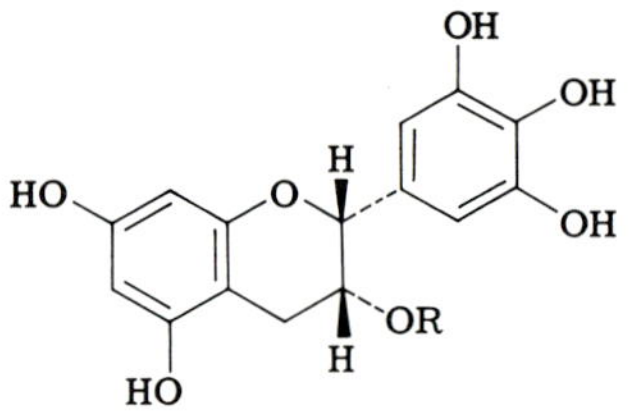

III (−)-Epigallocatechin; R = H
IV (−)-Epigallocatechin gallate; R = 3, 4, 5-trihydroxybenzoyl

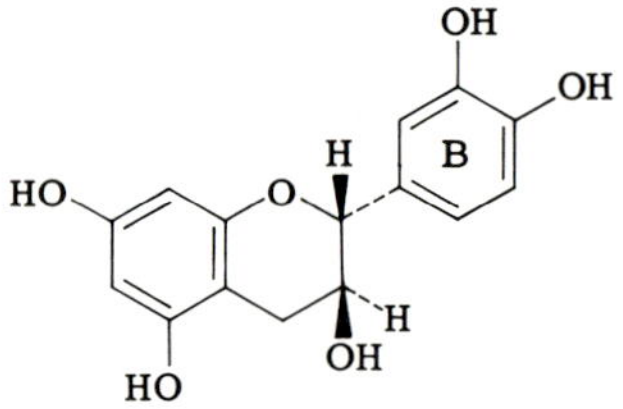

(+)-Catechin
V

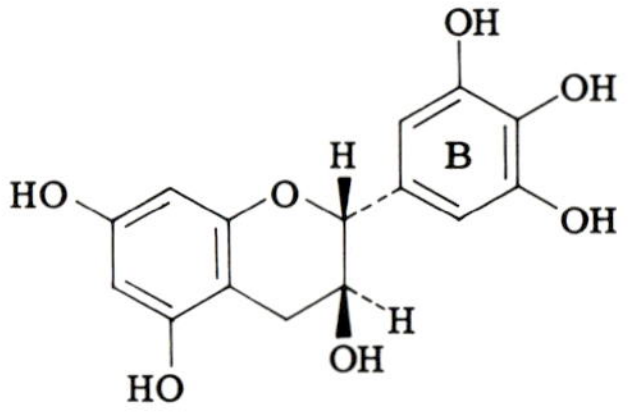

(+)-Gallocatechin
VI

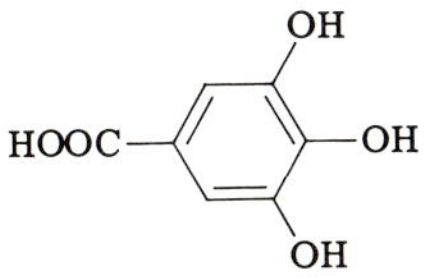

Gallic acid
VII

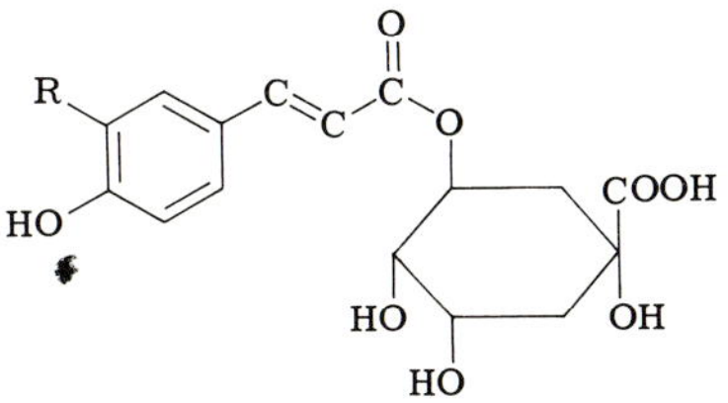

VIII Chlorogenic acid; R = OH
IX p-Coumarylquinic acid;
R = H

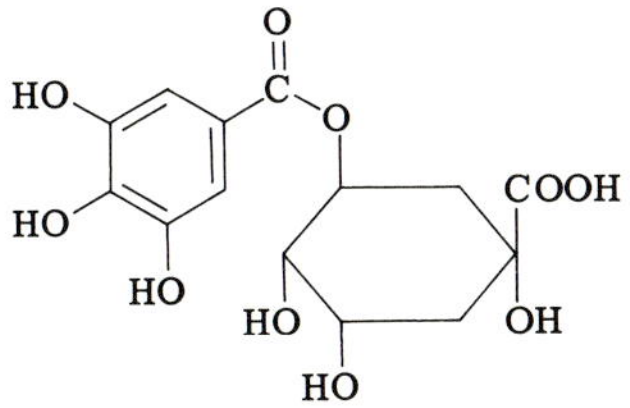

Theogallin
X

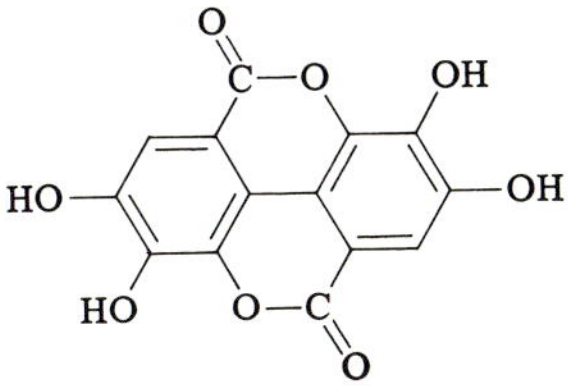

Ellagic acid
XI

3-gallate (IV), (+)-catechin (V), and (+)-gallocatechin (VI). Other epimers of these compounds have also been reported to be present in small amounts, but these may be artifacts resulting from the work-up procedures used (Roberts, 1961). It has been reported that the relative amounts of the various flavanols and their gallates may be genetically controlled and therefore a clonal characteristic (Bhatia and Chanda, 1959; Bhatia and Ullah, 1962). In view of the different oxidation products that are obtained from different flavanols, or different combination of flavanols, during tea fermentation (p. 285), this may be an important determinant of the biochemical potential of clones of tea plants to be converted into tea products with particular organoleptic characteristics.

Zaprometov (1962a, 1963) has studied the biosynthesis of catechins in detached tea shoot tips (flush) using precursors labeled with radioactive carbon-14. His results indicated that the B ring of catechin molecules (formulas I through VI) were formed by the shikimic acid pathway (Neish, 1960; Zaprometov, 1968; Grisebach and Barz, 1969) and that acetate units were favored precursors of the A ring; possibly by a polyketide biosynthetic route (Bu'Lock, 1965). Extraction and identification (Zaprometov, 1961) of quinic acid (about 1.5 percent on dry weight basis)

TABLE 2
Phenolic Compounds Found in Fresh Tea Flush

	Structure No. used in text	Molecular formula	Molecular weight	Amount found in flush (% dry wt.)
Flavanols				
(−)-Epicatechin	I	$C_{15}H_{14}O_6$	290	1–3
(−)-Epicatechin gallate	II	$C_{22}H_{18}O_{10}$	442	3–6
(−)-Epigallocatechin	III	$C_{15}H_{14}O_7$	306	3–6
(−)-Epigallocatechin gallate	IV	$C_{22}H_{18}O_{11}$	458	9–13
(+)-Catechin	V	$C_{15}H_{14}O_6$	290	1–2
(+)-Gallocatechin	VI	$C_{15}H_{14}O_7$	306	3–4
Flavonols and flavonol glucosides				
Quercetin	20	$C_{15}H_{10}O_7$	302	—[a]
Kaempferol	21	$C_{15}H_{10}O_6$	286	—[a]
Quercetin 3-rhamnoglucoside (rutin)	22	$C_{27}H_{30}O_{16}$	610	—[a]
Kaempferol 3-rhamnoglucoside	23	$C_{27}H_{30}O_{15}$	594	—[a]
Quercetin 3-rhamnodiglucoside	24	$C_{33}H_{40}O_{21}$	772	—[a]
Kaempferol 3-rhamnodiglucoside	25	$C_{33}H_{40}O_{20}$	756	—[a]
Leucoanthocyanins	—			2–3
Acids and depsides				∼5
Gallic acid	VII	$C_7H_6O_5$	170	—[a]
Chlorogenic acids (4 isomers)	VIII	$C_{16}H_{18}O_9$	354	—[a]
p-Coumarylquinic (4 isomers)	IX	$C_{16}H_{18}O_8$	338	—[a]
Theogallin	X	$C_{14}H_{15}O_{10}$	343	∼1
Ellagic acid	XI	$C_{14}H_6O_8$	302	—[a]
Total polyphenols				25–35

[a] Quantitative data not available.

and shikimic acid (about 0.035 percent on a dry weight basis) and demonstration of the presence of high levels of 5-dehydroshikimate reductase in tea flush (Sanderson, 1966a) also indicate the presence of an active shikimic acid pathway in tea flush. In fact, the specific activity of 5-dehydroshikimate reductase in crude preparations of tea flush was found (Sanderson, 1966a) to be 13–90 times greater than the level found in other organisms that had been studied; this is consistent with the unusually high level of flavonoid compounds in these tissues (Table 2). Patschke and Grisebach (1965) have shown that 4, 2′, 4′, 6′-tetrahydroxychalcone-2′-glucoside is a good precursor of (−)-epicatechin in tea leaves.

The largest part of the flavanols present in tea leaf tissues are esterified with gallic acid in the 3-position. Zaprometov and Bukhlaeva (1963) investigated the biochemistry of the formation of these esters, and their results suggested that esterification was a slow process compared to the formation of gallic acid. This problem needs more investigation.

It has been shown (Zaprometov, 1958; Sanderson and Sivapalan, 1966a) that the tea flavanols are biosynthesized in the leaves and that immature leaves are much more active in synthesizing flavanols than mature leaves (Sanderson and Sivapalan, 1966a). Histochemical studies carried out by Tambiah *et al.* (1966) and by Chalamberidze *et al.* (1969) have shown that the catechins are localized in the vacuoles of the palisade parenchyma of tea leaf tissues.

Another polyphenolic compound, which has been named theogallin (Cartwright and Roberts, 1954b, 1955; Roberts and Myers, 1958), is present in tea flush in rather large amounts (about 1 percent, see Table 2) and is of particular interest because of its unusual structure. Stagg and Swaine (1971) have recently confirmed that theogallin is a 3-galloyl quinic acid (X) as had been originally suggested by Roberts and Myers (1958).

As shown in Table 2, tea flush is known to contain several flavonols, flavonal glycosides, and other polyphenolic compounds in smaller quantities.

Accurate and rapid methods for determining the amount of the flavanols in tea extracts have recently been developed (Pierce *et al.*, 1969b; Collier and Mallows, 1971a). These methods are based on gas–liquid chromatography of trimethylsilyl derivatives of the flavanols, and it appears as though these methods could be extended to enable other polyphenolic compounds present in these extracts to be determined.

Enzymes

One of the most important fractions in tea flush is the proteins, because this fraction includes the enzymes required for the metabolism of the tea plant and for black tea manufacture. A list of the enzymes that have been found in tea plants to date is given in Table 3 together with a summary of their function in tea chemistry. Naturally, the myriad of enzymes required to sustain the life of a tea plant is present in the tea plant as part of its protein fraction, and any difficulty met in trying to detect a particular essential enzyme is most likely to be traceable to an inadequacy of the experimental procedures used, especially the extraction stage.

The extraction of enzymes in their native state from tea plant materials has been shown to be dependent on the inclusion of rather large amounts of an insoluble polyphenol absorbing material, such as Polyclar AT, Sephadex, powdered nylon, etc., in the maceration medium (Sanderson, 1965a, b, 1966a). It was shown in these studies (Fig. 2) that virtually no soluble protein could be extracted from tea leaf material unless one of

TABLE 3
ENZYMES DETECTED IN TEA LEAF

Name of enzyme	Function in tea chemistry	References
Catechol oxidase	Oxidation of tea catechins in tea fermentation	Sreerangacher (1939, 1943d), Sanderson (1964a), Bendall and Gregory (1963), Gregory and Bendall (1966), Takeo (1966b)
Peroxidase	Unknown	Bokuchava (1950), Roberts (1952)
Pectin methyl esterase	Demethylation of pectins during tea manufacture	Ramaswamy and Lamb (1958a,b)
Peptidase	Breakdown of proteins during withering	Sanderson and Roberts (1964)
Chlorophyllase	Unknown	Ogura and Takamiya (1966), Ogura (1969)
Phosphatases	Unknown	Tirimanna (1967)
5-Dehydroshikimate reductase	Biosynthesis of tea polyphenols	Sanderson (1966a)
Leucine transaminase	Biosynthesis of terpenes	Wickremasinghe et al. (1969)

these materials was used, and subsequent work showed that this was due to the protein precipitating activity of the polyphenolic compounds present in the tea flush (Sanderson, 1965a, b). Further application of the principles and practices developed by Loomis and Battaile (1964, 1966) and others (Anderson, 1968; Lam and Shaw, 1970) for the extraction of enzymes from plant tissues which are rich in polyphenolic compounds to investigations of the enzymology of tea biochemistry promises to be very fruitful.

The enzyme of most interest in tea flush is the tea leaf catechol oxidase because of its key role in tea fermentation (see p. 268 ff). The fact that tea fermentation is enzymatically catalyzed was apparent almost from the time that the nature of enzymes was understood because it was known from ancient times that steaming fresh tea leaves would destroy their ability to undergo fermentation (Bamber and Wright, 1902). However, the nature of the enzyme responsible for tea fermentation was later the subject of some controversy (Roberts and Sarma, 1938: Roberts, 1939a,b, 1940, 1952; Sreerangachar, 1939, 1943a,b,c), which was eventually resolved with the general acceptance (Roberts, 1958a, 1962) of Sreerangachar's identification of the enzyme as a copper-containing polyphenol oxidase (Sreerangachar, 1939, 1943c).

Early studies on the localization of the tea catechol oxidase in tea flush tissues led Li and Bonner (1947) to conclude that tea catechol oxidase was located in the chloroplasts on the basis of its ability to be precipitated by certain centrifugal forces. The insoluble nature of this enzyme was accepted by most investigators (Roberts, 1962; Bendall and Gregory, 1963; Gregory and Bendall, 1966) until methods were developed for the extraction of enzymes in their native state from tissues rich in tanninlike materials (Loomis and Battaile, 1964, 1966; Anderson, 1968). Application of these modern methods for handling plant enzymes produced results (Sanderson, 1964a; Buzyn et al., 1970) which suggest that the tea catechol oxidase is, in fact, soluble in its native state. Takeo (1966b) confirmed the finding that the use of a polyphenol adsorbent in the extraction of tea leaf catechol oxidase caused the enzyme to become disassociated from centrifugal precipitates obtained at <30 g (cellular debris) and at 30 g to 1400 g (chloroplasts). The major portion of the catechol oxidase activity was now found to be associated with the 1400 g to 15,000 g (mitochondrial) precipitate and the $>15,000$ g (soluble) fraction. Further work is required to ensure that neither (a) aggregation of enzyme molecules mediated by polyphenolic compounds nor (b) disruption of cell organelles occurred in

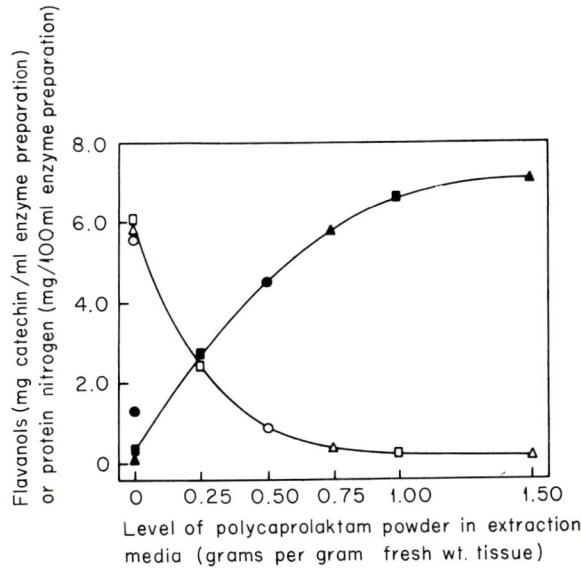

FIG. 2. Effect of polycaprolactam powder on the solubility of flavanols and protein in extracts of tea shoot tips. □ ○ △, total flavanols; ■ ● ▲, protein nitrogen; □ ■, Expt. I; ○ ●, Expt. II; △ ▲, Expt. III.

this study. It is interesting to note that Bamber and Wright (1902) reported that the enzyme of tea fermentation could be extracted from tea leaf tissues in a soluble form if hide powder were mixed with the tea leaf material during the extraction process.

Bendall and Gregory (1963; Gregory and Bendall, 1966) have purified the tea leaf catechol oxidase over 5000-fold on a dry weight basis. Four forms of the enzyme were found that appear to be at least partly interconvertible. The major fraction had an estimated molecular weight of 144,000 ± 16,000, and it contained 0.32 percent (w/w) copper or about 6.4–8.0 copper atoms per molecule.

Wickremasinghe et al. (1967) have used immunological techniques to determine the localization of the tea leaf catechol oxidase, and they found that the enzyme was located in the epidermis and around the vascular bundles of the tea leaf tissues. This contrasts with the intercellular localization of the catechins in these tissues (Tambiah et al., 1966; Chalamberidze et al., 1969) described above. This spatial separation of catechol oxidase and the tea catechins in the tissues of the tea flush effectively prevents tea fermentation from taking place until such time as the cell contents in these tissues are mixed, as by macerating the flush.

The supposed role of other enzymes listed in Table 3 in the chemistry of tea manufacture will be discussed in the following sections where appropriate.

Amino Acids

The following 19 amino acids usually associated with proteins have been reported (E. A. H. Roberts and Wood, 1951; G. R. Roberts and Sanderson, 1966) to be present in free form in fresh tea flush tissues: aspartic acid, glutamic acid, glycine, serine, asparagine, lysine, threonine, arginine, histidine, glutamine, alanine, β-alanine, tyrosine, proline, valine, tryptophan, leucine, isoleucine, and phenylalanine. However, the most abundant free amino acid present in tea flush is an unusual one which has been named theanine (Sakato et al., 1950). Theanine (XII) has been identified as 5-N-ethyl glutamine (Sakato et al., 1950; Cartwright et al., 1954),

Theanine
XII

Caffeine
XIII

and it usually accounts for 0.8–1.5 percent of the dry weight of these tissues.

γ-N-Ethylasparagine, a homolog of theanine, has been tentatively identified in extracts of tea flush (Wickremasinghe and Swain, 1965), but at much lower concentrations than theanine.

The amino acids are believed to play an important role in the development of tea aroma during the tea manufacture process, and this will be discussed later (see p. 295).

CARBOHYDRATES

Sanderson and Perera (1965, 1966) investigated the carbohydrates in tea flush and in tea roots, and their results are summarized in Table 4. The free sugars found were glucose, fructose, sucrose, raffinose, and stachyose. Polysaccharides were separated into crude fiber, pectin, starch (roots only), and another extractable polysaccharide fraction composed of eight different sugar residues, i.e., glucose, galactose, galacturonic acid, mannose, arabinose, xylose, ribose, and rhamnose. These results (Sanderson and Perera, 1965, 1966) agree well with the results of earlier qualitative investigations by Cartwright and Roberts (1954c), Torii and Kanazawa (1954, 1955), and Mizuno and Kimpyo (1955a,b, 1956a,b).

TABLE 4
The Carbohydrates Found in Tea Flush and in Tea Roots[a]

Carbohydrate	Level in flush (% dry wt.)	Level in roots (% dry wt.)
Free sugars		
Glucose		
Fructose	0.3–0.8	
Sucrose	0.9–2.3	0.7–0.9
Raffinose	Trace	
Stachyose	Trace	
Polysaccharides (sugar residues[b])		
Pectin (galacturonic acid, galactose, arabinose)	3.2–6.4	4.4–5.1
Starch (glucose)	None	0.4–19.4
Other extractable polysaccharide(s) (glucose, galactose, galacturonic acid, mannose, arabinose, xylose, ribose, rhamnose)	1.0–3.0	≤0.1
Crude fiber	9.6–11.6	4.1–7.4

[a] Adapted from data given by Sanderson and Perera (1965, 1966).
[b] Sugar residues obtained by hydrolyzing the polysaccharide.

The polysaccharides in tea plant tissues were fractionated and characterized for their component sugars by Mizuno et al. (1964). Fresh tea leaf was found to contain a total of about 20 percent (dry weight basis) extractable polysaccharides as compared to about 14 percent extractable polysaccharides in dried tea leaf. The polysaccharides were separated into eight fractions in both fresh leaf and in dried tea leaf, but the fractions all had different properties, suggesting that the polysaccharides were altered by the drying process. It is noteworthy that Mizuno et al. (1964; Mizuno and Kimpyo, 1963) reported about 4 percent (dry weight basis) starch in fresh tea leaves, but none in dried tea leaves. Other investigators (Shaw, 1935; Sanderson and Perera, 1965, 1966) have reported that they were unable to detect starch in tea leaves although starch was found readily in the woody parts of the tea plant.

Mizuno (1968; Mizuno and Harashima, 1970) has carried out some detailed studies on the structure of "theasaponin" from tea seeds and an arabinogalactan hemicellulose isolated from tea leaves. More studies of this kind are required before an adequate understanding of the nature of the tea polysaccharides will be achieved.

Radiochemical studies on the photosynthetic assimilation of carbon dioxide by tea leaves of different ages (Sanderson and Sivapalan, 1966a,b) have shown that both the total amount and the percentage of assimilated carbon incorporated into carbohydrates increase rapidly as the age of the tea leaves increases. Further, it was shown that assimilated carbon is readily translocated out of the mature leaves both to the immature buds and leaves and to the roots, whereas carbon assimilated by immature buds and leaves is not translocated out of the assimilating organ. These results showed clearly that the mature foliage on tea bushes contributes to the development of the flush and the root system, and to the food reserves of the tea plant under favorable conditions. The results also indicated that mature leaves do not become boarders on the remainder of the tea plant, even when these leaves cannot photosynthesize enough to attain their compensation point (such as under conditions of high shading).

Sugar phosphates were measured in two samples of tea leaves by Selvendran and Isherwood (1970), and the following results were obtained: glucose 1-phosphate, 1.25 and 0.63 μmoles/100 grams fresh weight; glucose 1,6-diphosphate, 0.2 and 0.07 μmoles/100 grams fresh weight; sucrose 6-phosphate, 0.05 and 0.01 μmoles/100 grams fresh weight.

Organic Acids

The organic acids present in Ceylon tea flush were studied by Sanderson and Selvendran (1965). These investigators were able to separate 10 organic acids, and they tentatively identified five of the acids present in greatest quantity. Oxalic acid, which is present as crystals in the vacuoles of many tea leaf cells (Wight and Barua, 1954; Barua, 1956; Green, 1971), was found to be the predominant acid in these tissues (about 0.26–0.50 percent on dry weight basis); malic and citric acids were present at about one-third the amount of oxalic acid; and succinic, isocitric, and five unidentified acids were present in much smaller quantities (Table 5).

Bokuchava and Soboleva (1958) studied the qualitative changes in organic acids which occur during the germination of tea seeds. Oxalic, citric, and malic acids were found in seeds and 40-day-old seedlings. Sixty-day-old seedlings contained quinic acid in their stems and an unidentified acid in their roots in addition to the three acids found in tea seeds. Leaves from 100-day-old seedlings were found to contain oxalic, quinic, citric, malic, and fumaric acids. The appearance of quinic acid in immature, aboveground tissues may be a manifestation of the development of metabolic factors necessary for the biosynthesis of tea flavanols and other polyphenolic compounds (see p. 252). It is noteworthy that quinic acid was not found in tea roots, which also appear to have little or no capacity to synthesize polyphenolic compounds.

Zaprometov (1961) demonstrated the presence of quinic and shikimic

TABLE 5
Organic Acids Found in Tea Flush

Acid[a]	Amount		Reference
	μEq/gram fresh wt.	mg/100 grams dry wt.	
Oxalic	14.7–27.8	2.6–5.0	Sanderson and Selvendran (1965)
Malic	4.1–10.3	1.1–2.8	Sanderson and Selvendran (1965)
Citric	4.1–6.1	1.0–1.5	Sanderson and Selvendran (1965)
Isocitric	Trace–0.9	Trace–0.2	Sanderson and Selvendran (1965)
Succinic	0.9–1.6	0.2–0.4	Sanderson and Selvendran (1965)
Unknown 1	0.7–1.2	—	Sanderson and Selvendran (1965)
Unknown 2	Trace–0.3	—	Sanderson and Selvendran (1965)
Unknown 3	Trace–0.3	—	Sanderson and Selvendran (1965)
Unknown 4	0.1	—	Sanderson and Selvendran (1965)
Unknown 5[b]	0.3–1.2	—	Sanderson and Selvendran (1965)
Quinic	—	15	Zaprometov (1961)
Shikimic	—	0.35	Zaprometov (1961)

[a] See Table 2 for phenolic acids.
[b] Probable identification = quinic acid.

acids in tea flush in amounts of 1.5 and 0.35 percent (dry weight basis), respectively, and pointed out their likely role in the biosynthesis of polyphenolic compounds in these tissues.

Chlorophylls and Carotenes

The main pigments in fresh tea flush are chlorophylls and carotenoid compounds. Chlorophylls a and b are the chlorophylls which have been reported to be present in fresh tea flush (Wickremasinghe et al., 1965; Wickremasinghe and Perera, 1966; Co and Sanderson, 1970). Co and Sanderson (1970) measured the amount of chlorophylls in a sample of fresh tea flush and found about 1.4 mg of chlorophylls per gram, dry weight, of tissue (Table 6). Further, it was shown that there is a large decrease in the amount of chlorophylls during the black tea manufacturing process and a concomitant appearance of the chlorophyll degradation products pheophytin a, pheophytin b, and pheophorbide (Table 6). These results corroborate the earlier findings of Wickremasinghe and co-workers (Wickremasinghe et al., 1965; Wickremasinghe and Perera, 1966).

Wickremasinghe and Perera (1966) found a positive correlation between the amounts of pheophytins and pheophorbide in black tea products and their blackness or brownness, respectively. If this suggested relationship

TABLE 6

Concentration of Chlorophylls in Tea Samples Taken at Various Stages of Black Tea Manufacture[a]

Samples	Chlorophyll b (μg/gram)	Chlorophyll a (μg/gram)	Pheophytin b (μg/gram)	Pheophytin a (μg/gram)	Pheophorbide (μg/gram)
Fresh tea flush	481	955	0	0	0
1-Hour fermented tea	364	794	0	0	0
2-Hour fermented tea	311	781	0	64	0
3-Hour fermented tea	326	727	0	124	0
Fired and dried tea	315	650	0	234	0
Hot water extract[b]	0	0	18	29	25
Spent leaves[c]	0	0	312	762	—
Commercial black tea (Lipton blend)	147	85	103	223	125

[a] From Co and Sanderson (unpublished data).
[b] Amounts corrected to whole tea basis. Extract solids = 43% of whole tea.
[c] Amounts corrected to whole tea basis. Spent leaves solids = 57% of whole tea.

TABLE 7
THE CAROTENES FOUND IN TEA FLUSH

Carotene	Reported by Tirimanna and Wickremasinghe (1965)	Reported by Sanderson et al. (1971a)
β-Carotene	Yes (ND)[b]	Yes (102 µg/gram dry weight)
γ-Carotene	Yes (trace only)	No
Lutein	Yes (ND)	Yes (260 µg/gram dry weight)
Neoxanthin	No	Yes (51 µg/gram dry weight)
Phytoene	Yes (trace only)	No
Phytofluene	Yes (trace only)	No
Lycopene	Yes (trace only)	No
Cryptoxanthin	Yes (trace only)	No
Violaxanthin	Yes (ND)	Yes (120 µg/gram dry weight)
Zeaxanthin	Yes (ND)	No
Unknowns 1, 2, 3,[a] 4[a]	Yes (ND)	No

[a] Two of the unknowns extracted by Tirimanna and Wickremasinghe (1965) were shown to be 5,6-epoxy carotenoids, and it is likely that one of these carotenoids is neoxanthin, found by Sanderson et al. (1971).

[b] ND, amount not determined.

does in fact hold, the factors affecting the nature of chlorophyll degradation—e.g., chlorophyllase activity, heavy metal ion (Fe^{3+}, Cu^{2+}, etc.) concentrations—must be important in determining the exact color of black tea.

Tirimanna and Wickremasinghe (1965) found 14 carotenoid compounds in tea flush and tentatively identified nine of them. Sanderson et al. (1971a) found only four of these carotenoid compounds in their investigation of tea carotenes, an observation suggesting that there may have been important qualitative differences between the tea materials used in these investigations. The analytical results from these investigations are summarized in Table 7. The role of the tea carotenes in the formation of black tea aroma will be discussed later (see p. 298).

CAFFEINE AND OTHER PURINES

Tea contains from 2.5 to 5.5 percent (dry weight basis) of caffeine (XIII) (Wood et al., 1964) along with much smaller amounts of the closely related methylated xanthines theobromine and theophylline. Michl and Haberler (1954) reported finding 2.5 percent caffeine, 0.17 percent theobromine, and 0.013 percent theophylline in black tea. It is likely that tea derives its wide acceptance because of the stimulating effects induced by ingestion of caffeine, although this is very difficult to prove. A good review of the pharmacological properties of caffeine in tea was made by Das et al. (1965).

Investigation of caffeine biosynthesis in detached tea leaves by Proiser and Serenkov (1963) and in detached tea shoot tips by Inone and Kawamura (1961) and by Zaprometov (1962b) indicated that incorporation of labeled acetate, shikimic acid, and sugars into caffeine in these tissues was slow and inefficient. These results suggested that there were several biosynthetic steps between the precursors tried and caffeine. Ogutuga and Northcote (1970a,b) have investigated the biosynthesis of caffeine in tea callus tissue produced in tissue culture, and they conclude that caffeine in tea is synthesized from purines produced by the breakdown of nucleic acids. The findings that ribonucleic acids decrease appreciably during withering (Bhattacharyya and Ghosh, 1968) and that caffeine increases appreciably during withering (Wood and Chanda, 1955; Sanderson, 1964d; Bhatia and Deb, 1965) appear to support the views of Ogutuga and Northcote (1970a,b).

Michl and Haberler (1954) have also reported the presence of the purines adenine (0.014 percent), guanine (<0.001 percent), xanthine (<0.001 percent), and hypoxanthine (<0.001 percent) in black tea.

Minerals

As shown in Table 1, the ash content of fresh tea flush is about 5 percent (dry weight basis). Of the several minerals in tea leaf (Eden, 1965), copper and aluminum perhaps deserve special mention. Copper is of particular importance in tea biochemistry because the tea catechol oxidase which is so important in the manufacture of black tea (see Fig. 12) is a copper-containing enzyme. Child (1955) reviewed the occurrence and role of copper in tea leaf, and he points out that, while there are no recorded instances of copper deficiency in tea plants which affects their growth, about 12–18 ppm (dry weight basis), copper in tea leaf appears to be required for enough catechol oxidase enzyme to be formed in these tissues to effect an adequate tea fermentation process. Spraying of tea plants with certain copper-containing formulations has been shown (Anonymous, 1969) to be effective in increasing the amount of copper and catechol oxidase activity in the tea leaf and in improving the tea fermentation properties of the tea leaf produced.

The aluminum status of the tea plant was reviewed by Chenery (1955). The data adequately establish the tea plant, and other members of the *Theaceae* family, as aluminum accumulators. The aluminum content of tea leaf tissue usually ranges from about 1000 to 10,000 ppm (dry weight basis) in contrast to aluminum contents of 10–50 ppm in most plant tissues used for food (Winton and Winton, 1935). The role of aluminum in the biochemistry of tea plants is obscure, but it may well be concerned with the metabolism or storage of the tea flavanols, with which aluminum readily complexes (Jurd, 1962).

Chemistry of Tea Manufacturing

Development of Black Tea Characteristics

Infusions of black tea are readily distinguished from infusions of freshly plucked green tea flush, or green tea, by their bright reddish brown color, their characteristic black tea aroma, and their brisk, slightly astringent taste. These black tea characteristics are developed during the black tea manufacturing process which is completed in the 6- to 24-hour period immediately following the harvesting of the tea flush. Therefore, an understanding of the chemistry of tea manufacture is important for developing an understanding of the chemical basis of black tea quality.

An approximate analysis of black tea is shown in Table 8. The corresponding analysis of the fresh tea flush is included for comparison. It

TABLE 8
Approximate General Analysis of Black Tea

Component	Fresh flush	Black tea[a]	Black tea brew[a,b]
Proteins	15[c]	15	Trace
Fiber	30	30	0
Pigments	5	5	Trace
Caffeine	4	4	~3.2
Polyphenols, simple	30	5	4.5
Polyphenols, oxidized	0	25	15
Amino acids	4	4	3.5
Ash	5	5	4.5
Carbohydrates	7	7	4
Volatile compounds	0.01	0.01	0.01

[a] Many of these components have been altered during the black tea manufacturing process, and they would not necessarily be recognizable as the component in the fresh flush from which they were derived.

[b] These are the substances extracted in a normal 5-minute extraction in a teapot.

[c] Values are expressed as percentage of whole tea dry weight.

should be noted that the substances present in black tea have, in many cases, undergone considerable change during the black tea manufacturing process, so that the substance in the fresh flush from which the black tea substances are formed is not always obvious. Further, there is some interaction of black tea components leading to the production of insoluble materials. This interaction is reflected in the poor extractability of some black tea components; this can be seen by comparing the last two columns in Table 8. The following discussion is largely an attempt to review current knowledge regarding the chemistry of black tea constituents and the chemistry of their formation from green tea constituents.

Biochemistry of the Withering Process

Withering, or partial desiccation, of tea flush is the first step in traditional tea-manufacturing processes (Eden, 1965; Keegel, 1958; Harler, 1963; Hainsworth, 1969). The withering physically conditions the tissues of the flush so that the flush will roll up and break up as desired in the subsequent macerating stage. However, besides the physical conditioning of the tea flush which results from withering, several biochemical changes are known to occur during withering which have important effects on subsequent stages of the black tea manufacturing process and the character of the finished black tea product (Sanderson, 1964d). These changes are summarized in Table 9.

It is noteworthy that only one of the biochemical changes known to occur in tea flush during withering, namely, the increase in cell membrane permeability, has been shown to be dependent on the loss of moisture itself (Table 9). As far as is known, the other changes—which are apparently involved in determining the level of aroma precursors (see p. 288ff), taste components, and enzymes—all take place on storage of the tea flush irrespective of whether moisture loss occurs or not. It has been pointed out (Sanderson, 1964d, 1968) that the increase in cell membrane permeability appears to be of considerable importance in orthodox black tea manufacturing processes in that it facilitates the mixing of cell contents during the maceration step. The importance of this mixing is clear when one considers that the enzyme catechol oxidase and its substrates, the tea flavanols, are spatially separated in the tissues of tea flush (see p. 258).

It is noteworthy that the biochemical changes which take place in tea flush as it withers are reminiscent of those described for senescing plant tissues in general (Varner, 1961). Recognition of this fact may help in further development of our understanding of the withering step.

TABLE 9
Summary of Biochemical Changes Known to Occur in Tea Flush during Withering Step of Black Tea Manufacture[a]

Description of change	Importance of change in black tea manufacture	Dependence of change on moisture loss	Reference
Loss of moisture	Conditions tea flush so that it will roll up and break up during subsequent maceration stage	—	Eden (1965), Harler (1963), Keegel (1958)
Increase in free amino acids	Precursors of aroma constituents	Independent	Roberts and Wood (1951), Roberts and Sanderson (1966)
Increase in caffeine	Increases physiological activity of black tea	Independent	Wood and Chanda (1955), Sanderson (1964d)
Increase in simple carbohydrates	Possibly contribute to taste and/or serve as aroma precursors	Unknown	Sanderson and Perera (1965)
Changes in level of organic acids	Possibly contribute to taste and/or serve as aroma precursors	Unknown	Sanderson and Selvendran (1965)
Changes in activity of catechol oxidase	Improves rate of tea fermentation	Independent	Sanderson (1964b), Takeo (1966a)
Increase in permeability of cell membranes	Facilitates even mixing of cell contents, required for satisfactory tea fermentation	Dependent	Sanderson (1968)

[a] Adapted from Sanderson (1964d).

FERMENTATION: TEA CATECHIN OXIDATION

Development of Roberts' (1962) Proposed Scheme for Fermentation

As mentioned above, the most dramatic change that occurs during the black tea manufacturing process is the enzymatic oxidation of the tea catechins. Further, the entire tea manufacturing process may be viewed as a process designed to provide conditions that promote this primary enzymatic oxidation and the secondary reactions that are coupled to the primary reaction. Roberts and his co-workers (Cartwright and Roberts, 1954a; Roberts et al., 1957; Roberts, 1958a,b, 1962; Roberts and Myers, 1959a,b,c, 1960b; Roberts and Williams, 1958) were among the first to study these reactions. They used a combination of solvent (ethyl acetate) extration/precipitation techniques, patterned after the earlier work of Bradfield and Penny (1944; Bradfield, 1946) on tea, and paper chromatography (Cartwright and Roberts, 1954a) to separate the pigments of aqueous black tea extracts (Roberts et al., 1957). The paper chromatography depended on a solvent system composed of first solvent = butanol:acetic acid:water (4:1:2.2), and second solvent = 2 percent acetic acid. Separations are shown diagrammatically in Fig. 3. This work showed that the black tea pigments could be separated into two main fractions, one of which was an orange neutral fraction and the other a brown acidic fraction. Neither of these fractions was homogeneous, but the colored pigments in these fractions were named (Roberts, 1958a) theaflavins and thearubigins, respectively.

The neutral theaflavins were separated by paper chromatography into two fractions called substance X and substance Y (Fig. 3). The acidic thearubigins appeared as a streak on these chromatograms, but they could be further fractionated by partitioning between solvents and solutions of varying polarity and acidity into S_I, S_{IA}, and S_{II} thearubigins. It was acknowledged that these fractions were clearly not homogeneous and that there were no sharp distinctions between the fractions.

Besides the theaflavins and the thearubigins, some other substances were also shown to be present in black tea infusions (Roberts et al., 1957; Roberts, 1958a). Colored substances in this category were called substances Z, P, and Q; and noncolored phenolic substances were called substances A, B, C, and D. These latter substances were colorless, but they gave a blue color on paper chromatograms when sprayed with a ferric chloride/potassium ferricyanide reagent.

The direct relationship between the oxidation of the tea catechins and the formation of the pigments responsible for the color of black tea infusions was established by Roberts (1958a) using a model tea fermentation system. It was shown in this investigation that theaflavins, gallic acid,

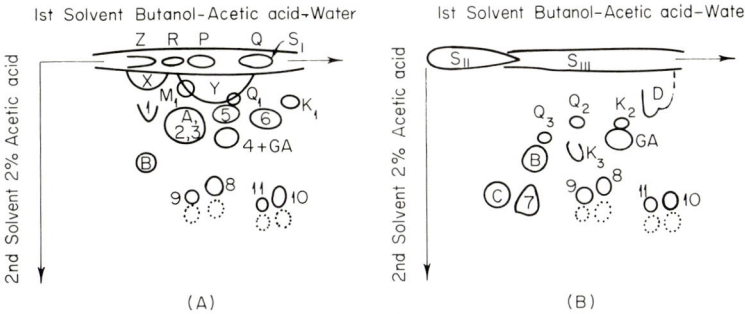

Fig. 3. Roberts' (1962) paper chromatograms of polyphenolic substances in black tea: (A) Polyphenolic substances soluble in ethyl acetate. (B) Polyphenolic substances insoluble in ethyl acetate. Key to abbreviations used: 1 = EGC = (−)-epigallocatechin (III); 2 = G Cat = (+)-gallocatechin (VI); 3 = EC = (−)-epicatechin (I); 4 = Cat = (+)-catechin (V); 5 = EGCG = (−)-epigallocatechin gallate (IV); 6 = ECG = (−)-epicatechin gallate (II); 7 = theogallin (X); 8, 9 = chlorogenic acids (VIII); 10, 11 = p-coumarylquinic acids (IX); GA = gallic acid (VII); M_1 = myricetin 3-glucoside; Q_1 = isoquercitrin; K_1 = kaempferol 3-glucoside; M_2 = myricetin 3-rhamnoglucoside; Q_2 = rutin; K_2 = kaempferol 3-rhamnoglucoside; Q_3 = quercetin 3-rhamnodiglucoside; K_3 = kaempferol 3-rhamnodiglucoside; A = bisflavanol A (XX); B = bisflavanol B (XXI); C = bisflavanol C (XXII); P = tricetinidin (XXV); R = flavanotropolone (?); S_I = thearubigins fraction I; S_{IA} = thearubigins fraction IA; S_{II} = thearubigins fraction II; Q = Epitheaflavic acids (XXIII, XXIV) + other unidentified compounds (?); X = theaflavin (XVI); Y = theaflavin gallates (XVII, XVIII, XIX); Z = uncharacterized substance with fluorescence similar to ellagic acid (XI).

and substances A, B, C, P, Q, and Z were formed in a system composed of an acetone-dried powder prepared from fresh tea flush (Roberts and Wood, 1950) for an enzyme source and a "mixed flavanol" preparation extracted from dried, unprocessed tea flush. The "mixed flavanol" preparation was shown to be composed mainly of the flavanols (−)-epigallocatechin, (+)-gallocatechin, (−)-epicatechin, (+)-catechin, (−)-epigallocatechin gallate, and (−)-epicatechin gallate together with small quantities of the 3-monoglucosides of myricetin, quercetin, and kaempferol, and traces of leucoanthocyanins, chlorogenic acids, and p-coumarylquinic acids, but no caffeine, sugars, or amino acids. Failure to detect thearubigin formation in this model system was attributed to the very low concentration of reactants in the model tea fermentation system as compared to the fermenting whole tea flush system. However, concurrent studies on the formation of theaflavins and thearubigins during fermentation of tea flush led Roberts (1958a) to postulate that thearubigins were formed from theaflavins.

Roberts and co-workers continued their studies on the chemistry of the formation of black tea pigments using individual tea catechins isolated from dried, unprocessed tea flush (Roberts and Myers, 1960a). Roberts (1958b, 1962) summarized the results and the conclusions drawn from these investigations (Roberts and Williams, 1958; Roberts and Myers, 1959a,b,c) in a proposed scheme for tea fermentation shown in Fig. 4. The salient features of this scheme will be discussed below in light of information developed since the publication of Roberts' last comprehensive report (Roberts, 1962).

The study of the oxidation potentials of tea catechins (Roberts, 1957), which showed that the gallocatechins had lower oxidation potentials than the other polyphenolic compounds present in tea flush, and visual observations of chromatograms prepared from unprocessed tea leaf and black tea, which were interpreted to show that only the gallocatechins underwent any appreciable oxidation, led Roberts (1958a) to conclude that most of the oxidation products formed during tea fermentation result from the oxidation of (−)-epigallocatechin and (−)-epigallocatechin gallate. However, Dzhemukhadze et al. (1957, 1964) showed clearly that all the tea catechins undergo appreciable decreases in amount during the conversion of fresh green flush to black tea, and the fact that (−)-epicatechin gallate decreases in amount during tea fermentation was confirmed by Vuataz and Brandenberger (1961) and by Bhatia and Ullah (1961, 1965). Investigations using improved methods for the analysis of

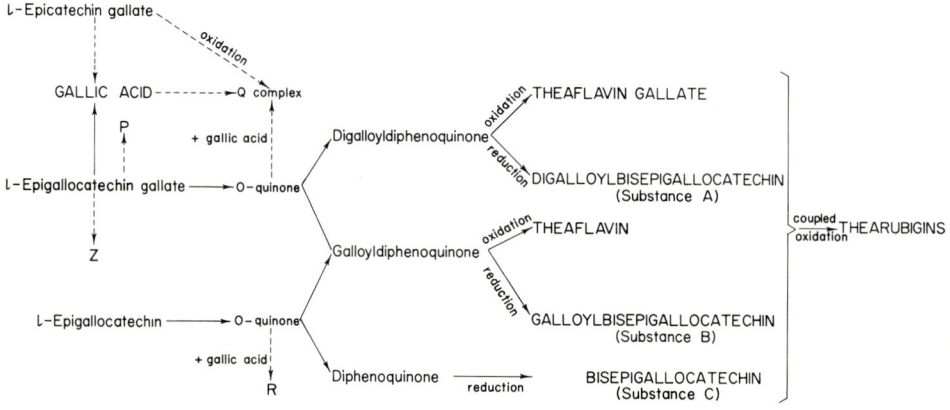

FIG. 4. Roberts' (1962) proposed reaction scheme for tea fermentation. (Reprinted with permission of Pergamon Press Ltd., Oxford.)

tea catechins (Pierce *et al.*, 1969b) have confirmed (Pierce *et al.*, 1969a) that all the tea catechins are oxidized during tea fermentation although the gallocatechins (III, IV, VI) are oxidized preferentially to the catechol-catechins (I, II, V) in agreement with data developed by Roberts (1957) regarding the relative oxidation potentials of these compounds. The extent to which the catechins are oxidized must be determined by the conditions of the tea fermentation process itself, which includes the efficiency of the maceration procedure used to initiate tea fermentation (Sanderson, 1964d, 1968), the length of the oxidation period, the activity of the catechol oxidase in the tea flush (Sanderson, 1964b; Takeo, 1966a), and the relative oxidation potentials of the catechins (Roberts, 1957).

Theaflavins

Theaflavins are reported (Roberts and Smith, 1963) to comprise 0.3–1.8 percent of the dry weight of black tea, which amounts to about 1.0–5.9 percent of the solids extracted in the brewing of a cup of tea. These compounds have a bright reddish color in solution, and to them has been attributed a major role in determining brightness in tea infusions and a contribution to the strength of the tea beverage (Roberts, 1962; Roberts and Smith, 1963).

Roberts *et al.* (1957) were the first to separate theaflavin from a black tea infusion. Model system studies (Roberts and Myers, 1959a) appeared to indicate that theaflavin was formed as a result of the mixed oxidation of (−)-epigallocatechin and (−)-epigallocatechin gallate. Further, it was found that theaflavin did not give rise to gallic acid when hydrolyzed by *Aspergillus niger* (Roberts and Myers, 1959b), that the absorption

XIV Proposed structure for theaflavin (Roberts, 1962); R = H

XV Proposed structure for theaflavin gallate (Roberts, 1962); R = 3, 4, 5-trihydroxybenzoyl spectra resembled that of purpurogallin (Roberts and Williams, 1958), and that the molecular weight and the elemental composition of theaflavin suggested a molecule derived from one molecule of (−)-epigallocatechin plus one molecule of (−)-epigallocatechin gallate (Roberts, (1958b). On the basis of these findings, Roberts (1958b, 1962) tentatively proposed structure (XIV) for theaflavin, and, on the basis of similar experimental evidence, structure (XV) for theaflavin gallate.

Takino et al. (1964) took up the study of the structure of theaflavin. These investigators found that a compound P_I, with paper chromatographic

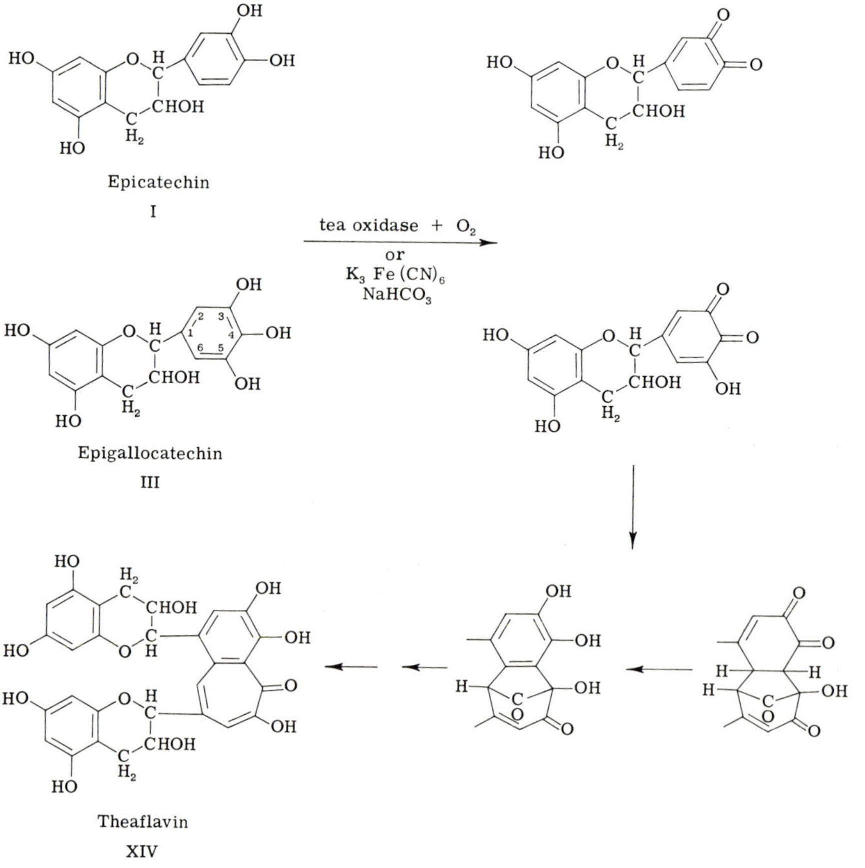

FIG. 5. Takino's proposed scheme for formation of theaflavin. From Takino et al. (1964).

and spectral properties that appeared to be the same as those reported for theaflavin by Roberts (1958b), was formed when (−)-epicatechin and (−)-epigallocatechin were oxidized together. The oxidation could be effected with either a tea catechol oxidase preparation or potassium ferricyanide in a sodium bicarbonate solution. The scheme shown in Fig. 5 was proposed for the formation of theaflavin, and structural formula (XVI), without configuration, was proposed for theaflavin. The proposed mechanism was supported by tentative mechanisms for the formation of purpurogallin suggested by Salfeld (1957) and by Horner and Durckheimer (1959). Further, elemental analysis of the purified oxidation product supported the proposed structure (XVI).

Later, Takino and Imagawa (1964) established the fact that the pigment formed from (−)-epicatechin and (−)-epigallocatechin was identical to theaflavin extracted from black tea by Roberts' method (Roberts, 1958b; Roberts and Myers, 1959c).

The structural formula proposed for theaflavin by Takino et al. (1964) was confirmed, and a configurational formula was established by Brown et al. (1966) and Takino et al. (1965, 1966), using a combination of mass spectrometry, ultraviolet, infrared, and nuclear magnetic resonance spectroscopy. Structure (XVI) is now accepted for theaflavin.

With the establishment of structure (XVI) for theaflavin and the recognition that theaflavin is formed by the oxidation and dimerization of (−)-epicatechin and (−)-epigallocatechin, it becomes apparent that analogous reactions involving the other possible combinations of tea catechins should lead to the formation of two different theaflavin gallates and one theaflavin digallate as follows:

$$(-)\text{-epicatechin (I)} + (-)\text{-epigallocatechin (III)} + O_2$$
$$\rightarrow \text{theaflavin (XVI)} + CO_2 + 2H_2O \quad (1)$$
$$(-)\text{-epicatechin (I)} + (-)\text{-epigallocatechin gallate (IV)} + O_2$$
$$\rightarrow \text{theaflavin gallate A (XVII)} + CO_2 + 2H_2O \quad (2)$$
$$(-)\text{-epicatechin gallate (II)} + (-)\text{-epigallocatechin (III)} + O_2$$
$$\rightarrow \text{theaflavin gallate B (XVIII)} + CO_2 + 2H_2O \quad (3)$$
$$(-)\text{-epicatechin gallate (II)} + (-)\text{-epigallocatechin gallate (IV)} + O_2$$
$$\rightarrow \text{theaflavin digallate (XIX)} + CO_2 + 2H_2O \quad (4)$$

Roberts (1958b, 1962) did recognize the presence of theaflavin gallate in black tea extracts in an amount which considerably exceeded theaflavin itself. However, his concept that theaflavin gallate was formed by the oxidative condensation of two molecules of (−)-epigallocatechin gallate to form a molecule with structure (XV) did not allow for more than one kind of theaflavin gallate. Also, Vuataz and Brandenberger (1961) had isolated two compounds (substances Y and Y_2) from black tea which were apparently two different theaflavin gallates. It remained for Nakagawa

XVI Theaflavin; $R_1 = R_2 = H$
XVII Theaflavin gallate A;
     $R_1 = H$; $R_2 = 3,4,5$-trihydroxybenzoyl
XVIII Theaflavin gallate B;
     $R_1 = 3,4,5$-trihydroxybenzoyl; $R_2 = H$
XIX Theaflavin digallate;
     $R_1 = R_2 = 3,4,5$-trihydroxybenzoyl

XX   Bisflavanol A; $R_1 = R_2 = 3,4,5$-trihydroxybenzoyl
XXI Bisflavanol B; $R_1 = 3,4,5$-trihydroxybenzoyl; $R_2 = H$
XXII Bisflavanol C; $R_1 = R_2 = H$

XXIII Epitheaflavic acid; $R = H$
XXIV 3-Galloylepitheaflavic acid; $R = 3,4,5$-trihydroxybenzoyl

and Torii (1965) in model tea fermentation system studies to show that several different compounds were formed in systems containing (a) (−)-epicatechin and (−)-epigallocatechin gallate, (b) epicatechin gallate and epigallocatechin, or (c) (−)-epicatechin gallate and (−)-epigallocatechin gallate (Table 10). The three oxidation products were only poorly resolved by paper chromatography with Roberts' solvent system (see Fig. 3). Indeed, their location on chromatograms corresponded to Roberts' substance Y (theaflavin gallate), and they gave the same color reactions as Roberts' theaflavins. However, Nakagawa and Torii (1965) showed (Fig. 6) that these compounds, which they called substances Y_2, Y_3, and Y, respectively, could be fully resolved on paper chromatograms using a solvent system composed of first solvent = phenol:water (3:1) and second solvent = butanol:acetic acid:water (4:1:2). Substances Y, Y_2, and Y_3

TABLE 10
Oxidation Products Detected in Model Tea Fermentation Systems Containing Selected Tea Catechins

Reaction mixture No.	Tea catechins present[a]	Oxidation products detected[a]	
		Nakagawa and Torii (1965)	Sanderson et al. (1972)
1	Cat	S, 16, 17, 18	S_B, 36, 37
2	EC	S, 11, 12, 13	S_A
3	ECG	S, Q, GA	S_C, Q, 32
4	EGC	C, 14, 15	C, 15, 30, R
5	EGCG	S, Y, A, GA	S_C, GA, A, B
6	Cat + EC	—	S_B, 37
7	Cat + ECG	—	S_E, Q
8	Cat + EGC	C, 22, 23, 24	S_D, C, 35, 37, 30, R
9	Cat + EGCG	S, C, 20, 21	S_B, A, B, 32, 37
10	EC + ECG	S, Q	S_C, GA, Q, 32
11	EC + EGC	X, C, 23	S_A, X, C, 35, 30, R
12	EC + EGCG	S, Y, C, GA	S_B, X, Y, GA
13	ECG + EGC	X, Y, C, GA	S_C, X, Y, Q, C, GA, 32, 34, 35, Z
14	ECG + EGCG	S, Y, Q, A, GA	S_C, Y, Q, GA, A, B, 32
15	EGC + EGCG	S, Y, C, GA	S_D, GA, A, B, C, 35, 33, 34, R
16	Cat + EC + ECG	—	S_C, Q, 32, 37
17	Cat + EC + EGC	—	S_B, C, 35, 37, R
18	Cat + EC + EGCG	—	S_B, X, 37
19	Cat + EGC + EGCG	C, 20, 22, 23	S_D, 35, 37, 33, 34, R, P
20	EC + ECG + EGC	—	S_F, X, Y, Q, C, GA, 32, 35, 33, Z, R
21	EC + ECG + EGCG	—	S_C, Y, Q, GA, 12, Z
22	ECG + EGC + EGCG	X, Y, C, GA	S_C, Y, Q, GA, C, 33, R
23	EGC + EGCG + EC	X, Y, C, GA	S_B, X, Y, GA, C, 35, 33, 34, R
24	Cat + ECG + EGCG	—	S_C, Y, Q, GA
25	Cat + EC + ECG + EGC	—	S_D, X, Y, Q, C, 35, 37
26	EC + ECG + EGC + EGCG	X, Y, C, GA	S_F, X, Y, Q, GA, 35, 33, 34, R
27	ECG + EGC + EGCG + Cat	—	S_F, X, Y, Q, C, 15, 33, 34, R
28	EGC + EGCG + EC + Cat	—	S_B, X, C, 35, 37, 33, 34, R
29	EGC + EGCG + ECG + Cat	—	S_F, Y, Q, A, B, C, 35, 33, 34, R, P
30	EGCG + EGC + ECG + EC + Cat	—	S_F, X, Y, Q, A, C, 33, 34, R, P

[a] Key to abbreviations used: EGC to Y, see Fig. 3; S, thearubigins; S_A–S_F, thearubigins with R_f in BAW of: SF_A = 0–0.5, S_B = 0–0.75, S_C = 0–1.0, S_D = 0.3–0.7, S_E = 0.5–1.0, S_F = 0.25–1.0; 12–37, unidentified substances.

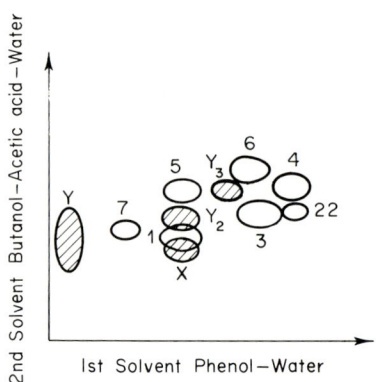

Fig. 6. Nakagawa and Torii's (1965) paper chromatogram of polyphenolic compounds formed in model tea fermentation systems. Key to abbreviations used: See Fig. 3 and Table 10.

must be theaflavin digallate (XIX), theaflavin gallate A (XVII), and theaflavin gallate B (XVIII), respectively.

The structures of theaflavin gallate A (XVII), theaflavin gallate B (XVIII), and theaflavin digallate (XIX) were established by Bryce et al. (1970) and by Coxon et al. (1970a) using mass spectrometry and nuclear magnetic resonance (NMR). The latter group of investigators (Coxon et al., 1970a) established that the configuration of these compounds was that expected from the mechanism Takino et al. (1964) for formation of theaflavin (Fig. 5) with retention of the stereochemistry of the flavanol substrates. Further, Bryce et al. (1970) and Coxon et al. (1970a) showed that these compounds are constituents of black tea infusions.

Coxon et al. (1970b) have reported the presence of a new compound, which they call isotheaflavin, in black tea. This compound is presumed to be formed from the oxidative coupling of (−)-epicatechin (I) and (+)-gallocatechin (VI) during tea fermentation.

An investigation of the substances formed from the different tea flavanols, or combination of flavanols, in model tea fermentation systems was carried out by Sanderson et al. (1972) using a crude soluble tea enzymes preparation (Co and Sanderson, 1970) as enzyme source. The results of this investigation corroborated and extended those of Nakagawa and and Torii (1965) as shown in Table 10.

Until recently, quantitative determination of theaflavins (and thearubigins; p. 280) was a spectrophotometric procedure developed by Roberts and Smith (1961, 1963). A more accurate gas–liquid chromatographic procedure which enables the individual theaflavins to be determined has been developed by Collier and Mallows (1971b).

Bisflavanols

Roberts and Myers (1959a) showed that among the oxidation products obtained from the oxidation of (−)-epigallocatechin and its gallate were some colorless compounds which they named substances A, B, and C. These compounds were formed as follows:

2 (−)-epigallocatechin gallate (IV) + $\frac{1}{2} O_2$
$\rightarrow$ substance A (bisflavanol A; XX) + H_2O (1)
(−)-epigallocatechin (III) + (−)-epigallocatechin gallate (IV) + $\frac{1}{2} O_2$
$\rightarrow$ substance B (bisflavanol B; XXI) + H_2O (2)
2 (−)-epigallocatechin (III) + $\frac{1}{2} O_2 \rightarrow$ substance C (bisflavanol C; XXII) + H_2O (3)

Roberts (1958b, 1961, 1962; Roberts and Myers, 1959a) assigned the structures XX, XXI, and XXII on the basis of ultraviolet absorption properties (Roberts and Williams, 1958), reactions with spray reagents (Roberts *et al.*, 1957), and the liberation of gallic acid upon incubation with *Aspergillus niger* (Roberts and Myers, 1959b).

Vuataz and Brandenberger (1961) isolated and crystallized substances A, B, and C from black tea. The physical data which these investigators obtained for these compounds agreed remarkably well with the structures proposed by Roberts and Myers (1959a).

Ferretti *et al.* (1968) obtained samples of substances A and B from Vuataz and Brandenberger and determined the configurational structure of these compounds by NMR. This work confirmed that the substances were bisflavanols as Roberts and Myers (1959a) had suggested and that the configuration of the catechins involved in their formation was maintained.

Roberts (1961, 1962) argued that the bisflavanols were intermediates in the formation of theaflavins. This argument is no longer tenable since it is known that the theaflavins are formed as a result of the coupled oxidation of (−)-epicatechin and (−)-epigallocatechin or their gallates (Brown *et al.*, 1966; Takino *et al.*, 1965, 1966), but, as Ferretti *et al.* (1968) pointed out, the small amounts of these compounds are as yet unexplained. It appears as though either the reaction sequence leading to the formation of the bisflavanols is of only minor significance or the bisflavanols are simply transient intermediates in a reaction sequence leading from the (−)-epigallocatechins to the thearubigins, or to some other black tea constituents.

Epitheaflavic Acids

Epitheaflavic acids (XXIII, XXIV) are bright red phenolic compounds which have just recently been shown to form during black tea fermentation

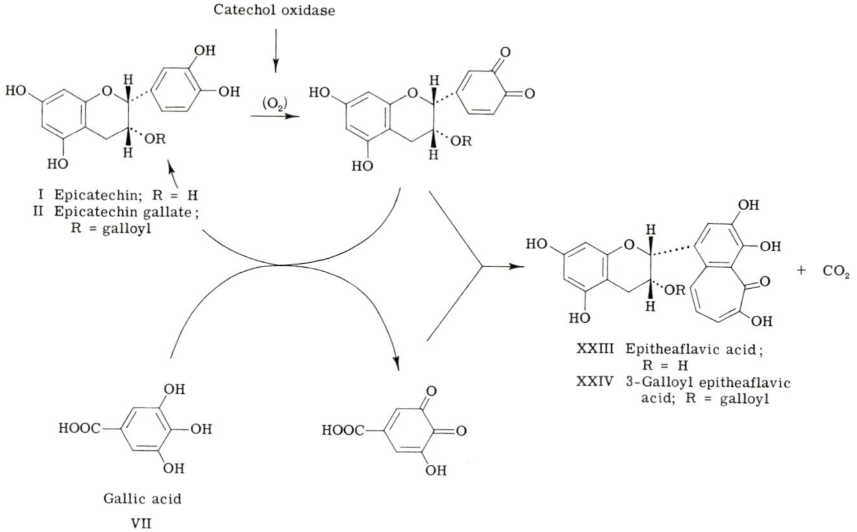

FIG. 7. Formation of epitheaflavic acids in tea fermentation system. Note dependence of gallic acid oxidation on (−)-epicatechin (gallate) oxidation, and formation of (3-galloyl)-epitheaflavic acid from oxidized (−)-epicatechin (gallate) and oxidized gallic acid (Berkowitz et al., 1971).

(Berkowitz et al., 1971). Epitheaflavic acid (XXIII) and 3-galloylepitheaflavic acid (XXIV) are formed from the coupled oxidation of (−)-epicatechin (I) and gallic acid (VII), or (−)-epicatechin gallate (II) and gallic acid (VII), respectively. Since gallic acid is not oxidized directly by the tea fermentation enzymes system (Roberts and Wood, 1950; Gregory and Bendall, 1966; Berkowitz et al., 1971), it must be the oxidized catechins, formed by the action of the tea catechol oxidase on the tea catechins, which act as oxidizing agents for the oxidation of gallic acid required for the formation of the epitheaflavic acids. These findings are summarized in Fig. 7.

Bryce et al. (1970) and Coxon et al. (1970c) have established that epitheaflavic acid has structure (XXIII) and the later investigators have confirmed the configuration of this compound. These investigators formed epitheaflavic acid by oxidizing mixtures of (−)-epicatechin (I) and gallic acid (VII) with potassium ferricyanide in aqueous sodium bicarbonate (Bryce et al., 1970; Coxon et al., 1970c). The preparation of an epimer of the epitheaflavic acid by oxidation of mixtures of (+)-catechin (V) and gallic acid (VII) with potassium ferricyanide in aqueous sodium bicarbonate was completed earlier by Horikawa (1969) and confirmed by Coxon et al. (1970c).

The epitheaflavic acids are present in black tea at very low levels even though they are the primary products formed from the coupled oxidation of the (−)-epicatechins and gallic acid, and this is apparently due to their reactivity in the fermenting tea system (Berkowitz et al., 1971). Because of their acidic nature and their solubility properties, the epitheaflavic acids themselves may be classified as thearubigins (Roberts, 1958a; Roberts and Smith, 1963). However, it has also been shown (Berkowitz et al., 1971) that the epitheaflavic acids are further oxidized in the tea fermentation process to other thearubigins through coupled oxidation with oxidizing tea catechins (Fig. 8). As with gallic acid, the tea enzymes system is not capable of directly oxidizing the epitheaflavic acids, but the oxidation of the epitheaflavic acids is efficiently accomplished in the presence of tea catechins undergoing oxidation in the tea fermentation system; presumably because of the high oxidation potentials of the oxidized tea catechins formed. Indeed, as will be pointed out below, the oxidation of gallic acid and epitheaflavic acid coupled with the enzymatic oxidation of the tea catechins appears to serve as a model for so many of the oxidative changes which occur during tea fermentation.

The epitheaflavic acids appear to be components of Roberts' (1962) "Q complex," a fact which Roberts must have suspected inasmuch as he did propose the correct structure for epitheaflavic acid in an earlier publication (Roberts, 1961). It has been shown (Berkowitz et al., 1971) that the epitheaflavic acids (XXIII, XXIV) cochromatograph with substance Q found in black tea in the paper chromatography system used by Roberts (Roberts et al., 1957). Also, Roberts and Myers (1959a, 1960b; Roberts, 1962) did recognize that an oxidation product of (−)-epicatechin gallate

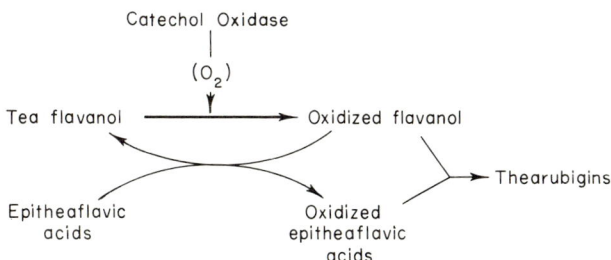

Fig. 8. Formation of thearubigins from epitheaflavic acids in tea fermentation system. Note dependence of epitheaflavic acids transformation on oxidation of tea flavanols (Berkowitz et al., 1971).

was a component of the "Q complex," and both epitheaflavic acid and 3-galloylepitheaflavic acid are formed from (−)-epicatechin gallate by the tea fermentation enzyme system (Berkowitz *et al.*, 1971); presumably because of some esterase activity in the tea fermentation enzymes system which liberates some gallic acid for participation in the formation of epitheaflavic acids.

It is noteworthy that gallic acid is formed characteristically when tea fermentation takes place (Roberts, 1962) and in model tea fermentation systems undergoing reaction when either (−)-epicatechin gallate or (−)-epigallocatechin gallate are present in the reaction mixture (Table 10). Further, substance Q (epitheaflavic acid) was formed in each model tea fermentation system which contained the tea catechins from which both (−)-epicatechin and gallic acid could be derived by deesterification (Table 10). The following reactions are illustrative:

(−)-epicatechin gallate → (−)-epicatechin + gallic acid + (−)-epicatechin gallate
  → epitheaflavic acids (XXIII and XXIV) + etc.   (1)
(−)-epigallocatechin gallate + (−)-epicatechin
  → (−)-epigallocatechin + gallic acid + (−)-epicatechin
  → epitheaflavic acid (XXIII) + etc.   (2)

Thearubigins

Thearubigins was the name given by Roberts (1958a) to all brown pigments in black tea with acidic properties. The thearubigins are a heterogeneous group of phenolic compounds which are perhaps best defined at the present time as the colored compounds separated from other colored black tea phenolic compounds by Roberts and Smith's (1963) procedure for their determination. That is, the thearubigins (the S_{II} type and part of the S_I type) in a tea infusion can be freed of other colored compounds by extracting the infusion with either ethyl acetate or isobutyl methyl ketone. The part of the S_I-type thearubigins which are extracted by these solvents can be recovered by extracting the solvent extract with a sodium bicarbonate solution.

The thearubigins are reported (Roberts and Smith, 1963) to comprise from about 9 percent to 19 percent of black tea leaf which is equivalent to about 30–60 percent of the solids in a black tea infusion. This group of compounds is responsible for much of the color of a black tea infusion, and it contributes significantly to the strength (Roberts and Smith, 1963), and to the mouth feel (Millin *et al.*, 1969a), of the tea beverage.

Molecular weight estimations (Roberts *et al.*, 1957), using ebullioscopic techniques, suggested that the thearubigins had an average molecular weight of about 700, and results of spectrophometric studies showed that the thearubigins lacked absorption maxima at about 380 nm and 455 nm, suggesting the absence of benzotropolone structures in the thearubigins (Roberts and Williams, 1958). Also, studies of the dynamics of the formation of theaflavins and thearubigins (Roberts, 1962) showed that, as tea fermentation progresses, theaflavins first increase and then begin to decrease whereas thearubigins increase continuously during the tea fermentation process. These results led Roberts (1958b, 1962) to propose that theaflavins (XVI–XIX) were converted to thearubigins by the oxidative opening of the benzotropolone structure producing a dicarboxylic acid. This would, of course, account for the acidic properties of the thearubigins. However, to date no one has reported such compounds to be present in tea. Roberts (1958c, 1962) concluded that thearubigins were formed by oxidation of the theaflavins and, possibly, the bisflavanols as shown in his scheme for tea fermentation (Fig. 4).

The results of model tea fermentation system studies (Nakagawa and Torii, 1965; Sanderson et al., 1972) have shown that thearubigins are formed by the oxidation of any one of the tea catechins or combination thereof (Table 10). Since theaflavins, or bisflavanols, or any other specific flavanol oxidation product are only formed as a result of the oxidation of specific tea catechins, or combinations of two specific catechins, it must be concluded that the thearubigins have no specific intermediate catechin oxidation product precursor. However, it was found (Sanderson et al., 1971b) that different thearubigins, which could be distinguished by their characteristic R_f's on paper chromatograms, were formed from different combinations of catechins. Further, it was shown in this latter investigation that the amount of thearubigins formed increased, and their solubility decreased, as the length of the oxidation period increased. This progressive decrease in solubility of the thearubigins must reflect an increase in the complexity and the molecular weight of the thearubigins brought about by oxidative polymerization reactions taking place between the polyphenolic compounds present in these model tea fermentation systems. These interrelationships will be summarized later (see p. 285).

Wickremasinghe (1967) was able to separate the thearubigins fraction of tea into five complexes of substances. The most colory complex was reported to contain mainly corilagin, theanine, and protein. Three other less highly colored thearubigin complexes were composed of corilagin, and theanine plus (a) theogallin, (b) chlorogenic acid, and (c) chlorogenic acid and epicatechins, respectively. The composition of the fifth thearubigin complex was not given. One wonders where the catechin oxidation products have gone in these analyses.

Millin et al. (1969c) studied the brown pigments in black tea infusions using a combination of solvent extraction and dialysis techniques. These investigators found that approximately 12 percent of the solids present in

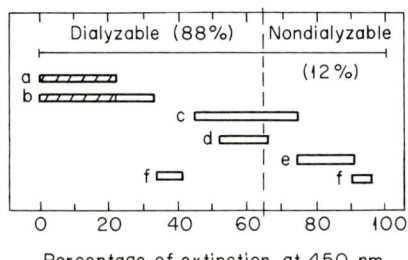

FIG. 9. Classification of colored substances in black tea liquors by dialysis and solvent extraction. (Figures in parentheses indicate percentage by weight of soluble tea solids in each fraction.) (a) Theaflavins; (b) ethyl acetate, pH 4.8 (19); (c) butanol, pH 4.8 (26); (d) ethyl acetate, pH 2.0 (15); (e) butanol, pH 2.0 (15); (f) insoluble in ethyl acetate or butanol, pH 2.0 (40). From Millin et al. (1969c).

a black tea infusion were nondialyzable and that these solids were responsible for approximately 35 percent of the color of the infusion (Fig. 9). Further, it was shown that the nondialyzable pigments behaved as Roberts' S_{II} type of thearubigins on paper chromatograms, suggesting that a portion of the thearubigins were, in fact, polymeric rather than simply dimeric as Roberts (1958b, 1962) had suggested. These results are reminiscent of the earlier results of Bradfield (1946).

The composition of the nondialyzable substances was investigated (Millin et al., 1969c), and it was shown that a major portion of this material was made up of polysaccharides and polyphenolic substances with smaller amounts of nitrogenous substances (Table 11). Most of the nitrogenous substances were classified as proteins because all the amino acids associated with proteins could be detected after hydrolyzing the nondialyzable material. A polysaccharide fraction free of polyphenolic and protein substances was obtained by passing the nondialyzable material through a column of polyvinylpyrrolidone. The amount of polysaccharides obtained from tea infusions by this procedure is shown in Table 11, but it must be recognized that additional polysaccharides which had become associated with polyphenolic substances may also have been held on the column. Certainly, there is a rather large amount of nondialyzable material which was not accounted for in these analyses. Finally, it was concluded that the protein was complexed with the polyphenolic substances present in the nondialyzable fraction although the evidence given was not conclusive. It is noteworthy that Vuataz and Brandenberger (1961) reported 0.55 percent nitrogen in their thearubigins preparation, and these investigators reported finding 14 different amino acids in acid hydrolyzates of their thearubigins fraction.

Millin et al. (1969b) studied the effects of aging black tea infusions on their color and their content of nondialyzable matter. It was found that appreciable darkening occurred in black tea infusions when aged for 1 or 2 hours and that these changes in color were directly related to increases in the amount of nondialyzable material. Again, these results are reminiscent of the earlier results of Bradfield and Penny (1944), which showed that boiling a solution of the red, ethyl acetate-extractable polyphenol fraction of tea infusions converted these materials into brownish, ethyl acetate-insoluble materials (thearubigins?).

Fractionation of aged tea infusions by chromatography on Sephadex LH–20 columns using 60 percent acetone as eluting agent produced results (Millin et al., 1969b) which indicated that there is an increase in the molecular weight of the black tea pigments on aging. These studies evidently form the basis for Millin and Rustidge's (1967) claim that the molecular weights of the thearubigins range from 700 to 40,000 since these investigators supposed (Crispin et al., 1968; Millin et al., 1969c) that black tea pigments were eluted in order of decreasing molecular weight from columns of Sephadex LH–20 developed with 60 percent acetone.

The most definitive study published to date on the chemistry of the thearubigins is the work of Ollis and his co-workers (Brown et al., 1969a,b). These investigators fractionated the thearubigins into five fractions by partitioning between solvents of different polarities. These fractions were then degraded by acid (hydrochloric acid in anhydrous isopropanol or methanol), or by reductive hydrolysis with aqueous sulfurous acid. The degradation products obtained were identified to be flavan-3-ols [(+)-catechin (V), (−)-epicatechin (I), (+)-gallocatechin (VI), and possibly

TABLE 11

Distribution and Composition of Nondialyzable Fraction of Tea Infusions[a]

Component	Centrifuged brew	Cream	Whole brew
Total solids (%)	7	21	11.7
$E_{450}^{1\%}$, nondialyzable material	8.8	17.5	14.8
Composition (%)			
Polysaccharide	60	7	34
Protein	8	8	6
Nucleic acid	1.2	1	0.8
Polyphenol (as catechin)	26	56	32
Unaccounted for	4.8	28	27.2

[a] From Millin et al. (1969c).

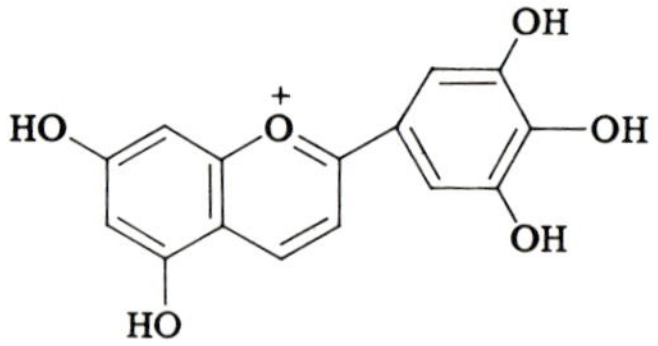

Tricetinidin
XXV

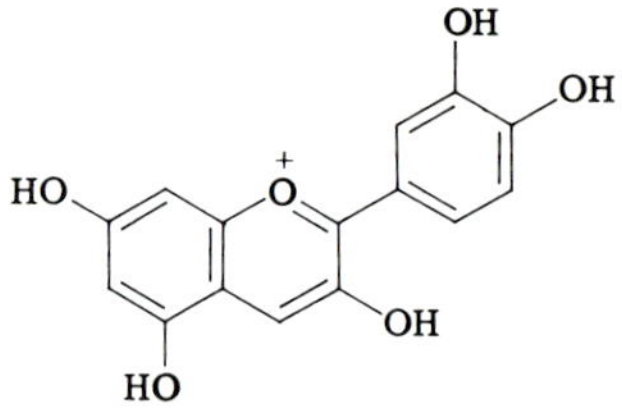

Cyanidin
XXVI

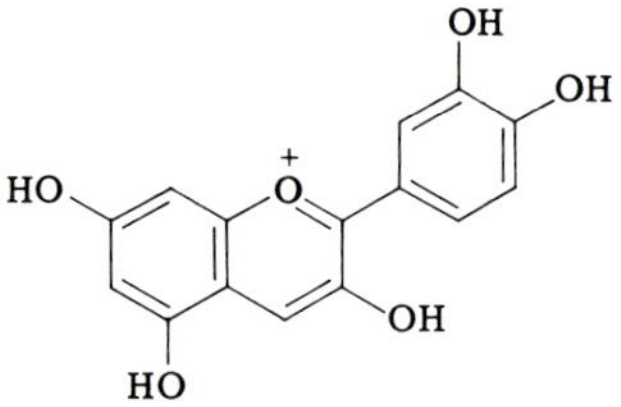

Delphinidin
XXVII

(−)-epigallocatechin (III) in one fraction; amounts not determined], flavan-3-ol gallates [(−)-epicatechin gallate (II) and (−)-epigallocatechin gallate (IV); amounts not determined], anthocyanidins [cyanidin (XXVI) and delphinidin (XXVII); about 20 percent of each fraction], and gallic acid (VII) (about 2–4 percent of each fraction). No nitrogen was found in any of the thearubigin fractions which could not be accounted for by caffeine, and no evidence for the presence of associated chebulinic acid, corilagin, or glucogallin as suggested by Wickremasinghe (1967) could be found. These results (Brown et al., 1969a,b) establish the thearubigins studied as polymeric proanthocyanidins (Seshadri, 1967; Weinges et al., 1969), although it was acknowledged (Brown et al., 1969a) that other thearubigin-type compounds exist in black tea infusions which were not extracted and studied in these investigations. Further, it was reported (Brown et al., 1969a) that it is probable that these latter substances are of the polyphenol-protein type (Calderon et al., 1968) in which the polyphenol is a polymeric proanthocyanidin.

No theaflavins were found as degradation products of the thearubigins (Brown et al., 1969a,b), but it is not certain that theaflavins would be recovered as such after the degradation procedures used even if they had been incorporated into the thearubigins. It was pointed out that more cyanidin (XXVI) was formed on degradation of the thearubigins than delphinidin (XXVII) in spite of the fact that their flavanoid precursors, i.e., epicatechins (I, II, and V) and epigallocatechins (III, IV, and VI), respectively, are present in fresh green tea leaf in relative proportions of about 1:3. Again, this suggests that the gallocatechins undergo more extensive alteration during the course of tea fermentation than the catechins.

The recovery of gallic acid upon hydrolysis of the thearubigins as carried out by Brown *et al.* (1969a,b) was also much lower than would be expected if the gallate groups in the average mixture of tea catechins found in fresh tea flush were present in the free state. As pointed out above, considerable free gallic acid is formed during tea fermentation (Sanderson *et al.*, 1972) (Table 10), and gallic acid is known to participate in certain oxidative condensations, such as the formation of epitheaflavic acids (XXIII, XXIV; Berkowitz *et al.*, 1971). Both these processes reduce the amount of gallic acid that would be expected to be attached to the thearubigins and that would therefore be recoverable by deesterification of the thearubigins.

Summary of Polyphenolic Oxidations Occurring during Tea Fermentation

The reactions which the polyphenolic compounds present in tea flush undergo during tea fermentation are summarized in Fig. 10. A summary of

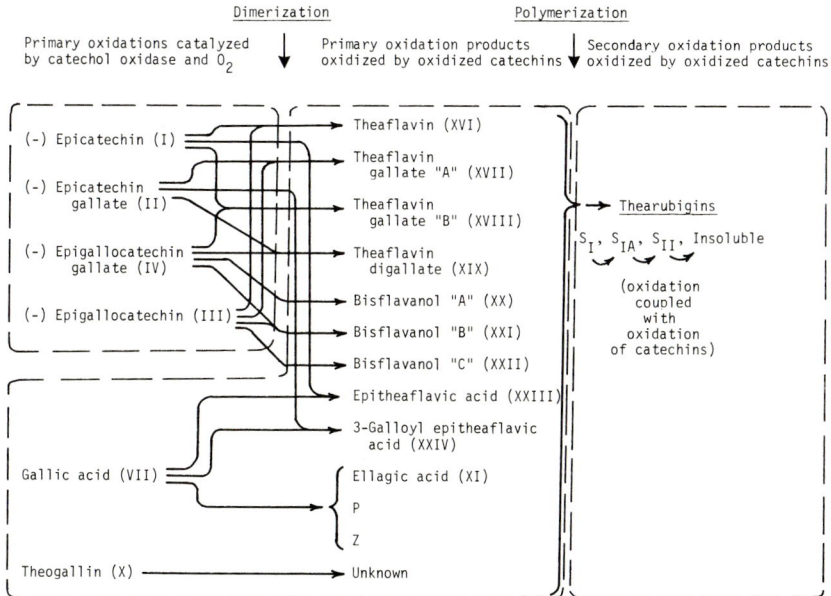

FIG. 10. Summary of reactions undergone by tea flavanols during tea fermentation.

TABLE 12
PHENOLIC COMPOUNDS KNOWN TO FORM DURING TEA FERMENTATION

Compound	Structure No. used in text	Molecular formula	Molecular weight	References[a] Original structure assignment	References[a] Best present structure assignment	Amount present in black tea (%)	Color
Theaflavin	XVI	$C_{29}H_{24}O_{12}$	564	(k)	(b,n)	0.28–1.63 (m)	Bright red
Theaflavin gallate A	XVII	$C_{36}H_{28}O_{16}$	716		(d, e)		
Theaflavin gallate B	XVIII	$C_{36}H_{28}O_{16}$	716				
Theaflavin digallate	XIX	$C_{43}H_{32}O_{20}$	868				
Thearubigins—S_I	Structure not known. (I, III, VII, XXVI, and XXVII are degradation products)	—	~700 to 40,000 (h)	(i, j)	(e)	5.1–14.8 (m)	Reddish brown
Thearubigins—S_{IA}							
Thearubigins—S_{II}							

Bisflavanol A	XX	$C_{44}H_{34}O_{22}$	914	(i)	Small amount (no data available) } Colorless
Bisflavanol B	XXI	$C_{37}H_{30}O_{18}$	762		
Bisflavanol C	XXII	$C_{30}H_{26}O_{14}$	610		
Epitheaflavic acid	XXIII	$C_{21}H_{16}O_{10}$	428	(d)	(d, f) Trace (a) } Bright red
3-Galloylepitheaflavic acid	XXIV	$C_{29}H_{20}O_{14}$	580	(a)	Unconfirmed
Tricetinidin	XXV	$C_{15}H_{11}O_6$	287	(l)	Unconfirmed Trace (j)
Ellagic acid	XI	$C_{14}H_6O_8$	302		Small amount (no data available) } Colorless
Unknown Q ("Q complex")	Thought to be a mixture of XXIII, XXIV, and others.			(k)	Unconfirmed Trace (j)
Unknown R				(i)	Unconfirmed Trace (j)
Unknown Z				(i)	Unconfirmed Trace (j)

[a] Key to references: (a) Berkowitz et al. (1971); (b) Brown et al. (1966); (c) Brown et al. (1969a,b); (d) Bryce et al. (1970); (e) Coxon et al. (1970a); (f) Coxon et al. (1970c); (g) Ferretti et al. (1968); (h) Millin and Rustidge (1967); (i) Roberts (1958b); (j) Roberts (1962) (k) Roberts and Myers (1959a,b); (l) Roberts and Williams (1958); (m) Roberts and Smith (1963); (n) Takino et al. (1965, 1966).

the properties of these black tea polyphenols is given in Table 12. It is known that compounds other than those shown in Fig. 10 are formed during tea fermentation (Table 10) (Sanderson *et al.*, 1972), but these other compounds have not yet been characterized. It is noteworthy that the only enzymatic oxidations affecting the tea catechins known to occur during tea fermentation are the oxidations of the tea catechins themselves catalyzed by the enzyme catechol oxidase. However, these primary enzymatically catalyzed oxidations appear to produce the driving force for the secondary oxidations which result in the formation of other dimeric oxidation products, such as the epitheaflavic acids, and the polymeric oxidation products, such as the thearubigins.

FERMENTATION: TEA AROMA AND ITS FORMATION

Composition of Tea Aroma

Tea, whether it be green tea, black tea, or any other type of tea, is a beverage that is prized the world over for its delicate exotic flavor. And the flavor of a tea is largely determined by its aroma. The dependent relationship between tea flavor and its aroma is readily demonstrated in the laboratory by tasting teas before and after removal of aroma; this can be accomplished by any one of several techniques, including solvent extraction, steam stripping, and vacuum distillation. Dearomatized tea infusions tend to be almost flavorless with only a slight astringency and body.

Interest in the chemical composition of tea aroma has a fairly long history, but until the advent of gas chromatography, the amount of information developed about the chemical nature of tea aroma was quite limited [see reviews of early work by Stahl (1962) and by Yamanishi (1968)]. The amount of essential oil found in tea flush and in finished tea products is only about 0.01 percent of their dry weight (Stahl, 1962), and it is now known that tea aroma contains over 100 components, thus making identification of individual components very difficult by classical analytical techniques.

A list of the compounds that have been identified as constituents of tea aroma is given in Table 13. Stahl (1962) reported that 41 tea aroma constituents have been identified at the time he wrote his review of tea chemistry, and since that time this list has grown to 140 identified components. As shown in Table 13, the principal contributors to our present knowledge of tea aroma are Yamanishi and co-workers in Japan, Gianturco and co-workers in the United States, and Egli and Brandenberger in Switzerland.

It is noteworthy that none of the compounds that have been identified as being components of tea aroma resemble tea aroma by itself (Yamanishi,

TABLE 13
Aroma Constituents of Tea[a]

Aroma constituent	Molecular formula	References[b] Found in fresh leaf	Found in green tea	Found in black tea
Hydrocarbons				
Limonene	$C_{10}H_{16}$		(a)	(c)
cis-β-Ocimene	$C_{10}H_{16}$			(c, g)
trans-β-Ocimene	$C_{10}H_{16}$			(g)
β-Myrcene	$C_{10}H_{16}$			(g)
Calamenene	$C_{15}H_{22}$		(a)	
α-Cubebene	$C_{15}H_{24}$		(a)	
α-Copaene	$C_{15}H_{24}$		(a)	
Caryophyllene	$C_{15}H_{24}$		(a)	
γ-Muurolene	$C_{15}H_{24}$		(a)	
α-Muurolene	$C_{15}H_{24}$		(a)	(a)
δ-Cadinene	$C_{15}H_{24}$		(a)	(a)
α-Humulene	$C_{15}H_{24}$		(a)	
β-Sesquiphellandrene	$C_{15}H_{24}$		(a)	
Alcohols				
Methanol	CH_4O			(b, c, d)
Ethanol	C_2H_6O			(c, d, j)
2-Propanol	C_3H_8O			(c)
2-Butanol	$C_4H_{10}O$			(c)
Isobutanol	$C_4H_{10}O$	(e)	(a)	(a, e)
2-Methylpropanol	$C_4H_{10}O$			(c)
n-Butanol	$C_4H_{10}O$	(e)	(a)	(a, e)
Furfuryl alcohol	$C_5H_6O_2$		(a)	(a)
2-Methyl-3-buten-2-ol	$C_5H_{10}O$			(c)
1-Penten-3-ol	$C_5H_{10}O$	(a)	(a)	(a, b, c)
cis-2-Penten-1-ol	$C_5H_{10}O$	(a)	(a)	(a, b, c)
trans-2-Penten-1-ol	$C_5H_{10}O$			(b)
2-Methylbutanol	$C_5H_{12}O$	(e)		(e)
Isoamyl alcohol	$C_5H_{12}O$	(a, e)	(a)	(a, b, c, e)
n-Amyl alcohol	$C_5H_{12}O$		(a)	(a, b, c)
cis-3-Hexen-1-ol	$C_6H_{12}O$	(e)	(a)	(a, b, c, e, f)
trans-3-Hexen-1-ol	$C_6H_{12}O$			(b, c)
trans-2-Hexen-1-ol	$C_6H_{12}O$		(a)	(a, b, c)
n-Hexanol	$C_6H_{14}O$	(e)	(a)	(a, b, c, e, f)
Benzyl alcohol	C_7H_8O	(e)	(a, e)	(a, b, c, e, f)
1-Heptanol	$C_7H_{16}O$		(a)	
Phenylethanol	$C_8H_{10}O$	(e)	(a, e)	(a, b, c, e, f)
Methylphenylmethanol	$C_8H_{10}O$			(a)

(Continued)

TABLE 13—Continued
AROMA CONSTITUENTS OF TEA[a]

Aroma constituent	Molecular formula	Found in fresh leaf	Found in green tea	Found in black tea
Linalool oxide				
cis, furanoid	$C_8H_{14}O_2$	(a)	(a)	(a, b, c)
trans, furanoid	$C_8H_{14}O_2$	(a)	(a)	(a, b, c)
cis, pyranoid	$C_8H_{14}O_2$	(a)	(a)	(a, b, c)
trans, pyranoid	$C_8H_{14}O_2$	(a)	(a)	(a, b, c)
1-Octen-3-ol	$C_8H_{16}O$			(b, c)
1-Octanol	$C_8H_{18}O$	(e)	(a, e)	(c, e, f)
4-Ethylguaiacol	$C_9H_{12}O_2$		(a)	
Nonanol	$C_9H_{20}O$		(a)	(c)
Linalool	$C_{10}H_8O$	(e)	(a, e)	(a, b, c, f)
3,4-Dimethyl-1,5,4-octatriene-3-ol	$C_{10}H_{16}O$		(a)	(a)
Nerol	$C_{10}H_{18}O$		(a)	(a, c)
Geraniol	$C_{10}H_{18}O$	(e)	(a, e)	(a, b, c, e, f)
α-Terpineol	$C_{10}H_{18}O$		(a)	(b, c)
Cadinenol	$C_{15}H_{24}O$			(a)
Nerolidol	$C_{15}H_{26}O$		(a)	(c)
α-Cadinol	$C_{15}H_{26}O$		(a)	
Cubenol	$C_{15}H_{26}O$		(a)	
Carbonyls				
Acetaldehyde	C_2H_4O	(a)		(b, c, d, j)
Acrolein	C_3H_4O			(d)
Propanal	C_3H_6O			(b, c, d)
Acetone	C_3H_6O			(b, c, d)
Diacetyl	$C_4H_6O_2$			(b, c, j)
Isobutanal	C_4H_8O	(a, e)		(a, b, c, d, e, f, j)
n-Butanal	C_4H_8O	(a, e)		(a, b, c, e, f)
Methyl ethyl ketone	C_4H_8O			(e)
2-Butanone	C_4H_8O			(c)
Furfural	$C_5H_4O_2$			(b, c, j)
trans-3-Penten-2-one	C_5H_8O			(c)
Acetylpropionyl	$C_5H_8O_2$			(c)
Isopentanal	$C_5H_{10}O$	(e)		(a, b, c, d, e, f)
2-Methylbutanal	$C_5H_{10}O$	(e)		(b, e, f)
n-Pentanal	$C_5H_{10}O$			(a, b, c)
5-Methylfurfural	$C_6H_6O_2$		(a)	(c)
trans,trans-2,4-Hexadienal	C_6H_8O			(b, c)
trans-2-Hexenal	$C_6H_{10}O$	(e)	(a)	(a, b, c, k)
4-Methyl-4-penten-2-one	$C_6H_{10}O$			(c)
4-Methyl-3-penten-2-one	$C_6H_{10}O$			(c)

TABLE 13—Continued
AROMA CONSTITUENTS OF TEA[a]

		References[b]		
Aroma constituent	Molecular formula	Found in fresh leaf	Found in green tea	Found in black tea
n-Hexanal	$C_6H_{12}O$			(a, b, c, f)
Benzaldehyde	C_7H_6O	(e)	(a, e)	(a, b, c, e, f)
trans,trans-2,4-Heptadienal	$C_7H_{10}O$			(b, c)
trans-2,cis-4-Heptadienal	$C_7H_{10}O$			(b)
n-Heptanal	$C_7H_{14}O$			(a, b, c)
Acetophenone	C_8H_8O		(a, e)	(j)
Phenylacetaldehyde	C_8H_8O			(a, b, c)
trans,trans-3,5-Octadien-2-one	$C_8H_{12}O$	(a)	(a)	(b, c)
trans-2-Octenal	$C_8H_{14}O$			(a, c)
2-Methyl-2-hepten-6-one	$C_8H_{14}O$			(b, c)
Octanal	$C_8H_{16}O$			(c)
2,2,6-Trimethylcyclohexanone	$C_9H_{16}O$			(c)
2,2,6-Trimethyl-6-hydroxy-cyclohexanone	$C_9H_{16}O_2$			(b)
1-Nonanal	$C_9H_{18}O$		(a)	
2-Phenylbutenal	$C_{10}H_{10}O$			(b)
trans,trans-2,4-Decadienal	$C_{10}H_{16}O$			(c)
Geranial	$C_{10}H_{16}O$			(a)
Citronellal	$C_{10}H_{18}O$			(f)
cis-Jasmone	$C_{11}H_{16}O$		(a)	(a, b, c)
Theaspirone (1-Oxa-8-oxo-2,6 10,10-tetramethylspiro [4,5]-6-decene)	$C_{12}H_{20}O_2$			(k, a)
α-Ionone	$C_{13}H_{20}O$		(a)	(a, b, c)
β-Ionone	$C_{13}H_{20}O$		(a)	(a, b, c)
5,6-Epoxyionone	$C_{13}H_{20}O_2$			(b)
2′,2″-Dihydroionone	$C_{13}H_{22}O$		(a)	
6,10,14-Trimethyl-2-pentadecanone	$C_{18}H_{36}O$		(a)	
Esters and lactones				
Ethyl formate	$C_3H_6O_2$			(c)
Ethyl acetate	$C_3H_8O_2$			(c, g)
γ-Butyrolactone	$C_4H_6O_2$			(c)
γ-Caprolactone	$C_6H_{10}O_2$			(b, c)
Isoamyl acetate	$C_7H_{14}O_2$			(a)
Methyl benzoate	$C_8H_8O_2$			(b)
Benzyl formate	$C_8H_8O_2$			(a)
Methyl salicylate	$C_8H_8O_2$	(a, e)	(a, e)	(a, b, c, f)
cis-3-Hexenyl acetate	$C_8H_{14}O_2$			(c)
Coumarin	$C_9H_6O_2$		(a)	
Benzyl acetate	$C_9H_{10}O_2$			(f)

(Continued)

TABLE 13—Continued
Aroma Constituents of Tea[a]

Aroma constituent	Molecular formula	Found in fresh leaf	Found in green tea	Found in black tea
Phenylethyl formate	$C_9H_{10}O_2$			(a)
Geranyl formate	$C_{11}H_8O_2$			(c)
Lactone of 2,6,6-trimethyl-2-hydroxycyclohexylideneacetic acid	$C_{11}H_{16}O_2$			(d, h)
Dihydroactinidiolide	$C_{11}H_{16}O_2$		(a)	
α-Terpinyl acetate	$C_{12}H_{20}O_2$		(a)	
cis-3-Hexenyl caproate	$C_{12}H_{22}O_2$		(a)	
cis-3-Hexenyl benzoate	$C_{13}H_{16}O_2$		(a)	(a)
Dibutyl phthalate	$C_{16}H_{22}O_4$		(a)	
Acids				
Formic	CHO_2	(a)		(a, i)
Acetic	$C_2H_2O_2$	(a, e)		(a, c, i)
Propionic	$C_3H_6O_2$	(a, e)	(a, c, f, i)	
Isobutyric	$C_4H_8O_2$	(a, e)		(a, c, i)
n-Butyric	$C_4H_8O_2$	(a, e)	(e)	(a, b, c, i)
Isovaleric	$C_5H_{10}O_2$	(a, e)		(a, c, f, i)
2-Methylbutyric	$C_5H_{10}O_2$			(c)
n-Valeric	$C_5H_{10}O_2$		(a)	(i)
cis-3-Hexenoic	$C_6H_{10}O_2$	(a)	(a)	(a)
trans-2-Hexenoic	$C_6H_{10}O_2$	(a)	(a)	(a, b)
Isocaproic	$C_6H_{12}O_2$			(i)
n-Caproic	$C_6H_{12}O_2$	(a, e)	(a, e)	(a, c, f, i)
Benzoic	$C_7H_6O_2$		(a)	
Salicylic	$C_7H_6O_3$	(e)	(e)	(a)
Isoheptanoic	$C_7H_{14}O_2$			(i)
n-Heptanoic	$C_7H_{14}O_2$			(i)
Isooctanoic	$C_8H_{16}O_2$			(i)
n-Octanoic	$C_8H_{16}O_2$			(i)
Isononanoic	$C_9H_{18}O_2$			(i)
n-Nonanoic	$C_9H_{18}O_2$			(i)
Isodecanoic	$C_{10}H_{20}O_2$			(i)
n-Decanoic	$C_{10}H_{20}O_2$			(i)
n-Lauric	$C_{12}H_{24}O_2$			(i)
Phenolic compounds				
o-Cresol	C_7H_8O	(e)	(e)	(a)
m-Cresol	C_7H_8O		(a)	(a)
p-Cresol	C_7H_8O		(a)	
Phenol	C_8H_6O	(e)	(a, e)	(a)
Sulfur compounds				
Hydrogen sulfide	H_2S		(a)	

TABLE 13—Continued

Aroma Constituents of Tea[a]

Aroma constituent	Molecular formula	References[b]		
		Found in fresh leaf	Found in green tea	Found in black tea
Methylmercaptan	CH_4S			(f)
Dimethyl sulfide	C_2H_6S		(a)	(c, d)
Nitrogen compounds				
Pyrrolealdehyde	C_5H_5NO			(a)
α-Methylbutyronitrile	C_5H_9N			(c)
Isovaleronitrile	C_5H_9N			(c)
Pyrryl methyl ketone	C_7H_7NO		(a)	(a)
N-Ethylformylpyrrole	C_7H_9NO		(a)	
1-Ethyl-2-formylpyrrole	$C_7H_{11}NO$			(b)
Indole	C_8H_7N		(a)	(a)
Phenylacetonitrile	C_8H_7N			(b)
Quinoline	C_9H_7N			(f)
Diphenylamine	$C_{12}H_{11}N$		(a)	
Miscellaneous oxygenated compounds				
1,1-Dimethoxyethane	$C_4H_{10}O_2$			(c)
2-Ethylfuran	C_6H_8O			(c)
Propionoin	$C_6H_{12}O_2$			(c)
n-Amylfuran	$C_9H_{14}O$			(c)
Halogenated compounds				
1,2,4-Trichlorobenzene	$C_6H_3Cl_3$		(a)	

[a] Adapted from Yamanishi (1968).
[b] Key to references: (a) Kiribuchi et al. (1963); Kobayashi et al. (1965a,b, 1966); Nakatani et al. (1969, 1970); Nose et al. (1971); Sato et al. (1970); Yamanishi et al. (1956, 1957, 1963, 1964, 1965a,b, 1966a,b, 1968a, 1970); (b) Bricout et al. (1967); Müggler-Chavan et al. (1966, 1969); Reymond et al. (1966); Viani et al. (1966); (c) Bondarovich et al. (1967); (d) Wickremasinghe and Swain (1965); (e) Takei and Sakato (1932); Takei et al. (1934a,b, 1935a,b,c, 1936, 1937a,b, 1938a,b); (f) Yamamoto et al. (1934, 1935, 1937, 1940a,b,c); (g) Saijo (1967); Saijo and Kuwabara (1967); (h) Ina, Sakato, and Fukami (1968); (i) Brandenberger and Muller (1962); (j) Bokuchava and Skobeleva (1957); (k) van Romburgh (1920).

1968). Further, Gianturco and co-workers (Bondarovich et al., 1967) have presented evidence that there are yet many unidentified constituents of the tea aroma complex, and these researchers suggest that some of the trace constituents may play an important role in determining the flavor of black tea. Finally, when one contemplates the chemical nature of the "essential" portion of tea aroma, one should recognize that there is no single tea aroma. Virtually every tea has a distinctive aroma, and this difference is so pronounced in teas from certain regions that an organoleptic assessment by a practiced taster is sufficient to identify the season and the geographical location in which the tea originated. Yamanishi *et al.* (1968a) compared the aromas of distinctive black teas obtained from the Dimbala and Uva districts of Ceylon, the Nilgiri and Darjeeling districts of India, and the Shizuoka district of Japan using gas chromatography and showed that the different teas had appreciable quantitative differences in the total and the relative amount of the several aroma constituents studied. However, no qualitative differences were detected in the aromas of these several different black teas. These researchers drew attention to the proportion of linalool (and its oxides) to geraniol and phenylethanol, and to the ratio of the total volatiles eluting from the gas chromatograph before and after linalool as being important distinguishing features of the aromas of these teas. Obviously, more information is required before a chemical description of the essential portion of tea aroma can be set down.

Formation of Tea Aroma

Green tea and black tea have aromas that are distinctive both from one another and from the fresh tea flush from which they originated. These distinctive aromas are developed during the tea manufacturing process. Certainly, in the case of black tea manufacture, the aroma of the tea flush material changes continuously during the manufacturing process, and only attains the characteristic black tea aroma after the final (firing) stage of manufacture. Harler (1963) has described the aroma of tea flush undergoing conversion to black tea as follows:

> The aroma of the leaf changes as fermentation proceeds. Withered leaf has the smell of apples. When rolling begins this changes to one of pears, which then fades and the acrid smell of green leaf returns. Later, a nutty aroma develops and finally a sweet smell, together with a flowery smell if flavour is present.

Yamanishi *et al.* (1966a) studied the chemical basis of the flavor changes which take place during black tea manufacture. These researchers used tea flush of the Benihomare variety grown in Japan to prepare black tea in their laboratory, and they reported changes in the amount of 47 aroma constituents to occur during the conversion of the fresh tea flush to black tea. The most notable changes found were as follows: During withering, hexyl alcohol, nerol, *trans*-2-hexenoic acid, *trans*-2-hexenol, linalool oxide (*cis*, furanoid), *n*-valeraldehyde, capronaldehyde, *n*-heptanal, *trans*-2-hexenal, *trans*-2-octenal, benzaldehyde, phenylacetaldehyde, *n*-butyric, isovaleric, *n*-caproic, *cis*-3-hexenoic, and salicylic acids, and *o*-cresol increased with particularly large increases in the first three named aroma constituents; while cis-2-pentenol, linalool, geraniol, benzyl alcohol, phenylethanol, and acetic acid markedly diminished. During fermentation, almost all aroma constituents increased, but increases were especially large for 1-penten-3-ol, cis-2-pentenol, benzyl alcohol, trans-2-hexenal, benzaldehyde, n-caproic, cis-3-hexenoic, and salicylic acids. On firing most alcohols, carbonyl and phenolic compounds were appreciably decreased whereas acetic, propionic, and isobutyric acids greatly increased.

Yamanishi et al. (1968b) carried out a gas chromatographic study of the aroma of Ceylon black teas, and they found that high-grown (better quality) black teas contained more linalool, linalool oxides, geraniol, and cis-jasmone than low-grown (lower quality) black teas.

Saijo and Kuwabara (1967) also studied the changes in the volatile compounds which take place during the manufacture of black tea. Fresh tea leaf was relatively rich in alcohols whereas the manufactured black tea had more aldehydes and acids. The amount of essential oil decreased during the black tea manufacturing process from 0.031 percent in the fresh tea leaf to 0.008 percent in the finished black tea. These investigators noted that black tea aroma did not develop in macerated tea leaf when the leaf was kept in a nitrogen atmosphere which would prevent any oxidative changes from taking place, including tea fermentation.

Amino Acids as Aroma Precursors

Bokuchava and Popov (1954) were perhaps the first researchers to investigate the chemistry of the formation of black tea aroma. These researchers reported that aromas are formed when certain amino acids are incubated with tea flavanols in hot aqueous solutions for 12 hours. The aromas obtained were characterized as being roselike (phenylalanine), flowery (glutamic acid and alanine), unpleasant (tryptophan and tyrosine), spicy (norleucine), and winy (threonine). Popov (1956) showed that carbon dioxide, ammonia, and aldehydes corresponding to the Strecker degradation products of the amino acids were formed in these systems, and he proposed a scheme to explain how oxidized tea catechins could cause these chemical changes. However, as yet no experimental evidence has been published to support this mechanism. In fact, operation of Popov's (1956) proposed scheme should lead to the production of some tea catechin-amino acid condensation products, but investigations utilizing radioactive amino acids have so far produced no results to suggest that such substances are formed (Berkowitz and Sanderson, unpublished observations).

Nakabayashi (1958) also showed that certain amino acids were converted to volatile carbonyl compounds corresponding to their Strecker degradation products in the fermenting tea leaf system. Further, it was shown that the formation of carbonyl compounds from amino acids was not enzymatic but depended on the enzymatic oxidation of the tea tannins. Nakabayshi (1958) also reported that neither sugars, alcohols, nor organic acids were precursors of volatile carbonyl compounds, but it may be worthwhile to reinvestigate this subject using the more sensitive analytical techniques now available.

Saijo and Takeo (1970a,b) and Co and Sanderson (1970) have recently studied the formation of carbonyl compounds from amino acids by the tea fermentation system in more detail with similar results (Table 14). In both investigations, model tea fermentation systems composed of preparations of crude soluble tea enzymes and purified tea catechins were used to show that aldehydes corresponding to Strecker degradation products were formed from glycine, alanine, valine, leucine, isoleucine, methionine, and phenylalanine when these amino acids were present in systems containing one or more tea catechins undergoing oxidation catalyzed by the soluble tea enzymes preparations. Omission of either the tea enzymes or the tea catechin(s) eliminated aldehyde formation although incorporation

TABLE 14

VOLATILE CARBONYL COMPOUNDS PRODUCED FROM AMINO ACIDS DURING TEA FERMENTATION[a]

Amino acid	Volatile carbonyl formed	Reported by
Glycine	Formaldehyde	(b)
Alanine	Acetaldehyde	(a, b)
Valine	Isobutyraldehyde	(a, b, c)
Leucine	Isovaleraldehyde	(a, b, c)
Isoleucine	2-Methylbutanal	(b, c)
Methionine	Methional	(a, b)
Phenylalanine	Phenylacetaldehyde	(b, c)
Aspartic acid	None	(c)
Glutamic acid	None[b]	(b, c)
Glutamine	None	(c)
Threonine	None	(c)
Serine	None	(c)
Tyrosine	None	(b)
Tryptophan	None	(b)
Theanine	None	(c)

[a] Summarized from results published by (a) Nakabayashi (1958); (b) Saijo and Takeo (1970a,b); and (c) Co and Sanderson (1970).

[b] The nonvolatile compound succinimide has been shown to form in this system by Moss, Coggon, and Sanderson, unpublished results.

of other oxidizing agents, such as dehydroascorbic acid (Co and Sanderson, 1970), β-naphthoquinone-4-sulfonate (Saijo and Takeo, 1970b), or oxidized D-catechin (Wickremasinghe and Swain, 1964), could also cause the formation of aldehydes from amino acids as expected from a Strecker degradation. Amino acids other than those listed above did not give rise to detectable levels of volatile compounds (Co and Sanderson, 1970), which may be due to the nonvolatile nature of the Strecker degradation products that were expected rather than to the lack of reactivity of these other amino acids. In both investigations (Saijo and Takeo, 1970a,b; Co and Sanderson, 1970), the results obtained in model tea fermentation systems were shown to take place during the manufacture of black tea by incorporating selected amino acids in whole tea flush systems undergoing tea fermentation and detecting an attendant increase in the amount of the expected carbonyl compounds as a result of tea fermentation and firing.

It is noteworthy that Finot *et al.* (1967) have shown that phenylacetaldehyde is formed from phenylalanine by reaction with black tea solids under conditions similar to the brewing of a cup of black tea, suggesting one more reason why the method used to brew tea may be important in determining the quality of the tea infusion. Negative results obtained in similar experiments by Mayuranathan and Gopalan (1965) appear to have been caused by the use of a fraction of the black tea extractable solids which were not reactive (Sanderson, 1966b).

Kiribuchi and Yamanishi (1963) have shown that methylmethionine sulfonium salt is present in green tea, and they have suggested that this is a precursor of dimethyl sulfide which is present in tea aroma (Table 13). A direct test of this supposition would be most valuable because it could provide the model for yet one more type of reaction involved in tea aroma formation.

Wickremasinghe (1967) has suggested that L-leucine may be an important precursor of some of the compounds that contribute to tea flavor, and these changes are accomplished by biosynthetic pathways. The finding that the leucine content of more flavory teas is lower than that of less flavory teas (Wickremasinghe and Swain, 1965) suggested that flavory teas contained a more active system for effecting this change. It was shown subsequently (Wickremasinghe *et al.*, 1969) that other biochemical factors needed for the biosynthetic conversion of leucine to terpenes, such as leucine transaminase, coenzyme A, and manganese, were present in fresh tea flush.

While a role for amino acids in contributing to the aroma of black tea is now well established, the number of volatile compounds known to be formed from amino acids is rather small (Table 14).

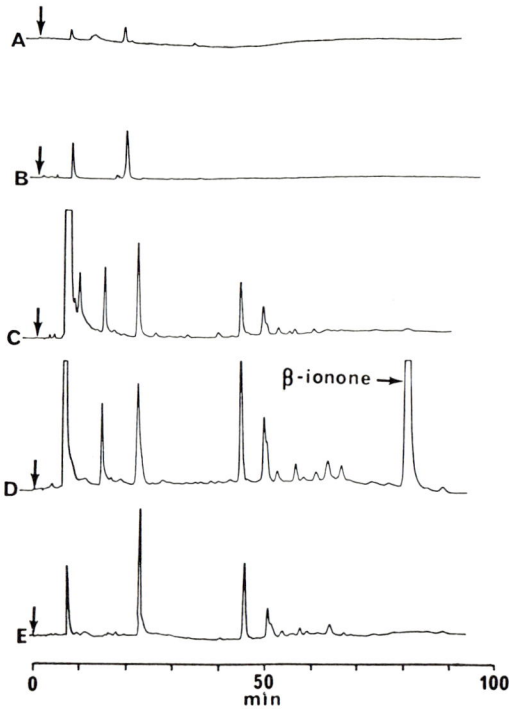

Fig. 11. Gas chromatograms of headspace volatiles produced in a model tea fermentation system containing β-carotene. (A) No fermentation; no drying; (B) 30-minute fermentation; no drying; (C) no fermentation; dried; (D) 30-minute fermentation; dried; (E) same as D, but no β-carotene. From Sanderson *et al.* (1971).

Carotenes as Aroma Precursors

It has been shown in several studies (Tirimanna and Wickremasinghe, 1965; Nikolaishvili and Adeishvili, 1966; Sanderson *et al.*, 1971) that the carotenoids present in fresh tea flush undergo appreciable decreases in concentration during the various stages of conversion to black tea. The data from the investigation carried out by Sanderson *et al.* (1971) shown in Table 15 illustrate the general finding that all the carotenoid compounds in tea flush undergo appreciable decrease in quantity during the course of black tea manufacture and that the major decrease occurs during the first part of the tea fermentation stage.

The suggestion that the degradation of carotenoid compounds in tea flush during black tea manufacture contributes to the formation of black tea aroma (Tirimanna and Wickremasinghe, 1965; Sanderson, 1965c;

TABLE 15

Concentration of Carotenoids in Tea Samples Taken at Various Stages of Black Tea Manufacture[a]

Samples	Neoxanthin		Violaxanthin		Lutein		β-Carotene	
	µg/gram dried sample	Percent of original	µg/gram dried sample	Percent of original	µg/gram dried sample	Percent of original	µg/gram dried sample	Percent of original
Fresh tea	51	100	120	100	260	100	102	100
1-Hour fermented tea	28	57	47	39	148	57	67	66
2-Hour fermented tea	30	58	53	43	154	59	64	62
3-Hour fermented tea	30	59	54	44	158	60	62	61
Fired and dried tea	23	46	36	30	154	59	61	60
Hot water extract[b]	0.6	1	0.7	0.5	8	3	3	3
Spent leaves[c]	11	22	18	15	147	57	58	57
Commercial black tea (Lipton blend)	18	—	0	—	141	—	33	—

[a] From Sanderson et al. (1971).
[b] Amounts corrected to whole tea basis. Extract solids = 43 percent of whole tea.
[c] Amounts corrected to whole tea basis. Spent leaves = 57 percent of whole tea.

TABLE 16

BLACK TEA AROMA CONSTITUENTS SUPPOSED TO BE DERIVED
FROM CAROTENOID COMPOUNDS[a,b]

Carotenoids found in tea leaves by Tirimanna and Wickremasinghe (1965)	Primary oxidation products	Secondary oxidation products
β-Carotene →	β-Ionone + TAK[c]	→ Dihydroactinidiole[d] → 2,2,6-Trimethylcyclohexanone → 5,6-Epoxyionone ↓ 2,2,6-Trimethyl-6-hydroxycyclohexanone Theaspirone[e]
Lutein →	[3-Hydroxy-β-ionone] — — — →Theaspirone[e] + [3-Hydroxy-α-ionone] + TAK	
Neoxanthin (d) →	3-Hydroxy-5,6-epoxyionone + 3,5-Dihydroxy-6-hydroionone + TAK	
Phytoene →	Linalool + TAK	
Phytofluene →	Linalool + TAK	
Lycopene →	Linalool + TAK	
γ-Carotene →	β-Ionone + TAK	
Cryptoxanthin →	β-Ionone + [3-Hydroxy-β-ionone] + TAK	
Violaxanthin →	[3-Hydroxy-5,6-epoxyionone] + TAK	
Zeaxanthin →	[3-Hydroxy-β-ionone] + TAK	

[a] From Sanderson *et al.* (1971a).

[b] Known reaction based on results of this investigation, →; highly probable reactions based on results of this investigation, →, probable reactions, --→. Compounds shown in brackets have not yet been identified in tea.

[c] TAK = Terpenoid-like aldehydes and ketones: oxidation products of all the carotenoids listed.

[d]

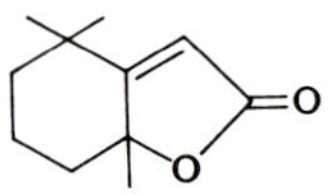

[e]

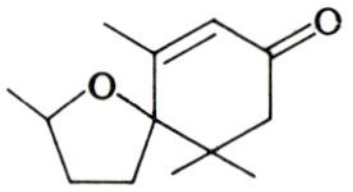

[f] Identified by Sanderson *et al.* (1971).

Müggler-Chavan et al., 1969) was investigated by Sanderson et al. (1971), and it was found that β-carotene is oxidatively degraded to β-ionone and several other unidentified volatile and nonvolatile compounds. It is probable that the degradation of β-carotene proceeds by a more or less random breaking of double bonds in the carotene molecule. These changes occurred readily during the tea fermentation process in macerated whole tea flush, but they would not occur in the more dilute model tea fermentation systems unless the oxidized reaction mixtures were taken to dryness. In any case, oxidation of tea catechins was required for the degradation of β-carotene. These results are summarized in Fig. 11, and they indicate that it is production of oxidized tea catechins which develops the oxidation potentials required to bring about the oxidative degradation of carotenoid compounds during the manufacture of black tea. The results obtained with β-carotene have been generalized (Sanderson et al., 1971) for all the carotenoid compounds reported to be in tea flush (Tirimanna and Wickremasinghe, 1965), and this generalization is summarized in Table 16. Examination of Table 16 reveals that it is most likely that many compounds known to be important in determining the character of black tea aroma are derived, at least in part, from the oxidative degradation of the tea carotenoids.

The compounds formed by these reactions are important enough to black tea aroma, and the development of oxidation potentials of sufficient magnitude to drive these reactions is critical enough, to suggest that the conditions which control the oxidative degradation of the tea carotenoids are of great importance in determining the character of the black tea produced. Factors involved in these reactions include (a) amount of tea catechins, (b) amount and makeup of tea carotenoids, (c) activity of tea catechol oxidase, (d) degree of mixing of cell contents achieved by tea flush maceration process, and (e) concentration of reactants possible as determined by degree of wither. Attention has been drawn to the probable importance of these factors before (Sanderson, 1964c,d, 1965c), but there is as yet very little definitive information on the effect of biochemical factors on the character of the finished black tea product.

Other Black Tea Aroma Precursors

It is likely that compounds other than amino acids and carotenes that are present in fresh tea leaves serve as precursors of black tea aroma constituents. Almost no work has been done as yet on tea lipids, but the report that the crude lipid content of fresh tea flush is negatively correlated with the quality of the black tea made from this flush (Sanderson and Kanapathipillai, 1964) may well be an indication of their role as precursors of aroma constituents. The role of lipids in flavor formation in other foods is well established (Forss, 1969).

The oxidation of alcohols to aldehydes to acids is also likely to occur

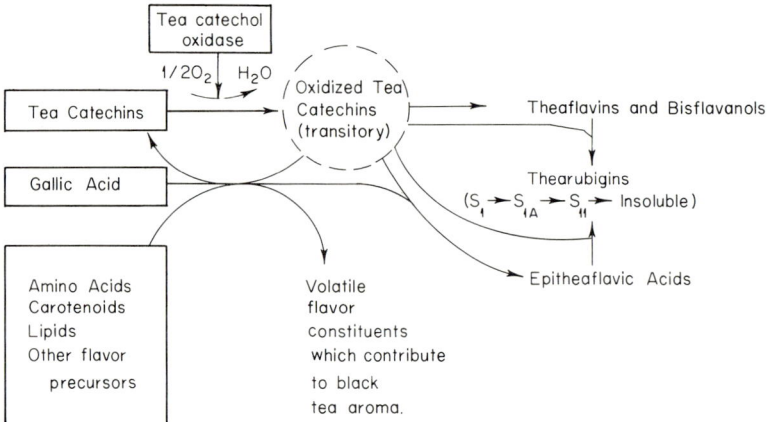

Fig. 12. Proposed scheme for tea fermentation. Materials present in fresh (withered) tea flush are enclosed in rectangles. The amount of any of the above materials present in black tea would be dependent on the conditions of black tea manufacture.

during tea manufacture, but such changes have not yet been studied in tea. Finally, much more is needed to be known about the ability of the firing, or drying, step to cause further chemical changes in the tea aroma constituents. Yamanishi *et al.* (1966a) have shown that firing tends to lower the overall level of black tea aroma constituents and to change their relative amounts, but the actual reactions taking place are virtually unknown as yet.

Mechanism of Tea Fermentation

The information regarding the chemistry of tea fermentation discussed above has been summarized in a proposed scheme for tea fermentation shown in Fig. 12. This scheme is a revision of a scheme proposed earlier (Sanderson, 1965c), and it incorporates such changes as appear to be required by the currently available information on the subject.

An important feature of the proposed scheme for tea fermentation (Fig. 12) is the essential function assigned to the tea enzyme catechol oxidase. The scheme indicates that tea fermentation is absolutely dependent on the activity of catechol oxidase since the oxidized tea flavanols either condense to form characteristic black tea constituents or serve as the oxidizing agents which drive the other reactions which are essential to the development of the black tea character.

The scheme points out the importance of at least three factors on the outcome of the tea fermentation. First, the composition of the fresh tea flush as modified by the withering treatment must be important in determining the potential of the tea flush to make finished tea products with any special characteristics. Next, the activity of the enzyme catechol oxidase must have an important effect on the outcome of the fermentation because it will determine the oxidation potentials developed during fermentation by determining the concentration of oxidized flavanols which are developed. This will, of course, determine which, and to what extent, secondary oxidations, which are so important to the development of the finished tea character, will take place. Finally, the importance of the maceration process which initiates tea fermentation can be deduced. As was pointed out above, catechol oxidase and its substrates, the flavanols, are spatially separated in the tea flush, and the secondary reactants are most likely also compartmentalized in these tissues. The extent to which the several reactants involved in tea fermentation are brought into contact must also be an important determinant of the outcome of tea fermentation. The effect of the maceration process on the physical form of the fermenting tea flush material as it affects the penetration of oxygen, an essential factor in tea fermentation, certainly should not be overlooked.

Tea Cream

The precipitate which forms naturally in a cup of tea when it cools is called tea cream (Roberts, 1962; Stahl, 1962). The amount of tea cream which forms in a tea brew is a measure of the strength and briskness (Roberts, 1962) of black teas and, generally speaking, the value of teas increases as the tendency of their infusions to form cream increases. While tea cream appears to be indicative of certain desirable attributes of black tea, it is one of the major problems in the manufacture of instant tea (Pintaro, 1970) due to the poor solubility of this material.

Roberts (1963) analyzed tea cream and found that it consisted essentially of thearubigins, theaflavins, and caffeine in a ratio of about 66:17:17. These results suggested that theaflavins are precipitated with the cream preferentially to the thearubigins (62 percent of the total theaflavins as compared to only 28 percent of the total thearubigins in this example). Finally, the amount of cream formed was shown to be directly related to the theaflavin, thearubigins, and caffeine contents of the infusion (Roberts, 1963), as well as to the overall concentration of tea solids in the infusion (Smith, 1968).

Wickremasinghe and Perera (1966) studied the composition of tea cream and compared it to the composition of the decreamed extracts. Their results (Table 17) indicated that there was some fractionation of tea infusion constituents on decreaming with the epicatechin (I), epigallocatechin (III),

TABLE 17

Composition of Tea Cream and Decreamed Extract[a]

Cream	Decreamed extract
Main constituents	p-Coumarylquinic acids
Caffeine	Chlorogenic acids
Theaflavin gallate	Theogallin
Thearubigins	Gallic acid
	Epicatechin
Secondary constituents	Epigallocatechin
Theobromine	Ellagic acid
Ellagic acid	Theaflavin
Epigallocatechin gallate	Phloroglucinol[b]
Epicatechin gallate	Unidentified—C and D[b]
Caffeic acid	
Gallic acid	
Triacetidin	
Unidentified—A and B	

[a] Adapted from Wickremasinghe and Perera (1966).
[b] Compounds not found in undecreamed extract.

and theaflavin (XVI) occurring exclusively in the decreamed extract; while the gallate esters of these compounds (II, IV, and XVII, XVIII, XIX, respectively) occur entirely in the cream. These results suggest that the more phenolic groups per molecule, the greater the tendency to participate in cream formation. Wickremasinghe and Perera (1966) also reported finding three phenolic compounds, i.e., phloroglucinol and two unidentified phenolic compounds, in decreamed extracts which were not found in undecreamed extracts. It was suggested that the appearance of these new compounds was due to chemical breakdown reactions which occur during cream formation. Certainly, cream formation is not entirely reversible (Sanderson, unpublished observations). The continuous formation of nondialyzable matter during aging of tea infusions (Millin et al., 1969b) is likely to be another manifestation of this phenomena.

Conclusions

The character and quality of black tea are determined by the color, taste, aroma, and reaction with milk of the tea infusion; and the peculiar stimulating properties of black tea are due to the caffeine content and the interaction of caffeine and the polyphenolic compounds present in the tea infusion. These quality and pysiological properties all have a chemical basis which is just beginning to be understood in some detail. The characteristic properties of black tea are largely dependent on chemical changes which take place during the tea fermentation step of black tea manufacture, and herein lies the source of, and the justification for, much of the research on the chemistry of tea fermentation.

Quantitatively, the major change taking place during tea fermentation is the oxidation of the tea flavanols, and accordingly, much of the research into tea chemistry has been concerned with studying the reactions involving these compounds. Our present knowledge of these reactions (Fig. 10) is considerably more detailed than it was at the time Dr. E. A. H. Roberts (1962) summarized the same information (Fig. 4) nine years ago. Further, our knowledge regarding the formation of black tea aroma, which is the main determinant of black tea flavor, has been developed almost entirely since Dr. Roberts' (1962) review. The overall summary of the chemistry of tea fermentation proposed in Fig. 12 attempts to show that the several chemical reactions taking place appear to be highly dependent on one another.

The proposed scheme for tea fermentation shown in Fig. 12 is likely to require revision in light of as yet unknown information, and it is sure to be found to be oversimplified. Much work needs to be done in almost every area of tea chemistry before a reasonably complete understanding of the subject is achieved, but major areas requiring elucidation at the present time are the following: (a) the chemistry of the thearubigins, (b) the chemistry of tea aroma formation, and (c) the chemistry of the firing step.

Finally, it deserves mentioning that the tea plant appears to be a plant which is particularly well suited for investigations of the biosynthesis of flavonoid compounds. The tea plant does accumulate unusually large amounts of flavonoid compounds, especially flavanols, and the single study carried out on the enzymes involved (Sanderson, 1966a) suggests that there is a particularly active biosynthetic system for these compounds. The recent development of methods for extracting active enzymes from plant tissues rich in polyphenolic compounds (Loomis and Battaile, 1964; Anderson, 1968; Lam and Shaw, 1970) promise to yield interesting results for the investigator who chooses to apply them to this problem.

ACKNOWLEDGMENTS

I wish to express my appreciation for the many contributions made to the experimental work which has gone into the development of the ideas expressed in this paper, and the development of the ideas themselves, by my several collaborators. I am particularly indebted to Dr. R. L. Wickremasinghe, Dr. G. R. Roberts, Dr. R. R. Selvendran, Mr. B. P. M. Perera, Mr. K. Sivapalan, Mr. T. S. Nathan, and Mr. V. Fernando of the Tea Research Institute of Ceylon, Talawakelle, Ceylon, and to Dr. H. N. Graham, Dr. H. Co, Mr. M. Gurkin, Miss J. E. Berkowitz, and Mr. J. G. Gonzalez of Thomas J. Lipton, Inc., Englewood Cliffs, New Jersey.

REFERENCES

Anderson, J. W. 1968. Extraction of enzymes and subcellular organelles from plant tissues. *Phytochemistry* **7**:1973–1988.
Anonymous. 1969. "Tea Growers Handbook." Tea Res. Inst. of East Afr., Kericho, Kenya.
Bamber, M. K., and H. Wright. 1902. The enzyme in tea: A preliminary note. *In* "Year Book of the Planters Association for 1901–1902." pp. 1–10 (special report). Planters Ass., Colombo, Ceylon.
Barua, D. N. 1956. Calcium oxalate crystals as an index of nutrient uptake in the tea plant. *Current Sci.* **8**:249–250.
Bendall, D. S., and R. P. F. Gregory. 1963. Purification of phenol oxidases. *In* "Enzyme Chemistry of Phenolic Compounds" (J. B. Pridham, ed.), pp. 7–24. Pergamon, London.
Berkowitz, J. E., P. Coggon, and G. W. Sanderson. 1971. Formation of epitheaflavic acid and its transformation to thearubigins during tea fermentation. *Phytochemistry* **10**:2271–2278.
Bhatia, I. S., and N. B. Chanda. 1959. Report of the biochemistry division. *Annual Report of the Tocklai Experiment Station for 1959.* p. 262.
Bhatia, I. S., and S. B. Deb. 1965. Nitrogen metabolism of detached tea shoots. I. Changes in amino acids and amides of tea shoots during withering. *J. Sci. Food Agr.* **16**:759–769.
Bhatia, I. S., and M. R. Ullah. 1961. Oxidation of l-epicatechin gallate during the processing of Assam tea leaf. *Chem. Ind. (London)* p. 1169.
Bhatia, I. S., and M. R. Ullah. 1962. Metabolism of polyphenols in the tea leaf. *Nature (London)* **193**:658–659.
Bhatia, I. S., and M. R. Ullah. 1965. Quantitative changes in the polyphenols during the processing of tea leaf and their relation to liquor characters of made tea. *J. Sci. Food Agr.* **16**:408–416.
Bhatia, I. S., and M. R. Ullah. 1968. Polyphenols of tea. IV. Qualitative and quantitative study of the polyphenols of different organs of some cultivated varieties of tea plant. *J. Sci. Food Agr.* **19**:535–542.
Bhattacharyya, A. K., and J. J. Ghosh. 1968. Studies on the ribonucleic acids of fresh and processed tea leaves. *Biochem. J.* **108**:121–124.
Bokuchava, M. A. 1950. Transformation of tannins under the action of polyphenoloxidase and peroxidase. *Biokhimiya* **6**:100–109.
Bokuchava, M. A., and V. R. Popov. 1954. The significance of amino acids in the formation of tea aroma by interactions with tannin substances under conditions of elevated temperature. *Dokl. Akad. Nauk SSSR* **99**:145–148. [*Chem. Abstr.* **29**:2329.]
Bokuchava, M. A., and N. I. Skobeleva. 1957. Study of volatile aldehydes of tea leaves. *Proc. Acad. Sci. USSR Biochem.* **112**:31–33.
Bokuchava, M. A., and G. A. Soboleva. 1958. Formation and transformation of organic acids in tea sprouts. *Fiziol. Rast.* **5**:70–74.
Bokuchava, M. A., and N. I. Skobeleva. 1969. The chemistry and biochemistry of tea and tea manufacture. *Advan. Food Res.* **17**:215–280.

Bondarovich, H. A., A. S. Giammarino, J. A. Renner, F. W. Shepard, A. J. Shingler, and M. A. Gianturco. 1967. Some aspects of the chemistry of tea. A contribution to the knowledge of the volatile constituents. *J. Agr. Food Chem.* **15**:36–47.

Bradfield, A. E. 1946. Some recent developments in the chemistry of tea. *Chem. Ind. (London)* pp. 242–244.

Bradfield, A. E., and M. Penny. 1944. The chemical composition of tea. The proximate composition of an infusion of black tea and its relation to quality. *J. Soc. Chem. Ind., London* **63**:306–310.

Brandenberger, H., and S. Muller. 1962. A gas chromatographic separation of the volatile fatty acids of black tea. *J. Chromatogr.* **7**:137–141.

Bricout, J., R. Viani, F. Muggler-Chavan, J. P. Marion, D. Reymond, and R. H. Egli. 1967. Sur la composition de l'arome de thé noir. II. *Helv. Chim. Acta* **50**:1517–1522.

Brown, A. G., C. P. Falshaw, E. Haslam, A. Holmes, and W. D. Ollis. 1966. The constitution of theaflavin. *Tetrahedron Lett.* pp. 1193–1204.

Brown, A. G., W. B. Eyton, A. Holmes, and W. D. Ollis. 1969a. Identification of the thearubigins as polymeric proanthocyanidins. *Nature (London)* **221**:742–744.

Brown, A. G., W. B. Eyton, A. Holmes, and W. D. Ollis. 1969b. The identification of the thearubigins as polymeric proanthocyanidins. *Phytochemistry* **8**:2333–2340.

Bryce, T., P. D. Collier, I. Fowlis, P. E. Thomas, D. Frost, and C. K. Wilkins. 1970. The structures of the theaflavins of black tea. *Tetrahedron Lett.* pp. 2789–2792.

Bu'Lock, J. D. 1965. "The Biosynthesis of Natural Products." McGraw-Hill, New York.

Buzyn, G. A., K. M. Dzhemukhadze, and L. F. Mileshko. 1970. Some properties of tea polyphenol oxidase. *Biokhimiya* **35**:1002–1006.

Calderon, P., J. Van Buren, and W. B. Robinson. 1968. Factors influencing the formation of precipitates and hazes by gelatin and condensed and hydrolyzable tannins. *J. Agr. Food Chem.* **16**:479–482.

Cartwright, R. A., and E. A. H. Roberts. 1954a. Paper chromatography of phenolic substances. *Chem. Ind. (London)* pp. 1389–1391.

Cartwright, R. A., and E. A. H. Roberts. 1954b. Theogallin, a polyphenol occurring in tea. *J. Sci. Food Agr.* **5**:593–597.

Cartwright, R. A., and E. A. H. Roberts. 1954c. The sugars of manufactured tea. *J. Sci. Food Agr.* **5**:600–601.

Cartwright, R. A., and E. A. H. Roberts. 1955. Theogallin as a galloyl ester of quinic acid. *Chem. Ind. (London)* pp. 230–231.

Cartwright, R. A., E. A. H. Roberts, and D. J. Wood. 1954. Theanine, an amino acid N-ethyl amide present in tea. *J. Sci. Food Agr.* **5**:597–599.

Chalamberidze, T. K., G. A. Soboleva, N. A. Pristupa, R. K. Petrova, and M. A. Bokuchava. 1969. Localization of catechins in tea leaf. *Soobshch. Akad. Nauk Gruz. SSR* **54**:697–700.

Chenery, E. M. 1955. A preliminary study of aluminium and the tea bush. *Plant Soil* **6**:174–200.

Child, R. 1955. Copper: Its occurrence and role in tea leaf. *Trop. Agr. (London)* **32**:100–106.

Co, H., and G. W. Sanderson. 1970. Biochemistry of tea fermentation: Conversion of amino acids to black tea aroma constituents. *J. Food Sci.* **35**:160–164.

Collier, P. D., and R. Mallows. 1971a. Estimation of flavanols in tea by gas chromatography of their trimethylsilyl derivatives. *J. Chromatogr.* **57**:29–45.

Collier, P. D., and R. Mallows. 1971b. Estimation of theaflavins in tea by gas-liquid chromatography of their trimethylsilyl esters. *J. Chromatogr.* **57**:19–27.

Coxon, D. T., A. Holmes, W. D. Ollis, and V. C. Vora. 1970a. The constitution and configuration of the theaflavin pigments of black tea. *Tetrahedron Lett.* pp. 5237–5240.

Coxon, D. T., A. Holmes, and W. D. Ollis. 1970b. Isotheaflavin. A new black tea pigment. *Tetrahedron Lett.* pp. 5241–5246.

Coxon, D. T., A. Holmes, and W. D. Ollis. 1970c. Theaflavic and epitheaflavic acids. *Tetrahedron Lett.* pp. 5247–5250.

Crispin, D. J., R. H. Payne, and D. Swaine. 1968. Analysis of the pigments of black tea extracts by chromatography on acetylated Sephadex. *J. Chromatogr.* **37**:118–119.

Das, D. N., J. J. Ghosh, K. C. Bhattacharyya, and B. C. Guha. 1965. Studies on tea. II. Pharmacological aspects. *Indian J. Appl. Chem.* **28**:15–40.

Dzhemukhadze, K. M., and F. G. Nakhmedov. 1964. Effect of watering on the catechin content of tea plants. *Dokl. Akad. Nauk SSSR* **155**:1447–1448.

Dzhemukhadze, K. M., G. A. Shalneva, and L. F. Milesko. 1957. Change in catechins during tea fermentation. *Biokhimiya* **22**:836–840.

Dzhemukhadze, K. M., G. A. Buzun, and L. F. Mileshko. 1964. Enzymic oxidation of catechols. *Biokhimiya* **29**:882–888.

Eden, T. 1965. "Tea," 2nd Ed. Longmans, Green, New York.

Ferretti, A., V. P. Flanagan, H. A. Bondarovich, and M. A. Gianturco. 1968. The chemistry of tea. Structures of compounds A and B of Roberts and some model compounds. *J. Agr. Food Chem.* **16**:756–761.

Finot, P. A., F. Müggler-Chavan, and L. Vuataz. 1967. La phénylalanine précurseur de la phénylacetaldehyde dans l'arome de thé noir. *Chimia* **21**:26–27.

Forss, D. A. 1969. Role of lipids in flavors. *J. Agr. Food Chem.* **17**:681–685.

Forsyth, W. G. C. 1964. Physiological aspects of curing plant products. *Ann. Rev. Plant Physiol.* **15**:443–450.

Gray, A. 1903. "The Little Tea Book." Baker & Taylor, New York.

Green, M. J. 1971. An evaluation of some criteria used in selecting large-yielding tea clones. *J. Agr. Sci.* **76**:143–156.

Gregory, R. P. F., and D. S. Bendall. 1966. The purification and some properties of the polyphenol oxidase from tea (*Camellia sinensis*, L.). *Biochem. J.* **101**:569–581.

Grisebach, H., and W. Barz. 1969. Biochemistry of flavonoids. *Naturwissenschaften* **56**:538–544.

Hainsworth, E. 1969. Tea. *In* "Encyclopedia of Chemical Technology" (A. Standen, ed.), 2nd Ed., Vol. 19, pp. 743–755. Wiley (Interscience), New York.

Harler, C. R. 1963. "Tea Manufacture." Oxford Univ. Press, London, and New York.

Horikawa, H. 1969. Benzotropolone pigment formed from (+) catechol and gallic acid. I. *Tokyo Kasei Daigaku Kenkyu Kiyo* **9**:1–8. [*Chem. Abstr.* **72**:80320d (1970).]

Horner, L., and W. Durckheimer. 1959. Zur Kenntnis der Ortho Chinone. XII. Ortho Chinone aus Brenzcatechin Derivaten. Pyro Catechin. *Z. Naturforsch. B* **14**:741.

Ina, K., Y. Sakato, and H. Fukami. 1968. Isolation and structure elucidation of theaspirone, a component of tea essential oil. *Tetrahedron Lett.* pp. 2777–2780.

Inone, T., and Y. Kawamura. 1961. Studies on biogenesis of tea components. II. Formation of caffeine in excised tea shoots. *Chem. Pharm. Bull.* **9**:236–238.

International Tea Committee. 1968. "Annual Bulletin of Statistics." Int. Tea Comm., London.

Jurd, L., 1962. Spectral properties of flavonoid compounds. *In* "Chemistry of Flavonoid Compounds" (T. A. Geissman, ed.), pp. 107–155. Pergamon, London.

Keegel, E. L. 1958. "Tea Manufacture in Ceylon." Tea Res. Inst., Talawakele, Ceylon.

Kiribuchi, T., and T. Yamanishi. 1963. Studies on the flavor of green tea. Part IV. Dimethyl sulfide and its precursor. *Agr. Biol. Chem.* **27**:56–59.

Kobayashi, A., H. Sato, and T. Yamanishi. 1965a. cis-2-Penten-1-ol in the essential oil from freshly plucked tea leaves and black tea. *Agr. Biol. Chem.* **29**:488–489.

Kobayashi, A., H. Sato, R. Arikawa, and T. Yamanishi. 1965b. Flavor of black tea. Part I. Volatile organic acids. *Agr. Biol. Chem.* **29**:902–907.

Kobayashi, A., H. Sato, H. Nakamura, K. Ohsawa, and T. Yamanishi. 1966. Flavor of black tea. Part III. Newly identified alcohols and aldehydes. *Agr. Biol. Chem.* **30**:779–783.

Kursanov, A. L. 1956. Tannins of the tea plant. *Kulturpflanze Beiheft* **1**:29–48.

Lam, T. H., and M. Shaw. 1970. Removal of phenolics from plant extracts by grinding with anion exchange resin. *Biochem. Biophys. Res. Commun.* **39**:965–968.

Li, L. P., and J. Bonner. 1947. Experiments on the localization and nature of tea oxidase. *Biochem. J.* **41**:105–110.

Loomis, W. D., and J. Battaile. 1964. Cell free enzyme from tannin containing plant tissues. *Plant Physiol.* **39**:Suppl., xxi.

Loomis, W. D., and J. Battaile. 1966. Plant phenolic compounds and the isolation of plant enzymes. *Phytochemistry* **5**:423–438.

Mayuranathan, P. S., and K. K. Gopalan. 1965. Reaction of amino acids and tea quinones. *J. Food Sci. Technol.* **2**:1–3.

Michl, H., and F. Haberler. 1954. Determination of purines in caffeine-containing drugs. *Monatsh. Chem.* **85**:779–795.

Millin, D. J., and D. W. Rustidge. 1967. Tea manufacture. *Process Biochem.* **2**:9–13.

Millin, D. J., D. J. Crispin, and D. Swaine. 1969a. Nonvolatile components of black tea and their contribution to the character of the beverage. *J. Agr. Food Chem.* **17**:717–722.

Millin, D. J., D. S. Sinclair, and D. Swaine. 1969b. Some effects of ageing on pigments of tea extracts. *J. Sci. Food Agr.* **20**:303–306.

Millin, D. J., D. Swaine, and P. L. Dix. 1969c. Separation and classification of the brown pigments of aqueous infusions of black tea. *J. Sci. Food Agr.* **20**:296–302.

Mizuno, T. 1968. Carbohydrates of tea. XII. The component sugars of theasaponin from the tea seed. *Nippon Nogei Kagaku Kaishi* **42**:291–501.

Mizuno, T., and E. Harashima. 1970. Studies on the carbohydrates of tea plants. Part XIV. On the hemicellulose "arabanogalactan" isolated from tea leaf. *Nippon Nogei Kagaku Kaishi* **44**:202–208.

Mizuno, T., and T. Kimpyo. 1955a. Carbohydrates of tea. I. The kinds of carbohydrates in black tea. *Nippon Nogei Kagaku Kaishi* **29**:847–852.

Mizuno, T., and T. Kimpyo. 1955b. Carbohydrates of tea. II Free sugars in black tea. *Nippon Nogei Kagaku Kaishi* **29**:852–856.

Mizuno, T., and T. Kimpyo. 1956a. Carbohydrates of tea. III. The kinds of carbohydrates in tea seed. *Nippon Nogei Kagaku Kaishi* **30**:474–477.

Mizuno, T., and T. Kimpyo. 1956b. Carbohydrates of tea. IV. The kinds of carbohydrates in the blossom of tea. *Nippon Nogei Kagaku Kaishi* **30**:477–479.

Mizuno, T., and T. Kimpyo. 1963. Quantitative determination of starch in tea leaves. *Nippon Shokuhin Kogyo Gakki-Shi.* **10**:216–223.

Mizuno, T., Y. Suzuki, H. Abe, and T. Kimpyo. 1964. Fractionation of tea leaf polysaccharides and identification of component sugars. *Nippon Shokuhin Kogyo Gakki-Shi.* **11**:146–152.

Müggler-Chavan, F., R. Viani, J. Bricout, D. Reymond, and R. H. Egli. 1966. Sur la composition de l'arome de thé. I. *Helv. Chim. Acta* **49**:1763–1767.

Müggler-Chavan, F., R. Viani, J. Bricout, J. P. Marion, H. Mechtler, D. Reymond, and R. H. Egli. 1969. Sur la composition de l'arôme de thé. III. Identification de deux cétones apparentées aux ionones. *Helv. Chim. Acta* **52**:549–550.

Nakabayashi, T. 1958. Studies on the formation mechanism of black tea aroma. Part V. On the precursor of the volatile carbonyl compounds of black tea. *Nippon Nogei Kagaku Kaishi* **32**:941–945. [*Chem. Abstr.* **55**:15772a.]

Nakagawa, M., and H. Torii. 1965. Studies on the flavanols in tea. Part IV. Enzymic oxidation of flavanols. *Agr. Biol. Chem.* **29**:278–284.

Nakatani, Y., S. Sato, and T. Yamanishi. 1969. 3S-(+)-3,7-Dimethyl-1,5,7-octatriene-3-ol in the essential oil of black tea. *Agr. Biol. Chem.* **33**:967–968.

Nakatani, Y., T. Yamanishi, N. Esumi, and T. Suzuki. 1970. On the structure of *cis*- and *trans*-theaspirone. *Agr. Biol. Chem.* **34**:152.

Neish, A. C. 1960. Biosynthetic pathways of aromatic compounds. *Annu. Rev. Plant Physiol.* **11**:55–80.

Nikolaishvili, D. K., and N. I. Adeishvili. 1966. Chromatographic study of the quantitative changes in tea leaf pigments during production of black tea. *Byull. Vses. Nauch.-Issled. Inst. Chai. Prom.* pp. 57–60.

Nose, M., Y. Nakatani, and T. Yamanishi. 1971. Studies on the flavor of green tea. Part IX. Identification and composition of intermediate and high boiling constituents in green tea flavor. *Agr. Biol. Chem.* **35**:261–271.

Ogura, N. 1969. Studies on chlorophyllase of tea leaves. II. Seasonal change of a soluble chlorophyllase. *Shokubutsugaku Zasshi* **82**:392–396.

Ogura, N., and A. Takamiya. 1966. Studies on chlorophyllase of tea leaves. *Shokubutsugaku Zasshi* **79**:588–594.

Ogutuga, D. B. A., and D. H. Northcote. 1970a. Caffeine formation in tea callus tissue. *J. Exp. Bot.* **21**:258–273.

Ogutuga, D. B. A., and D. H. Northcote. 1970b. Biosynthesis of caffeine in tea callus tissue. *Biochem. J.* **117**:715–720.

Patschke, L., and H. Griseback. 1965. 4,2',4',6'-Tetrahydroxychalkon-2'-glucoside-[β-^{14}C] als vorstufe für catechine im tee. *Z. Naturforsch.* **20b**:399.

Pierce, A. R., J. G. Gonzalez, and G. W. Sanderson. 1969a. Changes in tea flavanols during tea fermentation. Unpublished observations.

Pierce, A. R., H. N. Graham, S. Glassner, H. Madlin, and J. G. Gonzalez. 1969b. Analysis of tea flavanols by gas chromatography of their trimethylsilyl derivatives. *Anal. Chem.* **41**:298–302.

Pintaro, N. 1970. "Soluble Tea Processes." Noyes Data Corp., Park Ridge, New Jersey.

Popov, V. P. 1956. Oxidation of amino acids in the presence of tannins and polyphenoloxidase of tea. *Biokhimiya* **21**:383–387.

Proiser, E., and G. P. Serenkov. 1963. Caffeine biosynthesis in tea leaves. *Biokhimiya* **28**:857–861.

Ramaswamy, M. S. 1963. Chemical basis of liquoring characteristics of Ceylon tea. 2. Relationship between composition of the tea liquors and the valuations for the liquoring characteristics of black tea. *Tea Quart.* **34**:56–67.

Ramaswamy, M. S., and J. Lamb. 1958a. Studies on the fermentation of Ceylon tea. X. Pectic enzymes in tea leaf. *J. Sci. Food Agr.* **9**:46–51.

Ramaswamy, M. S., and J. Lamb. 1958b. Studies on the fermentation of Ceylon tea. XI. Relations between the polyphenol oxidase activity and pectin-methylesterase activity. *J. Sci. Food Agr.* **9**:51–56.

Reymond, D., F. Müggler-Chavan, R. Viani, L. Vuataz, and R. H. Egli. 1966. Gas chromatographic analysis of steam volatile aroma constituents: application to coffee, tea, and cocoa aromas. *J. Gas Chromatog.* **4**:28–31.

Roberts, E. A. H. 1939a. The fermentation process in tea manufacture. II. Some properties of tea peroxidase. III. The mechanism of fermentation. *Biochem. J.* **33:** 836–852.
Roberts, E. A. H. 1939b. Cytochrome oxidase in tea fermentation. *Nature (London)* **144:**867–868.
Roberts, E. A. H. 1940. The fermentation process in tea manufacture. 5. Cytochrome oxidase and its probable role. 6. The extent of dilution on the rate and extent of oxidations in fermenting tea leaf suspensions. *Biochem. J.* **34:**500–516.
Roberts, E. A. H. 1952. The chemistry of tea fermentation. *J. Sci. Food Agr.* **3:**193–198.
Roberts, E. A. H. 1957. Oxidation-reduction potentials in tea fermentation. *Chem. Ind. (London)* pp. 1354–1355.
Roberts, E. A. H. 1958a. The phenolic substances of manufactured tea. II. Their origin as enzymic oxidation products in fermentation. *J. Sci. Food Agr.* **9:**212–216.
Roberts, E. A. H. 1958b. The chemistry of tea manufacture. *J. Sci. Food Agr.* **9:**381–390.
Roberts, E. A. H. 1961. The nature of the phenolic oxidation products in manufactured black tea. *Tea Quart.* **32:**190–200.
Roberts, E. A. H. 1962. Economic importance of flavonoid substances: tea fermentation. *In* "Chemistry of Flavonoid Compounds" (T. A. Geissman, ed.), pp. 468–512. Pergamon, London.
Roberts, E. A. H. 1963. The phenolic substances of manufactured tea. X. The creaming down of tea liquors. *J. Sci. Food Agr.* **14:**700–705.
Roberts, E. A. H., and M. Myers. 1958. Theogallin, a polyphenol occurring in tea. II. Identification as a galloylquinic acid. *J. Sci. Food Agr.* **9:**701–705.
Roberts, E. A. H., and M. Myers. 1959a. The phenolic substances of manufactured tea. IV. Enzymic oxidations of individual substrates. *J. Sci. Food Agr.* **10:**167–179.
Roberts, E. A. H., and M. Myers. 1959b. The phenolic substances of manufactured tea. V. Hydrolysis of gallic acid esters by *Aspergillus niger*. *J. Sci. Food Agr.* **10:** 172–176.
Roberts, E. A. H., and M. Myers. 1959c. The phenolic substances of manufactured tea. VI. The preparation of theaflavin and theaflavin gallate. *J. Sci. Food Agr.* **10:**176–179.
Roberts, E. A. H., and M. Myers. 1960a. The phenolic substances of manufactured tea. VII. The preparation of individual flavanols. *J. Sci. Food Agr.* **11:**153–163.
Roberts, E. A. H., and M. Myers. 1960b. The phenolic substances of manufactured tea. VIII. Enzymic oxidations of polyphenolic mixtures. *J. Sci. Food Agr.* **11:** 158–163.
Roberts, E. A. H., and S. N. Sarma. 1938. The fermentation process in tea manufacture. I. The role of peroxidase. *Biochem. J.* **32:**1819–1828.
Roberts, E. A. H., and R. F. Smith. 1961. Spectrophotometric measurements of theaflavins and thearubigins in black tea liquors in assessments of quality in teas. *Analyst (London)* **86:**94–98.
Roberts, E. A. H., and R. F. Smith. 1963. The phenolic substances of manufactured tea. IX. The spectrophotometric evaluation of tea liquors. *J. Sci. Food Agr.* **14:** 689–700.
Roberts, E. A. H., and D. M. Williams. 1958. The phenolic substances of manufactured tea. III. Ultra-violet and visible absorption spectra. *J. Sci. Food Agr.* **9:**217–223.
Roberts, E. A. H., and D. J. Wood. 1950. The fermentation process in tea manufacture. II. Oxidation of substrates by tea oxidase. *Biochem. J.* **47:**175–186.
Roberts, E. A. H., and D. J. Wood. 1951. The amino acids and amides of fresh and withered tea leaf. *Curr. Sci.* **20:**151–153.
Roberts, E. A. H., R. A. Cartwright, and M. Oldschool. 1957. The phenolic substances

of manufactured tea. I. Fractionation and paper chromatography of water-soluble substances. *J. Sci. Food Agr.* **8**:72–80.

Roberts, G. R., and G. W. Sanderson. 1966. Changes undergone by free amino acids during the manufacture of black tea. *J. Sci. Food Agr.* **17**:182–188.

Saijo, R., and Y. Kuwabara. 1967. Volatile flavor of black tea. Part I. Formation of volatile components during black tea manufacture. *Agr. Biol. Chem.* **31**:389–396.

Saijo, R. 1967. Volatile flavor of black tea. Part II. Examination of the first fraction of effluents in gas chromatography. *Agr. Biol. Chem.* **31**:1265–1269.

Saijo, R., and T. Takeo. 1970a. The production of phenylacetaldehyde from L-phenylalanine in tea fermentation. *Agr. Biol. Chem.* **34**:222–226.

Saijo, R., and T. Takeo. 1970b. The formation of aldehydes from amino acids by tea leaf extracts. *Agr. Biol. Chem.* **34**:227–233.

Sakato, Y., T. Hahizume, and Y. Kishimoto. 1950. Studies on the chemical constituents of tea. Part V. Synthesis of theanine. *Nippon Nogei Kagaku Kaishi* **23**:269–271.

Salfeld, J. C. 1957. Zum Reaktionmechanismus der Purpurogallin-Bildung. *Angew. Chem.* **69**:723–724.

Sanderson, G. W. 1964a. Extraction of soluble catechol oxidase from tea shoot tips. *Biochim. Biophys. Acta* **92**:622–624.

Sanderson, G. W. 1964b. Changes in the level of polyphenol oxidase activity in tea flush in storage after plucking. *J. Sci. Food Agr.* **15**:634–639.

Sanderson, G. W. 1964c. The chemical composition of fresh tea flush as affected by clone and climate. *Tea Quart.* **35**:101–110.

Sanderson, G. W. 1964d. The theory of withering in tea manufacture. *Tea Quart.* **35**:146–163.

Sanderson, G. W. 1965a. The action of polyphenolic compounds on enzymes. *Biochem. J.* **95**:24P–25P.

Sanderson, G. W. 1965b. On the nature of the enzyme catechol oxidase in tea plants. *Tea Quart.* **36**:103–111.

Sanderson, G. W. 1965c. On the chemical basis of quality in black tea. *Tea Quart.* **36**:172–182.

Sanderson, G. W. 1966a. 5-Dehydroshikimate reductase in the tea plant (*Camellia sinensis* L.). *Biochem. J.* **98**:248–252.

Sanderson, G. W. 1966b. On the reaction of amino acids and tea quinones. *Curr. Sci.* **35**:392.

Sanderson, G. W. 1968. Change in cell membrane permeability in tea flush on storage after plucking and its effect on fermentation in tea manufacture. *J. Sci. Food Agr.* **19**:637–639.

Sanderson, G. W., and P. Kanapathipillai. 1964. Further studies on the effect of climate on the chemical composition of fresh tea flush. *Tea Quart.* **35**:222–229.

Sanderson, G. W., and B. P. M. Perera. 1965. Carbohydrates in tea plants. I. The carbohydrates of tea shoot tips. *Tea Quart.* **36**:6–13.

Sanderson, G. W., and B. P. M. Perera. 1966. Carbohydrates in tea plants. 2. The carbohydrates in tea roots. *Tea Quart.* **37**:86–92.

Sanderson, G. W., and G. R. Roberts. 1964. Peptidase activity in shoot tips of the tea plant (*Camellia sinensis* L.). *Biochem. J.* **93**:419–423.

Sanderson, G. W., and R. R. Selvendran. 1965. The organic acids in tea plants. A study of the non-volatile organic acids separated on silica gel. *J. Sci. Food Agr.* **16**:251–258.

Sanderson, G. W., and K. Sivapalan. 1966a. Effect of leaf age on photosynthetic assimilation of carbon dioxide in tea plants. *Tea Quart.* **37**:11–26.

Sanderson, G. W., and K. Sivapalan. 1966b. Translocation of photosynthetically assimilated carbon in tea plants. *Tea Quart.* **37**:140–153.
Sanderson, G. W., H. Co, and J. G. Gonzalez. 1971. Biochemistry of tea fermentation: The role of carotenes in black tea aroma formation. *J. Food Sci.* **36**:231–236.
Sanderson, G. W., J. E. Berkowitz, and H. Co. 1972. Biochemistry of tea fermentation: Products of the oxidation of tea flavanols in a model tea fermentation system. *J. Food Sci.* In press.
Sato, S., S. Sasakura, A. Kobayashi, Y. Nakatani, and T. Yamanishi. 1970. Flavor of black tea. Part VI. Intermediate and high boiling components of the neutral fraction. *Agr. Biol. Chem.* **34**:1355–1367.
Scott, J. M. 1964. "The Tea Story." Heinemann, London.
Selvendran, R. R., and F. A. Isherwood. 1970. Estimation of sucrose-6-phosphate and glucose-1,6-diphosphate in tea leaves and strawberry leaves. *Phytochemistry* **9**:533–536.
Seshadri, T. R. 1967. Present knowledge of the chemistry of proanthocyanidins. *J. Indian Chem. Soc.* **44**:628–636.
Shaw, W. S. 1935. Theotannin. I. Theotannin in relation to green leaf. *Bull. United Plant. Ass. South India* **4**:42–46.
Smith, R. F. 1968. Studies on the formation and composition of "cream" in tea infusions. *J. Sci. Food Agr.* **19**:530–534.
Sreerangachar, H. B. 1939. The endo-enzyme in tea fermentation. *Curr. Sci.* **8**:13.
Sreerangachar, H. B. 1943a. Studies on the fermentation of Ceylon tea. 4. Estimation of the oxidizing enzyme activity. *Biochem. J.* **37**:653–655.
Sreerangachar, H. B. 1943b. Studies on the fermentation of Ceylon tea. 5. The comparative rates of oxidation of different polyphenolic substrates by tea-oxidizing enzymes. *Biochem. J.* **37**:656–660.
Sreerangachar, H. B. 1943c. Studies on the fermentation of Ceylon tea. 6. The nature of the tea-oxidase system. *Biochem. J.* **37**:661–667.
Sreerangachar, H. B. 1943d. Studies on the fermentation of Ceylon tea. 7. The prosthetic group of tea oxidase. *Biochem. J.* **37**:667–674.
Stagg, G. V., and D. Swaine. 1971. The identification of theogallin as 3-galloylquinic acid. *Phytochemistry* **10**:1671–1673.
Stahl, W. H. 1962. The chemistry of tea and tea manufacturing. *Advan. Food Res.* **11**:201–262.
Takei, S., and Y. Sakato. 1932. The essential oil of green tea. I. *Rikagaku Kenkyusho Iho* **12**:13–21.
Takei, S., Y. Sakato, and M. Ono. 1934a. The essential oils of green tea. II. *Rikagaku Kenkyusho Iho* **13**:116–123.
Takei, S., Y. Sakato, and M. Ono. 1934b. The essential oil of green tea. III. Acids of raw tea oil. *Rikagaku Kenkyusho Iho* **13**:1561–1568.
Takei, S., Y. Sakato, and M. Ono. 1935a. The essential oil of green tea. IV. The flavor of black tea. *Rikagaku Kenkyusho Iho* **14**:303–307.
Takei, S., Y. Sakato, and M. Ono. 1935b. The essential oil of green tea. V. *Rikagaku Kenkyusho Iho* **14**:507–511.
Takei, S., Y. Sakato, and M. Ono. 1935c. Odorous substances of green tea. VI. Constituents of tea oil. *Rikagaku Kenkyusho Iho* **14**:1262–1274.
Takei, S., Y. Sakato, and M. Ono. 1936. The essential oil of green tea. VII. Constituents of tea oil. *Rikagaku Kenkyusho Iho* **15**:626–634.
Takei, S., Y. Sakato, and M. Ono. 1937a. The essential oil of green tea. VIII. *Rikagaku Kenkyusho Iho* **16**:7–16.

Takei, S., Y. Sakato, and M. Ono. 1937b. The essential oil of green tea. IX. Carbonyl compounds of black tea oil. *Rikagaku Kenkyusho Iho* **16**:773-782.

Takei, S., Y. Sakato, and M. Ono. 1938a. The essential oil of green tea. X. Synthesis of *p*-hydroxycinnamaldehyde. *Rikagaku Kenkyusho Iho* **17**:216-225.

Takei, S., Y. Sakato, and M. Ono. 1938b. Odoriferous principle of green tea. XI. Primary alcohols from the oil of black tea. *Rikagaku Kenkyusho Iho* **17**:871-879.

Takeo, T. 1966a. Tea leaf polyphenol oxidase. Part III. Studies on the changes of polyphenol oxidase activity during black tea manufacture. *Agr. Biol. Chem.* **30**:529-535.

Takeo, T. 1966b. Tea leaf polyphenol oxidase. Part IV. The localization of polyphenol oxidase in tea leaf cell. *Agr. Biol. Chem.* **30**:931-934.

Takino, Y., and H. Imagawa. 1964. Crystalline reddish orange pigment of manufactured black tea. *Agr. Biol. Chem.* **28**:255-256.

Takino, Y., H. Imagawa, H. Horikawa, and A. Tanaka. 1964. Studies on the mechanism of the oxidation of tea leaf catechins. Part III. Formation of a reddish orange pigment and its spectral relationship to some benzotropolone derivatives. *Agr. Biol. Chem.* **28**:64-71.

Takino, Y., A. Ferretti, V. Flanagan, M. A. Gianturco, and M. Vogel. 1965. The structure of theaflavin, a polyphenol of black tea. *Tetrahedron Lett.* pp. 4019-4025.

Takino, Y., A. Ferretti, V. Flanagan, M. A. Gianturco, and M. Vogel. 1966. The structure of theaflavin, a polyphenol of black tea (erratum). *Tetrahedron Lett.* p. 4024.

Tambiah, M. S., H. G. Nandadasa, and M. J. C. Amarasuriya. 1966. The localization and distribution of catechins, oils, and other ergastic substances in the leaves of *Camellia sinensis. Ceylon Ass. Advan. Sci., Proc. Annu. Sess.* p. 22.

Tirimanna, A. S. L. 1967. Acid phosphatases of the tea leaf. *Tea Quart.* **38**:331-334.

Tirimanna, A. S. L., and R. L. Wickremasinghe. 1965. Studies on the quality and flavor of tea. 2. The carotenoids. *Tea Quart.* **36**:115-121.

Torii, H., and J. Kanazawa. 1954. Studies on the carbohydrates in tea leaves. I. Detection of sugars by paper chromatography. *Nippon Nogei Kagaku Kaishi* **28**:34-38.

Torii, H., and J. Kanazawa. 1955. Studies on the carbohydrates in tea leaves. 2. Free oligosaccharides in the tea plant. *Nippon Nogei Kagaku Kaishi* **29**:620-622.

Ukers, W. H. 1935. "All about Tea," Vol. 1. Tea Coffee Trade J. Co., New York.

Ukers, W. H. 1936. "The Romance of Tea." Knopf, New York.

U.S. Department of Agriculture. 1969. World Tea Crop. *Foreign Agriculture Circular Tropical Products, FTEA 2-69*, October, 12 pp. U.S. Dep. Agr., Washington, D.C.

Van Romburgh, P. 1920. Unsaturated alcohol of essential oil of freshly fermented tea leaves. *Proc. Kon. Ned. Akad. Wetensch.* **22**:758-761.

Varner, J. E. 1961. Biochemistry of senescence. *Annu. Rev. Plant Physiol.* **12**:245-264.

Viani, R., F. Müggler-Chavan, and L. Vuataz. 1966. Phenylacetaldehyde, a constituent of the flavor of black tea. *Chimia* **20**:28-29.

Vuataz, L., and H. Brandenberger. 1961. Plant phenols. III. Separation of fermented and black tea polyphenols by cellulose column chromatography. *J. Chromatogr.* **5**:17-31.

Vuataz, L., H. Brandenberger, and R. H. Egii. 1959. Plant phenols. I. Separation of the tea leaf polyphenols by cellulose column chromatography. *J. Chromatogr.* **2**:173-187.

Weinges, D., W. Bahr, W. Ebert, K. Goritz, and H. D. Marx. 1969. Konstitution, Entstehung und Bedeutung der Flavonoid—Gerbstoffe. *Fortschr. Chem. Org. Naturst.* **27**:158-260.

Wickremasinghe, R. L. 1967. Fact and speculation in the chemistry and biochemistry of black tea manufacture. *Tea Quart.* **38**:205-209.

Wickremasinghe, R. L., and V. H. Perera. 1966. The blackness of tea and the colour of tip. *Tea Quart.* **37**:75–79.
Wickremasinghe, R. L., and T. Swain. 1964. The flavour of black tea. *Chem. Ind. (London)* pp. 1574–1575.
Wickremasinghe, R. L., and T. Swain. 1965. Studies of the quality on flavour of Ceylon tea. *J. Sci. Food Agr.* **16**:57–64.
Wickremasinghe, R. L., D. Kirthisinghe, K. P. W. C. Perera, and V. H. Perera. 1965. Effects of method of manufacture on the oxidation of polyphenols and chlorophylls. *Tea Quart.* **36**:167–171.
Wickremasinghe, R. L., G. R. Roberts, and B. P. M. Perera. 1967. The localization of the polyphenol oxidase of tea leaf. *Tea Quart.* **38**:309–310.
Wickremasinghe, R. L., B. P. M. Perera, and U. L. L. deSilva. 1969. Studies on the quality and flavour of tea. 4. Observations on the biosynthesis of volatile compounds. *Tea Quart.* **40**:26–30.
Wight, W., and D. N. Barua. 1954. Calcium oxalate crystals as an indicator of nutrient balance in the tea plant (*Camellia sinensis*). *Curr. Sci.* **23**:78–79.
Winton, A. L., and K. B. Winton. 1935. "The Structure and Composition of Foods," Vol. II. Wiley, New York.
Wood, D. J., I. S. Bhatia, S. Chakraborty, M. N. D. Choudbury, S. B. Deb, E. A. H. Roberts, and M. R. Ullah. 1964. The chemical basis of quality in tea. I. Analyses of freshly plucked shoots. *J. Sci. Food Agr.* **15**:8–14.
Wood, D. J., and N. B. Chanda. 1955. Report of the Biochemical Branch. *Annual Report of the Tocklai Experiment Station for 1954*. pp. 45–64.
Yamamoto, R., and Y. Kato. 1934. Study of the essential oil of black tea. I. The essential oil of fermented Formosan tea leaf. *Sci. Pap. Inst. Phys. Chem. Res. (Tokyo)* **27**: 122–129.
Yamamoto, R., and Y. Kato. 1935. Odor of black tea. II. The essential oil of Formosan black tea. *Nippon Nogei Kagaku Kaishi* **11**:639–644.
Yamamoto, R., and K. Ito. 1937. Essential oil of black tea. III. *Nippon Nogei Kagaku Kaishi* **13**:736–750.
Yamamoto, R., K. Ito, and H. Tin. 1940a. Essential oil of Formosan black tea. IV. *Nippon Nogei Kagaku Kaishi* **16**:781–790.
Yamamoto, R., K. Ito, and H. Tin. 1940b. Essential oil of Formosan black tea. V. *Nippon Nogei Kagaku Kaishi* **16**:791–799.
Yamamoto, R., K. Ito, and H. Tin. 1940c. Essential oil of Formosan black tea. VI. *Nippon Nogei Kagaku Kaishi* **16**:800–802.
Yamanishi, T. 1968. Tea aroma. *Eiyo To Shokuryo* **21**:227–235.
Yamanishi, T., J. Takagaki, and M. Tsujimura. 1956. Studies on the flavor of green tea. Part II. Changes in components of essential oil of tea leaves. *Bull. Agr. Chem. Soc. Jap.* **20**:127–130.
Yamanishi, T., J. Takagaki, H. Kurita, and M. Tsujimura. 1957. Studies on the flavor of green tea. Part III. Fatty acids in essential oils of fresh tea leaves and green tea. *Bull. Agr. Chem. Soc. Jap.* **21**:55–57.
Yamanishi, T., T. Kiribuchi, M. Sakai, N. Fugita, Y. Ikeda, and K. Sasa. 1963. Studies on the flavor of green tea. Part V. Examination of the essential oil of tea leaves by gas liquid chromatography. *Agr. Biol. Chem.* **27**:193–198.
Yamanishi, T., H. Sato, and A. Ohmura. 1964. Linalool epoxides in the essential oils from freshly plucked tea leaves and black tea. *Agr. Biol. Chem.* **28**:653–655.
Yamanishi, T., T. Kiribuchi, Y. Mikumo, H. Sato, A. Ohmura, A. Mine, and T. Kurata.

1965a. Studies on the flavor of green tea. Part VI. Neutral fraction of the essential oil of tea leaves. *Agr. Biol. Chem.* **29**:300–306.

Yamanishi, T., A. Kobayashi, H. Sato, A. Ohmura, and H. Nakamura. 1965b. Flavor of black tea. Part II. Alcohols and carbonyl compounds. *Agr. Biol. Chem.* **29**: 1016–1020.

Yamanishi, T., A. Kobayashi, H. Sato, H. Nakamura, K. Osawa, A. Uchida, S. Mori, and R. Saijo. 1966a. Flavor of black tea. Part IV. Changes in flavor constituents during the manufacture of black tea. *Agr. Biol. Chem.* **30**:784–792.

Yamanishi, T., A. Kobayashi, A. Uchida, and Y. Kawashima. 1966b. Studies on the flavor of green tea. Part VII. Flavor components of manufactured green tea. *Agr. Biol. Chem.* **30**:1102–1105.

Yamanishi, T., A. Kobayashi, H. Nakamura, A. Uchida, S. Mori, K. Ohsawa, and S. Sasakura. 1968a. Flavor of black tea. Part V. Comparison of aroma of various types of black tea. *Agr. Biol. Chem.* **32**:379–386.

Yamanishi, T., R. L. Wickremasinghe, and K. P. W. C. Perera. 1968b. Studies on the quality and flavor of tea. 3. Gas chromatographic analyses of the aroma complex. *Tea Quart.* **39**:81–86.

Yamanishi, T., M. Nose, and Y. Nakatani. 1970. Studies on the flavor of green tea. Part VIII. Further investigation of flavor constituents in manufactured green tea. *Agr. Biol. Chem.* **34**:599–608.

Zaprometov, M. N. 1958. The site of catechin synthesis in the tea plant. *Fiziol. Rast.* **5**:51–61.

Zaprometov, M. N. 1961. Isolation of quinic and shikimic acids from the shoots of the tea plants. *Biokhimiya* **26**:373–384.

Zaprometov, M. N. 1962a. The mechanism of biosynthesis of catechins. *Biokhimiya* **27**:366–377.

Zaprometov, M. N. 1962b. Formation of caffeine in tea plant shoots. *Biokhimiya* **27**:679–684.

Zaprometov, M. N. 1963. Catechin biosynthesis in tea shoots. *Fiziol. Rast.* **10**:73–78.

Zaprometov, M. N. 1968. Progress and prospects in the biochemistry of phenols. *In* "Phenolic Compounds and Their Biological Functions," Pap. 1st All-Union Symp. Phenolic Compounds, 1966, pp. 109–125. Nauka Publishing House, Moscow, Russia.

Zaprometov, M. N., and V. Y. Buklaeva. 1963. Gallic acid and the biosynthesis of catechins. *Biokhimiya* **28**:862–867.

A SPECULATIVE VIEW OF
TOBACCO ALKALOID BIOSYNTHESIS*

RAY F. DAWSON and THOMAS S. OSDENE

Philip Morris Research Center, Richmond, Virginia

Introduction	317
Correlative Information	318
Biochemical and Biophysical Properties of the Pyridine Ring	320
The Compartmentalization of Alkaloid Biosynthesis	322
Hormonal Interactions	325
Speculative Model Building	328
Probable Identity of the Macromolecule	329
The Pyridinium–Polynucleotide: General Considerations	329
The Pyridinium–tRNA Model	332
References	335

Introduction

The invitation to participate in this program requested a discussion of some aspect of tobacco chemistry or biochemistry. We have chosen to speculate upon a topic that does indeed relate to the tobacco plant but may relate also to a much broader question—that is, the significance of alkaloids in the lives of the plants that produce them. We shall confine our treatment

---

* This paper is dedicated to Professor Doctor Kurt Mothes, Director-Emeritus of the Institute for Plant Biochemistry in Halle, on the occasion of his 70th birthday.

to the *Nicotiana* alkaloids, for there is probably more factual information regarding the biosynthesis and physiological behavior of nicotine, nornicotine, and anabasine within the tobacco plant than can be assembled for any similar combination.

The term "speculate" is employed by necessity and without apology. Although much is known about the intermediates and pathways of alkaloid biosynthesis (cf. Leete, 1969, for review), efforts to deduce even plausible hypotheses concerning the biological significance of alkaloids have failed. We do not at present possess sufficient insight to relate alkaloid biosynthesis and metabolism to other phases of cell or plant activity. As a result the subject is necessarily ignored in textbooks. Nowhere is this situation more embarrassing than in the case of the tobacco alkaloids for the reason given above. There is required a reexamination of the problem in the hope that current accumulations of data may yield clues heretofore overlooked or unavailable.

We have attempted such a reexamination by bringing together the minutiae of tobacco science with the newer concepts of molecular biology and biophysics. It is hoped that the models thus developed, however vulnerable to criticism they may be, will initiate new trains of thought and thus give rise to valid hypotheses where none now exist in this baffling area of phytochemistry.

Correlative Information

The tobacco plant and other members of the genus *Nicotiana* accumulate a series of alkaloidal terminal metabolites, the common denominator among which is the presence of a 3-substituted pyridine ring. Nicotine (I) and anabasine (II) will concern us for present purposes (Fig. 1). Biosyntheti-

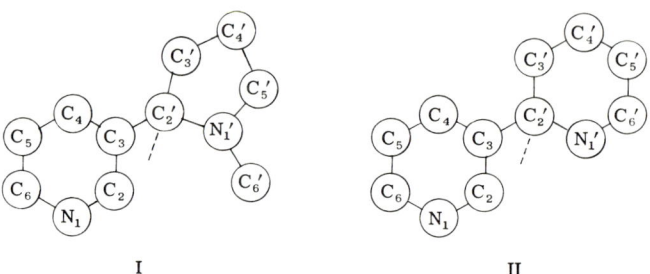

FIG. 1. The ring structures of nicotine (I) and of anabasine (II).

FIG. 2. Proposed intermediate in the biosynthesis of nicotine from nicotinic acid or nicotinamide. See Dawson (1962).

cally, the pyridine ring of both alkaloids originates with nicotinic acid (Dawson et al., 1960a, b) or one of its derivatives including nicotinamide adenine dinucleotide (Frost et al., 1967). The asymmetry of the ring is preserved during the transformations involved, but the carboxyl (or carboxamide) group at position 3 is lost by substitution, and the hydrogen at ring position 6 is made labile (Dawson et al., 1960a). Dawson (1962) has postulated the occurrence of a 3,6-dihydropyridinium intermediate which accounts satisfactorily for the exchange of hydrogen at position 6 only and for facilitated substitution of the carboxyl group (Fig. 2). Alternative hypotheses have been advanced to account for these changes (Wenkert, 1968; Leete, 1969), but we have rejected them for reasons of failure to conform to experimental fact (cf. Dawson et al., 1960a), to quantum mechanical estimates of electronic structure (Pullman and Pullman, 1959), and to molecular conformation requirements (cf. Arnott et al., 1969) of the models that we shall develop herein. It is helpful to recall, for purposes of this discussion, that the naturally occurring pyridine codehydrogenases, NAD and NADP, are reduced 1,4 rather than 3,6 (cf. Sund et al., 1964, for a broad review of this subject).

There have been noted in the preceding paragraph three items of basic importance for the discussion to be developed in the text to follow. The first of these is that the *Nicotiana* alkaloids accumulate within the plant body in the manner expected of terminal metabolites. That is, they may turn over very slowly or very rapidly, they may accumulate in high or low amount or not at all, and they may be introduced into the plant by artificial means, all with little or no visible consequence so far as the normal course of growth and morphogenesis is concerned. We interpret these facts to mean that, in our search for an answer to the question of biological significance, we must focus upon either the biological activity of the molecules at the moment of their formation within the cells of origin or, alternatively, the processes by which they are made in those cells. The second item of consequence is that nicotinic acid is the biosynthetic precursor of the element of structure that is common to all the alkaloids of this series. Therefore, we are obliged to examine the physical and chemical properties of nicotinic acid for clues to the answers we seek. Last, we have noted the occurrence of a reduced pyridine nucleus of unusual structure among the biosynthetic intermediates of these alkaloids. For lack of a better alternative, we assume that this intermediate occupies a pivotal position astride the pathways leading to alkaloids on the one hand or to the metabolic pools of nicotinic acid and its derivatives on the other. These three concepts are examined in somewhat more detail in the sections that follow.

BIOCHEMICAL AND BIOPHYSICAL PROPERTIES OF THE PYRIDINE RING

The pyridine ring of nicotinic acid is planar, but there is very likely potential freedom of rotation around the C(3)–C(7) bond. In the crystal, the carboxyl group and the pyridine ring of nicotinic acid are coplanar (Wright and King, 1953). Each molecule is linked to two others to form a zigzag chain parallel to the b axis. The carboxamide group of crystalline nicotinamide is tilted a few degrees out of the plane of the ring to accommodate a 2-dimensional network wherein each molecule is linked to four others (Wright and King, 1954). It is important to note that the linkages in these two instances are hydrogen bonds between carboxyl oxygen and ring nitrogen in the case of the acid and between carboxamide oxygen and nitrogen and ring nitrogen in nicotinamide. An instructive case is provided by the crystal structure of cytosine-5-acetic acid (Marsh et al., 1962). Here, the dihedral angle between the planes of the carboxyl and of the pyrimidine ring is close to 68°. Further, the molecules are held in a 3-dimensional network of hydrogen bonds involving all five available protons. Pairs of cytosine nuclei are arranged about a center of symmetry between two N(3) atoms in the same manner as the cytosine-guanine pairs of the Watson-Crick model of deoxyribonucleic acid (Watson and Crick, 1953). For our purposes, another matter of interest is the observation that, in one-half the molecules of the crystal, the proton on carboxyl O is transferred to ring nitrogen N(3) to form a zwitterion: the remaining half of the molecules are uncharged. The participation of carboxyl and carboxamide groupings in intermolecular hydrogen bonding is of special interest in the present context for the reason that resonance energy may contribute to the stability of the resulting complexes (Pauling, 1960).

Although the foregoing discussion relates to the properties of nicotinic acid and amide and related molecules in the solid state, there are indications that complex formation may occur also in solutions of these substances. Thus, hydrogen bonding between presumably coplanar molecules of size and structure similar to nicotinic acid (the purines and pyrimidines and their N-ribosides) has been shown to occur in nonaqueous solvents (Thomas and Kyogoku, 1967; Katz and Penman, 1966).

The concept of parallel plane stacking of aromatic and heteroaromatic molecules is also of importance in the present connection. Hydrophobic interactions between rings are considered to depend upon van der Waals attractions, and quantum mechanical calculations tell us that there are preferred conformations corresponding to more stable presentations and degrees of overlap (Pullman and Pullman, 1969). The tilted arrangement of individual molecules in crystals of polynuclear hydrocarbons and other planar compounds was used as an analogy for explaining the charge transfer properties of the chloroplast by Calvin (1959) over a decade ago.

The pyridine ring of nicotinic acid and its congeners falls within the purview of this discussion of molecular stacking and of the formation of charge transfer complexes (i.e., of semiconductors). Sarma *et al.* (1968, 1970; Sarma and Kaplan, 1970a,b) have investigated the intramolecular stacking of nicotinamide and adenine moieties of oxidized and reduced pyridine nucleotides and their analogs. Deranleau and Schwyzer (1970) describe intermolecular complex formation in 1 percent ethanolic-water solutions between pyridinium-type acceptors and indole-type donors by means of the appearance of new charge transfer bands in the corresponding absorption spectra. Estabrook *et al.* (1963) report a marked fluorescence enhancement when imidazole is added to nicotinamide adenine dinucleotide and its apoenzyme in solution. Nakano and Igarashi (1970) show that molecular interactions of a variety of bases in solution are almost entirely of the van der Waals type rather than hydrogen bonding. Some degree of dependence was noted upon the size of the π-electron system and the nature of the heteroaromatic ring structure and of ring substituents.

The ring nitrogen $N(1)$ of nicotinic acid and of nicotinamide may form glycosidic bonds with carbon $C(1)$ of ribosephosphate just as do the nitrogen atoms $N(9)$ and $N(1)$ of the purines and pyrimidines, respectively. In this regard, the ubiquitous dinucleotides of pyridine and adenine are homologous with the usual polynucleotides, and nicotinic acid and amide may be regarded as legitimate nucleotide bases. Further, the two mononucleotide residues of the nicotinamide adenine dinucleotides, NAD and NADP, are joined end-to-end in a phosphoric acid anhydride type of linkage, whereas the terminal phosphate residues of mononucleotide residues of the naturally occurring polynucleotides are joined in ester linkages to $C(3)$ of the next adjacent ribose moieties to form the backbone of the macromolecule. The anhydrophosphate linkage in NAD and NADP lengthens the molecular skeleton appreciably, with the result that a folded conformation may occur in which the nicotinamide and the adenine are stacked (Sarma *et al.*, 1968, 1970; Sarma and Kaplan, 1970a,b). It is tempting to suggest that the estimated interplanar distance of the stacked adenine and nicotinamide bases, namely, 3.90 Å, may suffice to allow the intercalation of a third planar molecule and thus to permit a stacking interaction between NAD and/or NADP and other molecular species of comparable geometry. Obviously, this kind of complex formation would represent an approach to the solid state (i.e., semicrystallinity) and to the properties that such structure confers.

THE COMPARTMENTALIZATION OF ALKALOID BIOSYNTHESIS

We have directed attention thus far to those properties of the pyridine ring precursor of nicotine and of anabasine, namely, nicotinic acid and amide, which relate principally to the ordered arrangement of pyridine or pyridinium compounds in solid structures. The reasons for this emphasis upon solid rather than liquid state phenomena are now to be developed by considering what is known about the compartmentalization of alkaloid biosynthesis within the tobacco plant (Dawson, 1948; Dawson and Solt, 1959; Solt and Dawson, 1958).

An explanation of the special importance of the root system for nicotine production has never been achieved. One can think of many possibilities in addition to the disparity in numbers of apical meristems mentioned earlier. For example, the complete machinery for nicotine synthesis may be allocated to roots in common, or perhaps in linkage, with other genetically controlled characters by differential transcription of the genome. Such transcription appears to be under hormonal control, and we shall discuss this aspect of the subject at a later point. Still a third possibility is an involvement with the reductive assimilation of nitrate (cf. Dawson, 1946). The latter is an obvious functional difference between root and shoot, the realization of which may depend more upon environment than upon the genetic code, although our ignorance of the details of the process is so great that even this possibility is not assured. If nicotine synthesis were to be related to the reductive assimilation of nitrate, an attractive hypothesis would include a 1,6- or 3,6-dihydronicotinamide derivative as the reductant. Nicotinamide adenine dinucleotide reduced 1,4- is a well recognized reductant in many organisms for nitrite. However, there are some organisms, such as the alga *Dunaliella* (Grant, 1970), which employ other, still unidentified reductants. The tobacco root may fall into this general category, but in the absence of even the most elementary information we cannot pursue this inviting prospect further.

The formation of anabasine contrasts sharply with that of nicotine in that the capability is much more evenly distributed between root and shoot (Dawson, 1944) and along the root axis (Solt *et al.*, 1960). The significance of this distribution will be discussed at a later point.

Much has been learned about the production of nicotine and of anabasine by the use of microbially sterile cultures of detached tobacco roots (Dawson, 1942, 1962; Solt, 1957). This experimental system eliminates, or at least minimizes, the complex interactions between different plant organs in respect to the production and transport of foodstuffs, the absorption and distribution of mineral salts and the interplay of different hormonal systems of root and shoot. Thus, the amount of nicotine synthesized by tobacco root tips in culture is directly proportional to the mass of new tissues produced during steady-state growth (Solt, 1957). That is, the root tips may be regarded as uniformly moving point sources for both new cells and new alkaloid molecules. Furthermore, the proportionality constant has not differed significantly between different clones of a Samsun Turkish variety and between these and three different cigar and cigarette tobaccos (Dawson, 1962). These findings have been interpreted to mean that a fixed number of molecules of nicotine is produced on the average for each new cell produced by the root apical meristem.

In developing an explanation of such stoichiometry, it is useful to note the extraordinary stability of nicotine production rates in excised root culture. In general, those environmental factors which affect growth rates of the roots affect nicotine production rates proportionally. Also, additions to the excised root cultures of precursors of the pyridine or pyrrolidine rings or of both do not affect either rates or yields of nicotine production. The same is true of additions of precursors of anabasine (Solt et al., 1960). This stability of rate may result from dependence upon an event of biological nature which proceeds from a definite starting point to an equally definite conclusion more or less independently of ordinary kinetic influences. Such events would include the replication of biological units (e.g., the cell or its organelles) and the functioning of one or more of these during one phase or another of the development of the unit. For example, Schmid and Serrano (1948) reported a close parallelism between total protein and total nicotine contents of various types of tobacco throughout the growth cycle; and in one case they found a mole-for-mole relationship between nicotine and protein amino acid residues. This report is of sufficient importance to require verification and elaboration (cf. Mothes et al., 1957). We suggest that nicotine synthesis may be associated with the formation or with the specific activity of cell-structure-related macromolecules and that it is, therefore, more nearly a phenomenon of the solid state rather than of the liquid milieu in which the organelles of the cell are suspended. We think it likely that nicotinic acid is bound to one of these macromolecules during the transformation to nicotine and to anabasine. Should this be so, it may be more readily understood why efforts to identify significant enzyme participation in alkaloid biosynthesis have never achieved more than peripheral success (cf. Leete, 1969, for review).

Not only is the process of transformation of nicotinic acid to nicotine characterized by a stoichiometry to growth and a high degree of rate stability, it is also a highly compartmentalized process. For example, additions to the excised roots of *Nicotiana glauca* of precursors of the pyrrolidine ring of nicotine do not divert alkaloid production away from anabasine and toward nicotine as preferred product. The reverse situation is also true. Although we know that isotopically labeled nicotinic acid and precursors of the reduced rings penetrate the compartments where respective syntheses occur, it is impossible to divert one process to the other or to affect the rate of one with respect to the other by additions of exogenous materials. This leads us to conclude that the processes by which nicotinic acid is converted to nicotine and to anabasine are not only sharply confined but also highly specific.

Clues to the identity of the intracellular locus (loci) of these processes are not yielded by knowledge of the intermediates in the biosynthesis of nicotinic acid. The pyridine ring originates with one product of the Krebs tricarboxylic acid cycle, aspartic acid, and another of the Embden–Meyerhof–Parnas pathway, phosphoglyceraldehyde (for reviews, cf. Dawson *et al.*, 1971; Leete, 1969). The situation with respect to the origin of the pyrrolidine ring of nicotine is more illuminating. This ring is produced from intermediates of the Krebs tricarboxylic acid cycle while the latter is functioning in a predominantly biosynthetic rather than in an oxidative mode (Wu *et al.*, 1962; Christman and Dawson, 1963). That is, patterns of labeling indicate that phosphoenolpyruvate and oxaloacetate are being fed into the cycle in essentially equimolar amounts. Leete (1969) has raised minor objections to this interpretation, but these objections may be removed by pointing out that the probable occurrence of the hexosemonophosphate pathway in the root tissues would favor a differential incorporation of label from glucose-1- and glucose-6-^{14}C. Further, the failure of citric-1-^{14}C acid to yield the same labeling patterns in the pyrrolidine ring of nicotine as malic, succinic, and acetic acids, among others, is probably due to its failure to penetrate the alkaloid-producing compartments as readily as the latter acids, as judged by the low radiochemical yield observed (Christman and Dawson, 1963). In such cases, randomization of label through a greater or lesser number of undefined pathways is virtually certain to occur.

The information just adduced permits us to locate the nicotinic acid transforming compartment in close physical proximity to the mitochondria of cells which are in a high state of anabolic activity. If this were not so, label distribution patterns within the pyrrolidine ring, as noted above, would likely have been randomized in greater degree. The cells of the meristems are in high states of anabolic activity, for they must synthesize all the components required for the production of organelles, walls, membranes, and so on. Brown and Broadbent (1950) have provided an exceptionally clear picture of the localization and the properties of those cells within the root tip which are actively producing protein. Such cells are also in an active state of expansion (growth in volume), and we infer that the production of oxaloacetic and α-ketoglutaric acids for the synthesis of aspartic and glutamic acids, respectively, is a principal activity of at least a portion of the mitochondrial complement during this developmental phase. It is tempting to speculate, therefore, that nicotine may be produced principally in cells that are in active growth and that such production may accompany or result from the synthesis of protein, perhaps a specific protein, which makes up the structure of one or more of the organelles of the young cell. It should be noted, however, that an association with mitotic phenomena or with cell differentiation instead cannot be excluded.

Hormonal Interactions

The question why some tissues make substantial amounts of nicotine, while others of homologous nature do not, may relate to the uneven distribution of hormones within the plant body. One of the first indications of such a relationship was the report of Scott et al. (1964) that free cells of tobacco in culture could be induced to produce nicotine in a sustained fashion by growing them in a medium containing auxin and a cytokinin supplemented with a variety of other materials including amino acids. Earlier it was reported (Dawson, 1962) that tobacco callus tissues in culture upon a medium to which, incidentally, hormones were not deliberately added gradually lost their ability to synthesize the alkaloid during successive transfers. On a unit dry weight basis the actual production of nicotine in the report of Scott et al. amounted to approximately one-fourth the production by excised tobacco root tips.

Examples of the differential inhibition of nicotine production with respect to growth are more abundant. Very small amounts of an auxin added to excised tobacco root tips in culture gave such a result (Solt, 1957). Applications of 2,4-dichlorophenoxyacetic acid (an auxinlike material in low-level dosages) to intact tobacco plants either through the roots by hydroponic culture or through the leaves by spraying, substantially reduced nicotine content (Yasumatsu and Murayama, 1969). Gibberellic acid reduces nicotine yields of excised tobacco roots without notably affecting growth rates (Parups, 1960; Solt et al., 1961), and helminthosporic acid is reported to behave similarly (Yasumatsu et al., 1967).

There is abundant literature to suggest that auxins and cytokinins interact within the plant body to establish hormonal balances favorable to one stage or another of growth and development (Leopold, 1964). Less is known about the mode of action of the gibberellins and the helminthosporins. With the fragments of available information suggesting that hormonal balance may be a decisive factor for the occurrence or nonoccurrence of nicotine biosynthesis in cells of stalk origin, we may ask whether known or inferred patterns of hormone distribution can be correlated with patterns of alkaloid production within the intact plant.

Auxin is produced in aerial buds and in young leaves and is moved downward in the stalk to stimulate cell growth and to inhibit lateral bud development. There is a pronounced downward gradient of auxin concentration in the shoot such that the lower portions contain much less of this hormone than the upper ones. Thimann (1937) showed that growth of roots is only moderately stimulated by auxin at any concentration and is inhibited entirely by concentrations well below the stimulatory range for stems. It is thus tempting to suggest a perhaps simplistic reason for the low level of nicotine synthesis in shoots, namely, that the concentrations of growth hormone may be so high in the shoot as to be essentially inhibitory to the process.

Much of our knowledge of plant hormonal balance has been obtained by the use of the callus tissue which forms upon tobacco pith explants in sterile culture. The failure of such explants to produce nicotine beyond the first few transfers without added auxin and kinetin (Dawson, 1962) has been noted already. Lavee and Galston (1968) have clarified this situation by showing that cell size, RNA, and protein content of cells, and the hardness and compactness as well as the growth potential of such explants increase as the initial inoculum is obtained consecutively from apex to base of the tobacco stalk. This conforms to our experience (Dawson, 1962) in that the loss of nicotine synthesizing ability of the callus tissue in culture was associated with a progressive shift of the cultured mass from a hard, compact state to a loose, friable condition. Although the exact causes remain unknown, this observation may represent still another example of the complex interrelationships between hormones, chemical composition, and morphogenetic processes.

It is reported variously that additions of auxins alone to tobacco pith cells may cause the latter to enlarge and to increase their ribosomal, ribonucleic acid and protein contents. Additions of cytokinins alone have led to the initiation of endomitotic activity and to some increased production of protein. Additions of auxin to nuclei or to chromatin isolated from tobacco pith cells in culture gave increased rates of synthesis of ribonucleic acid when a specific protein mediator was also present. Apparently, the effect of the auxin-protein combination is on the chromatin and involves an activation of the genome for transcription (Mathysse and Phillips, 1969).

Additions of both auxin and cytokinin to tobacco pith cells led, in another study, to exponential rates of cell division and of protein synthesis. It is interesting to note that the ratio of molar concentrations of the two hormones yielding maximum rates of these processes was close to unity (Jouanneau and Peaud-Lenoel, 1967). It has been suggested that the proteins and the ribonucleic acid thus synthesized may be specific in their structure and properties. If this were so, it may help to explain some of the results of Zwar and Rijven (1967), who could not discern strikingly different effects upon the incorporation of several different amino acids into protein when auxin and cytokinin alone were supplied to tobacco tissues.

If cytokinins do indeed direct the transcription of a part of the auxin-activated genome into the synthesis of ribonucleic acids and proteins that are specific for different phases or facets of cell growth and development, then it may be of the greatest importance that cytokinins have been shown to occupy an anticodon-adjacent position in the nucleotide sequence of certain amino acid transfer ribonucleic acid molecules (Hall et al., 1966; Burrows et al., 1970). The specific molecules involved are 6-(3-methyl-2-butenylamino)-9-β-D-ribofuranosylpurine, the 4-hydroxy-2-butenylamino derivative, and the corresponding 2-methylthio-9-β-D-ribofuranosyl analogs. Although we have not constructed models of these molecules within an RNA strand, it seems possible that the bulky side chains may hinder base pairing, in the Watson-Crick sense, of folded portions of the strand and thus serve to stabilize the anticodon loop (cf. Cramer, 1969, for general details).

Wood and Braun (1967; Wood, 1964; Wood et al., 1969) describe a low-molecular-weight (ca. 400-gram molecules) cytokinin of undetermined structure which was believed to contain nicotinamide, a methyl group, glucose, and a sulfate residue. Considering the methods of isolation, it does not seem likely that this material was a component of a polynucleotide, although the possibility cannot be excluded. The material, which was detected in tobacco pith parenchyma tissue, is considered to be the native cytokinin of the species tested. Biosynthesis of the nicotinamide-type cell division factor was induced by additions of kinetin (6-furfurylaminopurine) to cultures of this cell type, but the evidence suggests that the two substances do not replace one another in the naturally occurring cell division-promoting system. It is tempting, therefore, to speculate that both types of cytokinins may be essential for controlling the biochemistry of cell multiplication in *Nicotiana* and that both may be incorporated into molecules, such as tRNA, which are actively involved in the process. In the discussions that follow, we shall describe the incorporation of nicotinic acid or amide into polynucleotides. Although we should feel more confident of our position if the pyridine structure of the Wood and Braun cytokinin were to be confirmed by other methods in addition to the infrared spectra provided by these authors, we shall make the assumption in the discussion to follow that the cytokinin of Wood and Braun may be identical to the pyridinium precursor of nicotine and of anabasine.

Braun (for review, see 1970) and co-workers have made much of the hormonal balance concept for explaining the origin of the tumor cell. They report that the Kostoff genetic tumors arise in hybrids between *Nicotiana glauca* and *N. langsdorfii* in response to high auxin to cytokinin ratios. The reverse ratios lead to normal morphogenetic development. In our experience with root cultures of *N. glauca*, bud regeneration occurred more frequently than in the case of cultures of roots of *N. tabacum*. The obvious suggestion is that a different hormonal balance may explain the contrasting distributions of nicotine- and of anabasine-synthesizing capacities within the plant bodies of these two species that we have described earlier.

It is curious that the hormonal balances of the root and shoot should differ so markedly if the functions of auxin and cytokinin in protein synthesis are indeed as outlined above. It has been mentioned earlier that the reductive assimilation of nitrate is a characteristic of protein synthesis in roots. Possibly this process could be related to the hormonal picture as well as to the synthesis of nicotine. However, we are unaware of any experimental data which implicate either auxins or cytokinins in the reductive assimilation of nitrate, and we have been unable to adduce a connection between the latter and nicotine formation as noted earlier.

Speculative Model Building

The chains of reasoning developed in the preceding sections may now be drawn together tentatively to formulate models of alkaloid production by the tobacco plant. These models are required to conform to the following specifications.

1. Nicotinic acid or amide, in the role of pyridine ring precursor, is bound to a macromolecule at the moment of alkaloid synthesis. Such bonding may be by covalent, hydrogen bonding or even van der Waals forces.

2. The macromolecule in question must be either a component of a cell structure or a catalyst involved in the formation or functioning of such a structure. However, those functions related principally to energy metabolism are excluded on kinetic grounds.

3. An interaction of regulatory nature must be accommodated between alkaloid synthesis and the known activities of auxins and cytokinins.

4. The model must relate specifically to cell growth.

5. The model must allow for known phylogenetic and ontogenetic patterns and trends in respect to alkaloid production in the genus *Nicotiana* (cf. Dawson, 1962).

Probable Identity of the Macromolecule

Nicotinic acid is known to be bound to a variety of macromolecules by one or another of the bond types listed above. For example, it occurs widely in cereals in complex, alkali-labile association with carbohydrates, amino acids, and perhaps other molecular species (Christianson *et al.*, 1968). It occurs in especially large amounts in the kernels of maize that have been bred and selected for high content of sugar and water-soluble polysaccharides (Richey and Dawson, 1951), and an interaction between the synthesis, or at least the accumulation, of this compound and the metabolism of carbohydrate has been proposed. However, it is not possible at present to relate either the heterogeneous macromolecular complex of the cereals or the concept of a relation to polysaccharide (starch) formation in maize to so distant a process as alkaloid formation.

Except for the reports already mentioned regarding formation of charge transfer complexes between pyridinium compounds and imidazoles (e.g., histidine) or various indoles including 3-indoleacetic acid and amide, there is little reason also to associate tobacco alkaloid production with the activities of proteins as such. It may be necessary to revise this conclusion at a later date, but at present, there is little ground in this sector upon which to speculate. Therefore, we turn to the polynucleotides for ideas which may lead to fruitful experimentation.

The Pyridinium–Polynucleotide: General Considerations

There are two ways in which we may visualize the structure of a pyridinium–polynucleotide type of intermediate in tobacco alkaloid biosynthesis. First, nicotinic acid or amide mononucleotide may substitute for guanylic acid, for example, in a polynucleotide. In this case, the pyridinium component would be considered a "foreign base" albeit of highly unconventional character. The second is the intercalation of the pyridinium moiety of a molecule of unspecified size between the stacked purine and pyrimidine rings of a single-stranded polynucleotide. In both cases, the possibilities for establishing a charge transfer complex are present with opportunities for base tilt provided by the pyridinium $C(3)$–$C(7)$ single bond. It seems unnecessary to defend the first of these two models on structural grounds, for the universal occurrence of the nicotinamide adenine dinucleotides affords adequate conceptual precedent. Indeed, it is not difficult to imagine an enzyme-catalyzed reaction in which guanylic or adenylic acid may be exchanged for the nicotinamide mononucleotide moiety of nicotinamide adenine dinucleotide (NAD). Olivera et al. (1968) have described a reasonable analogy.

The fact that pyridines have yet to be reported among the purines and pyrimidines of polynucleotide hydrolyzates may be of little moment. In the first place, it is doubtful whether anyone has looked for them, especially in preparations obtained from tobacco root apices. Second, it is not likely that they would ever be present in sufficient quantities in such hydrolyzates to be detected accidentally and without deliberate search. Third, their ultraviolet absorption properties do not distinguish them readily from purines and pyrimidines. Fourth, to detect them on chromatograms or in eluates would require the use of reagents which are not within the usual "bag of tricks" of the nucleic acid analyst.

The conversion of the pyridinium residue to nicotine is viewed essentially as described by Dawson (1962) with appropriate corrections since applied by Leete (1969). However, it is likely that a considerably stronger reductant than those which convert NAD to NADH would be required to produce the 3,6-dihydropyridinium intermediate of this pathway (cf. Pullman and Pullman, 1959). The identity of this reductant cannot be guessed, but it is possible that the reductant may be located at some distance from the pyridinium acceptor if a charge transfer complex is indeed set up among the stacked bases of the polynucleotide.

Where the pyridinium precursor is covalently bonded to a polynucleotide, substitution of the carboxyl or carboxamide grouping by the incoming precursor of the pyrrolidyl or piperidyl rings would be expected to result in a local increase in steric blocking as well as in a charge inversion. Thus, the incipient nicotine molecule may prevent a short section of the polynucleotide chain from engaging in perfect helix formation of the double-stranded, Watson-Crick type and indeed might initiate the helix-coil transition (Poland and Scheraga, 1970). Owing to the fact that rates of nicotine turnover are usually very slow, it is likely that the nicotine-containing polynucleotide would eventually lyse to yield both free nicotine and conventional nucleotide monomers.

There is little more that can be conjectured with profit about the case in which polynucleotides are involved only as base stacking partners with the alkaloid precursor. A charge-transfer complex might be envisaged in the mitochondrial polynucleotides as a means of channeling electrons between the components of the pyridine dehydrogenase–flavoprotein–cytochrome–oxygen system. However, purines and pyrimidines do not appear to be ideal structures for composing semiconductors, and, in any case, the atypical 3,6-reduction of the pyridinium acceptor would have to be explained.

The version of the polynucleotide model which engenders greatest interest is that in which the pyridinium intermediate is an integral part of the macromolecule. Proceeding on the assumption that the proposed nicotine and anabasine synthesizing scheme may involve DNA structure, wherein the pyridinium component serves as a "recognition point," it is useful to explore the consequences of the possible occurrence of a pyridine nucleotide at infrequent intervals along the backbone of a DNA or of an RNA chain. First, we visualize the pyridine nucleotide as a suitable substitute for guanylic acid, although substitution for adenylic acid cannot be excluded. Utilizing bond lengths and angles listed by Arnott et al. (1969) and by Wright and King (1953, 1954) (the latter are somewhat less reliable), a planar projection on paper (for rationale, cf. Nash and Bradley, 1966) of the nicotinic acid:cytosine and of the guanine:cytosine base pairs was prepared in such a manner that the carboxyl oxygens of nicotinic acid were superimposed over ring atoms N(1) and C(6) of guanine in the Watson-Crick hydrogen-bonded conformation (Fig. 3). The in-plane position of the pentose C(1) attached to pyridine N(1) is shifted about 0.4 Å, or about one-third the length of the average covalent bond, from the position occupied by this atom when attached to guanine N(9). The angular difference between the corresponding glycosidic bonds of these two bases becomes about −18°. The possibilities for accommodating such

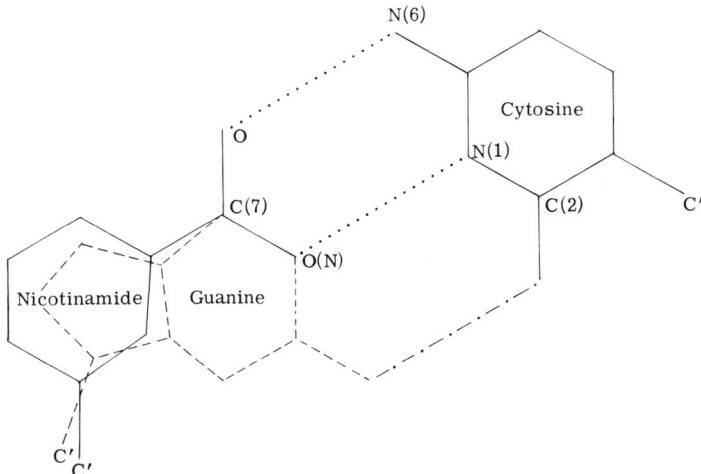

FIG. 3. Watson–Crick type planar projection of nicotinamide substituting for guanine in two-stranded polynucleotide helix.

geometric nonconformities within a perfect helix of the Watson-Crick type may be assessed quite readily by the construction of molecular models. The question to be answered is whether small, angular adjustments of several bonds in the mononucleotide chain can yield the desired accommodation. In case they cannot, then the pyridine nucleotide may still function in tobacco root RNA as an occasional small loop stabilizer. However, it is difficult to visualize a role for pyridine nucleotides in DNA that could account quantitatively for the approximately 10^{12} molecules of nicotine that are produced on the average for each new cell generated by the root apical meristem.

THE PYRIDINIUM–tRNA MODEL

A much more satisfying model is afforded by amino acid transfer ribonucleic acid, tRNA, which is known to be a repository of foreign bases in any case. Using the model described by Cramer (1969), we note that six of the eight loop-adjacent positions on the polynucleotide chain are composed of guanine:cytosine pairs (Fig. 4). It seems likely that the 3-fold greater stability of the hydrogen bond complex of this base pair compared with that of the adenine:uracil pair (Nash and Bradley, 1966) may contribute in an important manner to the stability of loop structure. Now, if one or more of these three-bonded guanine:cytosine pairs were to be substituted by a two-bonded nicotinic acid:cytosine or a nicotinamide:cytosine pair, we should have a situation in which sacrifice of intermolecular bonding strength is probably not great, because of resonance and ionic interactions as noted earlier. If there should be proton transfer from nicotinic acid carboxyl to cytosine ring N, it would probably occur exactly as in the case of cytosine acetic acid, i.e., between carboxyl O and cytosine N(3) (Marsh *et al.*, 1962). It may be of interest that pK values for cytosine are 4.60 and 12.16, while that for nicotinic acid is 4.76. Finally, this model would be far less exacting than the DNA case, for hydrogen bonding of the Watson-Crick type between bases at a loop-adjacent site should be subject to fewer conformational restraints.

It is now generally accepted that changes in tRNA populations may play an important part in controlling developmental changes of the cell. An attractive feature of the present model is that it affords a simple (if wasteful of nitrogen) mechanism for achieving the turnover of this type of polynucleotide. Once the pyridinium polynucleotide is transformed into a nicotine polynucleotide, the bulk of the pyrrolidine ring and the localized charge inversion should effectively prevent loop reclosure at the critical point. This mechanism could be especially conducive to inactivation of the tRNA molecule if it were to involve the anticodon loop. Thus, the produc-

Tobacco Alkaloid Biosynthesis

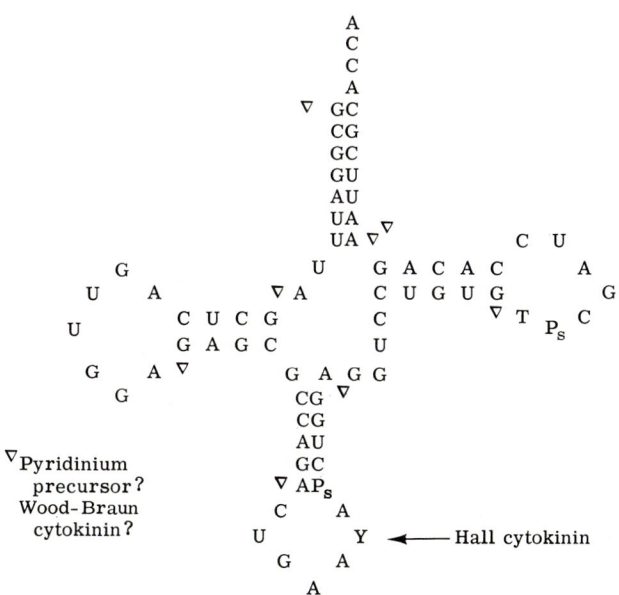

FIG. 4. First- and second-order structure of tRNA^{phe} (and others) (cf. Cramer, 1969) showing the location of the Hall cytokinin and the loop-adjacent locations of the hypothetical pyridinium precursor of nicotine as well as of the Wood–Braun cytokinin, assuming the identity of the last two.

tion of nicotine may initiate a shift in the populational characteristics of tRNA in a fashion somewhat analogous to the modification of rat liver tRNA by N-2-acetylaminofluorene (Agarwal and Weinstein, 1970). In the latter case, the carcinogen combines with guanine covalently rather than with ribose, but the net effect is to interfere with amino acid transfer to protein.

An estimate of tRNA turnover in the tobacco root tip may be made if it is assumed that, on the average, one molecule of nicotine is produced for each molecule of tRNA to be metabolized. Crude calculations show that approximately 20 amino acid residues are incorporated into protein for each nicotine molecule produced. Therefore, the tRNA molecule turns over after 20 amino acid transfers, a figure that would very likely be increased substantially if we were able to specify which tRNAs of the total population were to follow this pathway. The figure is not an unreasonable one, however, and we are led to propose that the special significance of alkaloid biosynthesis for the tobacco plant is precisely to initiate the turnover of at least a portion of its tRNA.

Hormonal relationships are also built into this tRNA model. Hall et al. (1966), Armstrong et al. (1969a, b), Burrows et al. (1970), and Fittler et al. (1968) have shown that cytokinins occupy an anticodon-adjacent position in tRNA and have demonstrated some of the properties and consequences of such. In particular, the work of Leonard et al. (1969) indicates that these modified bases attached to the 3' end of the anticodon triplet may inhibit base pairing in that portion of the loop and may also stabilize stacking of bases belonging to the five adjacent nucleotides of the single-stranded anticodon loop. If we assume that auxin plays a role in the production of messenger RNA, while cytokinin stabilizes tRNA structure, then the relationship of these regulators to protein synthesis on the one hand and to alkaloid production on the other becomes relatively clear. The role of the pyridinium component would be similar to that of the purine cytokinin, namely, the stabilization of the third-order structure of tRNA. The Wood and Braun claim (Wood, 1964) of a pyridine structure for tobacco cytokinin has been noted earlier. In passing, the work of Nozawa and Mizuno (1969) and of Yamomoto and Ohyama (1962) indicate that further relations between pyridine biochemistry and plant growth may eventually materialize.

The role of the 3,6-dihydropyridinium intermediate is also clarified by this model; it is the step that triggers tRNA turnover. However, the fact that the step is energetically somewhat more difficult than the reduction of NAD makes one wonder whether it may not have even more significance than the initiation of turnover. Either the reduction goes in this way because of steric factors induced by base stacking or there is indeed some significant charge transfer process occurring in the postulated region of binding to tRNA. The nature and function of such a transfer cannot be imagined at the present time.

The uncoiling of the tRNA strand subsequent to ring reduction and carboxyl substitution requires no special discussion (cf. theoretical treatment by Poland and Scheraga, 1970), nor do the fragmentation of the polynucleotide strand and the elimination of nicotine or anabasine. The freed alkaloids would follow pathways of transport already delineated and accumulate in leaves and other organs in already familiar patterns.

There is required a final mention of the unequal distribution of nicotine synthesis within the plant body. Thus far, we have mentioned unequal distribution of hormones and of the reductive assimilation of nitrate as correlatives. There is still a third possibility. One of us has reasoned elsewhere (Dawson, 1962) that the process of nicotine production in various groups of the Plant Kingdom is close to elimination in the evolutionary sense. We find it not too difficult to imagine that the specific 3,6- or 1,6-reduction of the pyridinium intermediate may be the factor that is being eliminated within not only the phylogeny of *Nicotiana* but also the ontogenetic development of the species *Nicotiana tabacum* and *N. rustica* as well. Thus, the root system may still retain the proposed pathway involving the pyridinium polynucleotide intermediate, whereas the shoot system may virtually have eliminated this pathway from the biochemistry of its meristems. Alternatives that may bring the species into the mainstream of evolution include either the removal of nicotinic acid or amide from the polynucleotide directly followed by incorporation of the latter into the pool of nicotinic acid *N*-glucoside reported by Mizusaki *et al.* (1970) or the even more simple substitution of guanine as the preferred base for stabilizing loop structure. Indeed, if this interpretation should be correct, we may glimpse here an evolutionarily primitive system in process of transition to a more advanced state the driving force for which would be the profligate waste of nitrogen, and hence the pressure upon survival, that tobacco alkaloid production represents. Both *N. tabacum* and *N. rustica* survive at present because they were rescued from extinction by primitive man who, like his present-day successor, sought release from the realities of a world to which he may have been only marginally better adapted than the plants that gave him solace.

REFERENCES

Agarwal, M. K., and I. B. Weinstein. 1970. Modifications of ribonucleic acid by chemical carcinogens. II. *In vivo* reactions of *N*-2-acetylaminofluorene with rat liver ribonucleic acid. *Biochemistry* **9**:503–507.
Armstrong, D. J., F. Skoog, L. H. Kirkegaard, A. C. Hampel, R. M. Bock, I. Gillam, and G. M. Tener. 1969a. Cytokinins: Distribution in species of yeast transfer RNA. *Proc. Nat. Acad. Sci. U.S.* **63**:504–511.
Armstrong, D. J., W. J. Burrows, F. Skoog, K. L. Roy, and D. Söll. 1969b. Cytokinins: Distribution in transfer RNA species of *Escherichia coli*. *Proc. Nat. Acad. Sci. U.S.* **63**:834–841.
Arnott, S., D. S. Dover, and A. J. Wonacott. 1969. Least-squares refinement of the crystal and molecular structures of DNA and RNA from X-ray data and standard bond lengths and angles. *Acta Crystallogr., Sect. B* **25**:2192–2206.
Braun, A. C. 1970. On the origin of the cancer cell. *Amer. Sci.* **58**:307–320.
Brown, R., and D. Broadbent. 1950. The development of cells in the growing zones of the root. *J. Exp. Bot.* **1**:249–263.
Burrows, W. J., D. J. Armstrong, M. Kanimek, F. Skoog, R. M. Bock, S. M. Hecht, L. G. Dammann, N. J. Leonard, and J. Occolowitz. 1970. Isolation and identification of four cytokinins from wheat germ transfer ribonucleic acid. *Biochemistry* **9**:1867–1872.
Calvin, M. 1959. From microstructure to macrostructure and function in the photochemical apparatus. *Brookhaven Symp. Biol.* **11**:160–180.

Christianson, D. A., J. S. Wall, R. J. Dimber, and A. N. Booth. 1968. Nutritionally unavailable niacin in corn. Isolation and biological activity. *J. Agr. Food Chem.* **16**:100–104.

Christman, D. R., and R. F. Dawson. 1963. An isotopic study of nicotine biosynthesis in relation to the Krebs tricarboxylic acid cycle. *Biochemistry* **2**:182–186.

Cramer, F. 1969. Some modified nucleotides as tools in nculeic acid research. *Accounts Chem. Res.* **2**:338–344.

Dawson, R. F. 1942. Nicotine synthesis in excised tobacco roots. *Amer. J. Bot.* **29**: 813–815.

Dawson, R. F. 1944. Accumulation of anabasine in reciprocal grafts of *Nicotiana glauca* and tomato. *Amer. J. Bot.* **31**:351–355.

Dawson, R. F. 1946. The Ninth Stephen Hales Lecture: Development of some recent concepts in the physiological chemistry of the tobacco alkaloids. *Plant Physiol.* **21**:115–130.

Dawson, R. F. 1948. Alkaloid biogenesis. *Advan. Enzymol. Relat. Subj. Biochem.* **8**: 203–251.

Dawson, R. F. 1962. Biosynthesis of the *Nicotiana* alkaloids. *In* "Science in Progress" (W. R. Brode, ed.), pp. 117–143. Yale Univ. Press, New Haven, Connecticut.

Dawson, R. F., and M. L. Solt. 1959. Estimated contributions of root and shoot to the nicotine content of the tobacco plant. *Plant Physiol.* **34**:656–661.

Dawson, R. F., D. R. Christman, A. D'Adamo, M. L. Solt, and A. P. Wolf, 1960a. The biosynthesis of nicotine from isotopically labeled nicotinic acids. *J. Amer. Chem. Soc.* **82**:2628–2633.

Dawson, R. F., D. R. Christman, M. L. Solt, and A. P. Wolf. 1960b. The biosynthesis of nicotine from nicotinic acid: chemical and radiochemical yields. *Arch. Biochem. Biophys.* **91**:144–150.

Dawson, R. F., D. R. Christman, and R. U. Byerrum. 1971. Biosynthesis of nicotinic acid in plants and microbes. *Methods Enzymol.* **18**:90–113.

Deranleau, D., and R. Schwyzer. 1970. Charge transfer as a molecular probe in systems of biological interest. Intermolecular interactions of the indole-pyridinium type. *Biochemistry* **9**:126–134.

Estabrook, R. W., J. Gonze, and S. P. Nissley. 1963. A possible role for pyridine nucleotides in coupling mechanisms of oxidative phosphorylation. *Fed. Proc. Fed. Amer. Soc. Exp. Biol.* **22**:1071–1075.

Fittler, F., L. K. Kline, and R. H. Hall. 1968. N^6-(Δ^2-Isopentenyl) adenosine: Biosynthesis in vitro by an enzyme extract from yeast and rat liver. *Biochem. Biophys. Res. Commun.* **31**:571–576.

Frost, G. M., K. S. Yang, and G. R. Waller. 1967. Nicotinamide adenine dinucleotide as a precursor of nicotine in *Nicotiana rustica*. *J. Biol. Chem.* **242**:887–888.

Grant, B. R. 1970. Nitrite reductase in *Dunaliella tertiolecta*: Isolation and properties. *Plant Cell Physiol.* **11**:55–64.

Hall, R. H., M. J. Robins, L. Stasink, and R. Thedford. 1966. Isolation of N^6-(γ,γ-dimethylallyl) adenosine from soluble ribonucleic acid. *J. Amer. Chem. Soc.* **88**: 2614–2615.

Jouanneau, J.-P., and C. Peaud-Lenoel. 1967. Croissance et synthèse des protéines de suspensions cellulaires de tabac sensibles à la kinétine. *Physiol. Plant.* **20**:834–850.

Katz, L., and S. Penman. 1966. Association by hydrogen bonding of free nucleotides in non-aqueous solution. *J. Mol. Biol.* **15**: 220–231.

Lavee, S., and A. W. Galston. 1968. Structural, physiological and biochemical gradients in tobacco pith tissue. *Plant Physiol.* **43**:1760–1768.

Leete, E. 1969. Alkaloid biosynthesis. *Advan. Enzymol. Relat. Areas Mol. Biol.* **32:** 373–422.
Leonard, N. J., H. Iwamura, and J. Eisinger. 1969. Synthetic spectroscopic models related to coenzymes and base pairs. IV. Stacking interactions in tRNA; the anticodon-adjacent base. *Proc. Nat. Acad. Sci. U.S.* **64:**352–359.
Leopold, A. C. 1964. "Plant Growth and Development," 466 pp. McGraw-Hill, New York.
Marsh, R. E., R. Bierstedt, and E. L. Eichhorn. 1962. The crystal structure of cytosine-5-acetic acid. *Acta Crystallogr.* **15:**310–316.
Matthysse, A. G., and C. Phillips. 1969. Protein intermediary in the interaction of a hormone with the genome. *Proc. Nat. Acad. Sci. U.S.* **63:**897–903.
Mizusaki, S., Y. Tanabe, T. Kisaki, and E. Tamaki. 1970. Metabolism of nicotinic acid in tobacco plants. *Phytochemistry* **9:**549–554.
Mothes, K., L. Engelbrecht, K.-H. Tachope, and G. Hutschenreuter-Trefftz. 1957. Wurzelaktivität und Nikotinbildung. *Flora (Jena)* **144:**518–536.
Nakano, N. J., and S. J. Igarashi. 1970. Molecular interactions of pyrimidines, purines and some other hetero-aromatic compounds in aqueous media. *Biochemistry* **9:** 577–583.
Nash, H. A., and D. F. Bradley. 1966. Calculation of the lowest energy configurations of nucleotide base pairs on the basis of an electrostatic model. *J. Chem. Phys.* **45:** 1380–1386.
Nozawa, R., and D. Mizuno. 1969. Replication and properties of DNA in nicotinamide adenine dinucleotide deficiency of *Escherichia coli* cells. *Proc. Nat. Acad. Sci. U.S.* **63:**904–910.
Olivera, B. M., Z. W. Hall, and I. R. Lehman. 1968. Enzymatic joining of polynucleotides. V. A DNA-adenylate intermediate in the polynucleotide joining reaction. *Proc. Nat. Acad. Sci. U.S.* **61:**237–244.
Parups, E. V. 1960. Effect of gibberellic acid applications to leaves of *Nicotiana* on nornicotine, anabasine, metanicotine, oxynicotine, and nicotinic acid content. *Tobacco* **151:**20–22 (*Tob. Sci.* **4:**163–165).
Pauling, L. 1960. "Nature of the Chemical Bond," 3d. Ed. Cornell Univ. Press, Ithaca, New York.
Poland, D., and H. Scheraga. 1970. "Theory of Helix-Coil Transitions in Biopolymers." Academic Press, New York.
Pullman, B., and A. Pullman. 1959. The electronic structure of the respiratory coenzymes. *Proc. Nat. Acad. Sci. U.S.* **45:**136–144.
Pullman, B., and A. Pullman. 1969. Quantum-mechanical investigations of the electronic structure of nucleic acids and their constituents. *Progr. Nucl. Acid Res. Mol. Biol.* **9:**327–402.
Richey, F. D., and R. F. Dawson. 1951. Experiments on the inheritance of niacin in corn (maize). *Plant Physiol.* **26:**475–495.
Sarma, R. H., and N. O. Kaplan. 1970a. High frequency nuclear magnetic resonance study of the M and P helices of reduced pyridine dinucleotides. *Biochemistry* **9:**539–548.
Sarma, R. H., and N. O. Kaplan. 1970b. High frequency nuclear magnetic resonance investigation of the backbone of oxidized and reduced pyridine nucleotides. *Biochemistry* **9:**557–564.
Sarma, R. H., V. Ross, and N. O. Kaplan. 1968. Investigation of the conformation of β-diphosphopyridine nucleotide (β-nicotinamide-adenine dinucleotide) and pyridine dinucleotide analogs by proton magnetic resonance. *Biochemistry* **7:**3052–3062.

Sarma, R. H., M. Moore, and N. O. Kaplan. 1970. Investigation of the configuration and conformation of N-methyl-N-ethylnicotinamide-adenine dinucleotide by nuclear magnetic resonance spectroscopy. *Biochemistry* **9**:549–556.

Schmid, H., and M. Serrano. 1948. Untersuchungen über die Nikotinbildung des Tabaks. *Experientia* **4**:311–312.

Scott, T. A., H. Hussey, T. Speaks, P. McCloskey, and W. K. Smith. 1964. Isolation of nicotine from cell cultures of *Nicotiana tabacum*. *Nature (London)* **201**:614–615.

Solt, M. L. 1957. Nicotine production and the growth of tobacco scions on tomato rootstocks. *Plant Physiol.* **32**:484–490.

Solt, M. L., and R. F. Dawson. 1958. Production, translocation and accumulation of alkaloids in tobacco scions grafted to tomato rootstocks. *Plant Physiol.* **33**:375–381.

Solt, M. L., R. F. Dawson, and D. R. Christman. 1960. Biosynthesis of anabasine and of nicotine by excised root cultures of *Nicotiana glauca*. *Plant Physiol.* **35**:887–894.

Solt, M. L., R. F. Dawson, and D. R. Christman. 1961. The gibberellic acid inhibition of nicotine biosynthesis. *Tobacco* **153**:20–23 (*Tob. Sci.* **5**:95–98).

Sund, H., H. Dickmann, and K. Wallenfels. 1964. Die Wasserstoff-übertragung mit Pyridinnucleotiden. *Advan. Enzymol. Relat. Subj. Biochem.* **26**:115–191.

Thimann, K. V. 1937. On the nature of inhibition caused by auxin. *Amer. J. Bot.* **24**:407–412.

Thomas, G. J., Jr., and Y. Kyogoku. 1967. Hypochromism accompanying purine-pyrimidine association interactions. *J. Amer. Chem. Soc.* **89**:4170–4175.

Watson, J. D., and F. H. C. Crick. 1953. Genetical Implications of the structure of deoxyribonucleic acid. *Nature (London)* **171**:346–349.

Wenkert, E. 1968. Alkaloid synthesis. *Accounts Chem. Res.* **1**:78–81.

Wood, H. N. 1964. The characterization of naturally occurring kinins from crown gall tumor cells of *Vinca rosea* L. *Colloq. Int. Cent. Nat. Rech. Sci.* **123**:97–102.

Wood, H. N., and A. C. Braun. 1967. The role of kinetin (6-furfurylaminopurine) in promoting division of cells of *Vinca rosea* L. *Ann. N.Y. Acad. Sci.* **144**:244–250.

Wood, H. N., A. C. Braun, H. Brandes, and H. Kende. 1969. Studies on the distribution and properties of a new class of cell division-promoting substances from higher plant species. *Proc. Nat. Acad. Sci. U.S.* **62**:349–356.

Wright, W. B., and G. S. D. King. 1953. The crystal structure of nicotinic acid. *Acta Crystallogr.* **6**:305–317.

Wright, W. B., and G. S. D. King. 1954. The crystal structure of nicotinamide. *Acta Crystallogr.* **7**:283–288.

Wu, P. L., T. Griffith, and R. U. Byerrum. 1962. Synthesis of pyrrolidine ring of nicotine from several C^{14}-labeled metabolites by *Nicotiana rustica*. *J. Biol. Chem.* **237**:887–890.

Yamomoto, Y., and H. Ohyama. 1962. Effect of kinetin on pyridine nucleotide level in leaves. *Shokubutsugaku Zasshi* **75**:368–370.

Yasumatsu, N., and T. Murayama. 1969. Studies on the chemical regulation of alkaloid biosynthesis on tobacco plants. VI. Effects of simultaneous application of auxin and gibberellin A_3 on nicotine content of flue-cured tobacco. *Hatano Tob. Exp. Sta., Bull.* **63**:69–74.

Yasumatsu, N., A. Sakurai, and S. Tamura. 1967. Studies on the chemical regulation of alkaloid biosynthesis in tobacco plants. Part I. Inhibition of nicotine biosynthesis by helminthosporol and helminthosporic acid. *Agr. Biol. Chem.* **31**:1061–1065.

Zwar, J. A., and A. H. G. C. Rijven. 1967. Effect of growth substances on amino acid incorporation by tobacco tissues. *Aust. Commonw. Sci. Ind. Res. Organ., Div. Plant Ind., Annu. Rep.* pp. 79–80.

SUBJECT INDEX

A

Acetate, 253, 264
Acetic acid, 295, 324
N-2-Acetylaminofluorene, 333
Acetylcholine, 52, 81–101
  identification and measurement, 88
  interactions with respiration, 95
  role of calcium, 94
  synthesis, 101
  topographical distribution of, 91
Acetyl cholinesterase, 88
Acetyl-CoA, 101
Acetylenic ketoester, 176
Achatocarpaceae, 121
Acids, 301
  in tea manufacture, 301
Actinic light, 7
Actinomycin D, 52
Adenine, 264, 332
Adenylic acid, 331
ADP, 4, 5, 18, 21, 27
Adventive oxidation, see Oxidation
Aerobic, 55
  relating to leaflet closure, 55
Agdestidaceae, 121
Aglycons, phenolic, 186
Alachlor, 243
Alanine, 212, 258, 295, 296
β-Alanine, 258
Albizzia, 53, 55, 68, 77
Albizzia julibrissin, 52
Alcohols, 296, 301
Aldehyde, 169, 186, 295, 296, 301
Alkaloid, 106, 209
  localization, 174
  production, 334
  tobacco alkaloid biosynthesis, 317–335
Alkylating agents, 242

Alkylcarboxylic, 71
S-Alkyltransferase, 243
Allyl isothiocyanate, 198
Aluminum, 264
Amide, 328, 329
Amino acids, 151, 174, 205, 209, 258, 282, 295, 325, 329, 333
4-Amino-6-t-butyl-3-(methylthio)-as-triazin-5-($4H$)-one, 243
2-Amino-5-nitrophenol, 243
Aminophenols, 167, 185
Ammonia, 295
AMO-1618, 93
AMP, 18
Amylases, 170
Anabasine, 318, 322–324, 331, 334
Anaerobic, 55
  relating to leaflet closure, 55
Anilide, 237, 243
Anilines, 233
Anthocyanidins, 158, 161, 186, 284
Anthocyanin, 107, 132, 135, 143, 188, 209
Anthocyanin–copigment complexes, 147
Anthraquinones, 172
Anthrols, 171, 172
Antibiotic substances, 166, 167
Antibody, 39
Antigen, 39
Apigenin, 136
Apigeninidin, 136
Arabinogalactan hemicellulose, 260
Arabinose, 259
Araliaceae, 176
Arbutin, 170, 187
Arginine, 258
Aroma, 249, 266, 297, 304, 305
Arylalkyl compounds, 242
S-Arylalkyltransferase, 243
Ascorbic acid, 148, 154

Asparagine, 258
Aspartate, 44
Aspartic acid, 258, 324
Aspergillus niger, 271, 277
ATP, 4, 5, 10, 11, 18, 21, 27, 55, 73, 74, 76, 94
Atrazine, 231, 241
Atrazine glutathione conjugation, 241
Atropine, 92
Aurovertin, 27
Autoxidation, 172
Auxin, 325, 326, 328, 334
Azodrin, 231

B

Bacteria, 166, 167, 175
  photosynthetic, 4
Bacterial photosynthesis, *see* Photosynthesis
Bacteriochlorophyll, 7
Baeyer-Villiger mechanism, 149
Barban, 242, 243
Barbeuia, 121
Barbeuiaceae, 121
Bark, 166
Basellaceae, 121
Batidaceae, 121, 132
Batis argillicola, 121
Batis maritima, 121
Benzaldehyde, 294, 295
Benzidine, 167
Benzyl alcohol, 295
Benzyl chloride, 242, 243
(+)-β Bisabolol, 199
Betacyanins, 107
Betalains, 105–132
  betacyanins, 108
  betaxanthins, 109
  biogenesis, 121
  synthesis of betalains, 130
Betalains, 115
  functions of, 131
  genetics of biogenesis, 130
  structure determination and distribution, 115
Betalamic acid, 130
Betaxanthins, 107
Berberine, 175
Bidrin, 231

Bioelectric potential, 52, 53, 82, 101
Bisflavanols, 277, 281
Black shank, 184
Black tea, 263, 265, 288, 304
Bleaching, oxidative, 7
Boll weevil (*Anthonomus grandis*), 199
Boraginaceae, 172
Bovine serum albumin, 40
Brevicoryne brassicae, 205
Bromine, 263
Brown pigments, 280
Browning, 175, 185
Bruchophagus roddi, 199
Bud differentiation, 76
Bundle sheath, 184
  cells of, 169
n-Butyric acid, 294

C

C-Benzyl product, 155
Caffeine, 213, 263, 284, 303, 304
  biosynthesis of, 264
Calcium, 63, 94
  intracellular, 87
Callus tissue, 264
Calvin cycle, 3
Camellia sinensis, 248
Caproate, 71
n-Caproic acid, 294, 295
Capronaldehyde, 294
Carajurin, 136
Carbamates, 231
Carbanilate, 243
Carbohydrate, 2, 10, 170, 259, 260, 329
Carbon assimilation, 3
Carbon dioxide, 2, 3, 11, 295
Carbon monoxide, 229
Carbonyl, 186, 295
Carboxamide, 320
Carboxin, 237
Cardenolides, 216
β-Carotene, 301
Carotenes, 262
  as aroma precursors, 298
  factors affecting, 301
Carotenoids, 6, 263, 298, 301
Caryophyllaceae, 107, 121
Casparian band, 166, 176

Subject Index

Catechin, 152, 158, 253, 255, 258, 268–271, 278, 281, 288, 295, 297, 301
   biosynthesis of, 253
   oxidation of, 268
Catechol, 174
Catecholcatechins, 271
Catechol oxidase, 256, 257, 264, 266, 271, 273, 288, 301, 302
CCC, 93
Cell membrane permeability, 266
Centrospermae, 107, 121, 131
Chalcone, 140, 146
Chebulinic acid, 284
Chelating agents, 229
Chemical barrier, 166
Chemical free energy, 5, 14
Chemoreceptors, 209
Chenopodiaceae, 121
Chlorella, 5
3-(3-Chloro-4-bromophenyl)-1-methoxy-1-methylurea, 237
Chlorogenic acids, 269, 281
3-(4-Chlorophenyl)-1-hydroxymethylurea, 239
4-Chlorophenylurea, 238
Chlorophyll, 5, 6, 262
Chlorophyll *a* and *b* molecules, 10, 17, 262
Chlorophyllase, 263
Chlorophyllide *a*, 16, 17
Chlorophyllin *a*, 13, 16, 17
Chloroplasts, 3, 4, 8, 11, 18, 26, 27, 52, 76, 95, 228, 233, 257
   movements of, 53
Chloropropham, 237
1-Chloro-3-tosylamido-7-amino-L-2-heptanone (HCl), 242
2-Chloro-s-triazine, 239
   N-dealkylated, 243
Chloroxuron, 237
Choline, 101
Choline acetyl transferase, 101
Cholinesterase, 101
Chromatin, 326
Chromatophores, 7
Chromenols, 138, 143
Chromen-4-sulfonic acid, 146
Chromophore, 16, 48, 186
Chromoprotein, 36
Chromosomes, 166, 175

Chymotrypsin, 45, 243
Cicatrix, 176
o-Cinnamylphenol, 159
Citrate, 94
Citric acid, 261
Coenzyme A, 297
Colchicine, 175
Color, 263, 304
Compartmentalization, 322
Compositae, 176
Copper, 258, 264
   deficiency, 264
Coreopsis, 177
   C. saxicola, 177–179
   C. grandiflora, 177, 178
Corilagin, 281, 284
Cork, 166
Cortex, 166, 170, 175
Cortical cell zone, 175, 182
Coumarin, 188
Coumarin 3,5-diglucoside, 149
p-Coumarylquinic acids, 269
Coumestrol, 151
o-Cresol, 294
Crude fiber, 259
Cyanidin, 135, 284
Cyanidin mannoside, 209
Cyclic photophosphorylation, 4, 9, 11
Cyclodopa, 130
Cysteine, 242
Cytochrome *c*, 13
Cytochrome oxidase, 169
Cytochromes, 9, 20, 229
Cytokinin, 325, 327, 328, 334
Cytoplasm, 169, 233
   fibrillar, 53
Cytosine, 331, 332
Cytosine-5-acetic acid, 320

D

Dark reaction 2, 3
N-Dealkylation, 231
Dehydroascorbic acid, 297
Dehydrogenases, 183
   localization and disease resistance, 183
5-Dehydroshikimate reductase, 254
Delphinidin, 135, 284

N-Demethylase, 227, 228
  difference between plant and animal systems, 230
  inhibitors, 229, 231, 237
  specificity, 231
N-Demethylation, 226, 231
  oxidative, 238
3-Deoxyanthocyanidins, 136
Dephlogisticated air, 2
Depolarizer, 10
Dermal cell, 174
Detergents, 229
Detoxication, 226
Dichlorodicyanobenzoquinone, 159
1,2-Dichloro-4-nitrobenzene, 243
2,4-Dichlorophenoxyacetic acid, 325
4-(3,4-Dichlorophenyl)-1,1-dimethylsemicarbazide, 232
4-(3,4-Dichlorophenyl)-2-methylsemicarbazide, 232
4-(3,4-Dichlorophenyl)-1,1,2-trimethylsemicarbazide, 232
Dicryl, 237
Diene ketones, 186
3,5-Diglucosides, 149
Dihydrochalcones, 157
3,6-Dihydronicotinamide, 322
3,6-Dihydropyridinium, 319, 330, 334
3,4-Dihydroxybenzoic, 145
Dihydroxyphenylalanine, 130, 185
α-Diketones, 143
β-Diketones, 151
Dimedone, 154
2,3-Dimercaptopropanol, 242
N,N-Dimethylacetamide, 18
2,4-Dimethylimidazole, 23, 24
2:6-Dimethylocta-1:3:5:7-tetraene, 181
Dimethylphenylurea, 238
Dimethyl sulfide, 297
1,1-Dimethyl-3-[3-N-t-butylcarbamoyloxy)phenyl]urea, 237
2,4-Dinitrofluorobenzene, 242
2,4-Dinitrophenol, 55
Diphenamide, 231, 238
Diphenylcarbamyl chloride, 242
Diphenylpicrylhydrazyl, 13
Disease resistance, degree of, 166
Dithiol compounds, 242
1,4-Dithiothreitol, 242
Diurnal cycle, 55

Diuron, 230, 232, 233
DNA, 107, 166, 320, 331
DNA-RNA hybridization, 107, 115
Dorsal cells, 71
Dracorubin, 136
Dyrene, 243
Dysphania, 121
Dysphaniaceae, 121

E

Ecothiopate iodide, 92
Ecotype, 170
Electric potentials, 169
Electron acceptor, 3, 8, 11, 14, 229
Electron spin resonance (ESR), 13, 23
Electron transfer, 10, 14, 21, 26
Electron transport, 8, 12, 18, 19
Empoasca fabae, 205
Endodermis, 166, 167, 169, 171, 174, 175, 181, 185
Endomitotic activity, 326
Endopeptidase, 42, 47
Endoplasmic reticulum, 228
Endopolyploidy, 175
  induction of, by wounding, 175
Energy transfer, 5
Enolic benzoates, 149
Enzymes, 167, 175, 255, 266, see also specific enzymes
  activities, 52
  induction, 53
  localization in healthy and diseased plants, 182
  relating to disease resistance, 165
  oxidation of tea tannins, 296
Epicatechin, 252, 254, 269, 273, 274, 276, 278, 281, 283, 303
Epicatechin-gallate, 252, 253, 269, 270, 274, 284
Epidermis, 166, 172, 258
Epigallocatechin, 252, 269–271, 273, 274, 277, 284, 303
Epigallocatechin-gallate, 252, 269–271, 273, 274, 277
Epitheaflavic acids, 277, 285, 288
S-Epoxidetransferase, 243
Ericaceae, 172, 188
Eserine, 93
Eserine sulfate, 88

Subject Index

Esterases, 170, 174
Esters, 170
Ethyl acetate, 268
γ-N-Ethylasparagine, 259
Ethyl chlorophyllide a, 16
Ethylene, 169
5-N-Ethyl glutamine, 258

F

Far-red light, 35, 55, 65, 68, 77, 81, 91, 101
Fatty acids, 186
Fenuron, 230
Fermentation, 248, 268, 271, 295, 305
Ferredoxin, 13, 17
    clostridial, 12
Ferric oxalate, 3
Ferricyanide, 3, 7, 11
Ferrohemochrome, 18, 19
Firing, 305
Flavan-3,4-diols, 148, 160
Flavan-3-ol gallates, 284
Flavan-3-ols [(+)-catechin], 283
Flavanols, 252, 253, 255, 261, 265, 266, 276, 295, 302, 305
Flavanones, 136
Flav-2-ene, 152, 155
Flav-3-enes, 157
Flavin, 13, 20
    oxidized, 10
Flavonal glycosides, 255
Flavones, 136, 186
Flavonoid, 107, 254, 305
Flavylium–catechin condensation, 151
Flavylium–flavan condensation, 153
Flavylium salts, 135–161
    effect of light, 140
    of pH, 138
    electron deficient, 138
    hydrolytic reactions, 138
    peroxide oxidation, 148
    polyphenol and, 151
    xanthylium salt formation, 160
Flavylium salt–polyphenol condensation reactions, 151
Flower initiation, 53
Flowering, 76
Fluometuron, 230, 231
    N-demethylation of, 233
Fluorescence, 7, 321

FMN, 4
Formaldehyde, 227, 238
Formaldomethone, 238
Formate, 71
Free sugars, 259
Fructose, 259
Fumaric acid, 261
Fungi, 166, 167, 174, 186
Fusarium oxysporum, 184, 185, 188

G

Galactose, 259
Galacturonic acid, 259
Gallic acid, 271, 278, 280, 284, 285
Gallocatechins, 186, 253, 269–271, 276, 283, 285
3-Galloylepitheaflavic acid, 278, 280
Galloyl-3-quinic acid, 255
Gene–enzyme system of resistance, 183
Genes, 166
Genetic tumors, 328
Geraniol, 294, 295
Germination of tea seeds, 261
Gibberellic acid, 325
Globin, 185
Globular protein, 44
β-1,4-Glucan(s), 229
Glucogallin, 284
Gluconasturtiin, 213
Glucose, 229, 259, 324, 327
Glucose 1-phosphate, 260
Glucose 1,6-diphosphate, 260
β-Glucosidase, 170, 186, 187
Glucoside, 187
Glutamate, 44, 94
Glutamic acid, 258, 295
Glutamine, 258
Glutathione, 12, 13, 242
    conjugation, 239
Glutathione S-transferase, 239, 243
    activity
        distribution, 241
        endogenous inhibitors, 241
        genetic control, 241
        inhibition, 242
        localization, 241
        purification, 240
        specificity, 242
Glycine, 258, 296

Glycolate, 76
Glyconapin, 213
O-Glycosides, 239
Glycosidic bond, 321
Glyoxysomes, 76
Gossypol, 199
Grasshopper (*Poekilocerus bufonius*), 216
Green algae, 8
Green tea, 265, 288
Guanine, 264, 331, 332, 335
Guanylic acid, 329, 331
Gyrostemonaceae, 121

H

Halophytaceae, 121
Halophytum, 121
Heavy metal, 263
Helminthosporic acid, 325
Hematoporphyrin, 12, 13, 21, 22, 26
Hematoporphyrin dihydrochloride, 185
Heme enzymes, 185
Hemicellulose, 260
Hemin, 185
Hemoglobin, 185
n-Heptanal, 294
Herbicide metabolism, 225–244
  N-demethylation, 226
  Glutathione conjugation, 239
Hexadecatrienoic acid, 181
cis-3-Hexenoic, 294, 295
trans-2-Hexenoic acid, 294
trans-2-Hexenol, 294, 295
Hexosemonophosphate, 324
Hexyl alcohol, 294
Histidine, 258, 329
Histochemistry of plants, 165–190
  development of chemical barrier, 166
  enzyme localization, 182
  localization of compounds, 187
Hormonal function, 101
Hormonal interactions on alkaloid biosynthesis, 325
Hormone, 76, 82, 325, 334
Host–pathogen interactions, 183
Hydrogen donor, 167
Hydroxyapatite, 44
4-Hydroxy-2-butenylamino derivative, 327
2'-Hydroxychalcones, 157, 158

3-Hydroxyflavylium, 148
N-Hydroxymethyl glycoside, 239
2-Hydroxy-s-triazines, 243
trans-2-Hydroxy-4,6,4'-trimethoxychalcone, 140
Hypericin, 216
Hypocotyls, 229
Hypodermal, 182
Hypodermis, 174
Hypoxanthine, 264

I

Imidazole, 18, 23, 26
Imidazolyl radical, 19, 26
Immunity, 165, 166, 186
3-Indoleacetic acid, 329
Indolequinone, 185
Infection, 167
Infusion, 303
Inhibition, 242, 243, 325
  of demethylation, 231
Inorganic orthophosphate, 4, 26
Insects, 167
Intracellular locus, 324
Ion exchange, 166
Ion flux, 59
Ion selection, 166
β-Ionone, 301
Ipazine, 243
Irradiation
  red light, 55
  far-red light, 55
Isobutyric acid, 295
Isocitric acid, 261
Isoenzymes, 169, 184, 185
  gravity effect, 169
  light effect, 169
  of peroxidase, 185
Isoflavones, 186
Isoleucine, 258, 296
Isotheaflavin, 276
Isovaleric acid, 294

J

cis-Jasmone, 295
Juglandaceae, 172

Subject Index

K

Kaempferol, 269
Ketodiols, 148
α-Ketoglutaric acid, 325
Kinetin, 326, 327
Krebs tricarboxylic acid cycle, 324

L

Lecithin, 186
Leucine, 258, 296, 297
Leucine transaminase, 297
Leucoanthocyanins, 269
Light, 2, 169
 red and far-red irradiation, 35
Limonene, 199
Linalool, 294, 295
Linalool oxide (cis-furanoid), 294
Linoleic acid, 186
Linolenate acid, 186
Linolenic, 186
Linoxyn, 166
Linuron, 230, 233
Lipase, 170
Lipids, 169, 171, 175, 176, 181, 205, 301
 genetics of endodermal, 176
Lipophilic microsomal membrane system, 233
Lithium aluminum hydride, 155
Lumiflavin, 16, 17
Luteolinidin, 136
Lysine, 258

M

Mahonia, 175
Malic acid, 261, 324
Malvin, 148, 209
Malvone, 148
Mammals, 167
Manganese, 10, 297
Mannose, 259
Membrane permeability, 76
Membrane theory of phytochrome action, 101
2-Mercaptoethanol, 242
Meristems, 185
Methionine, 296

3'Methoxy-4',7-dihydroxy-flavylium chloride, 138
8-Methoxy-4'-hydroxy-3-methylflavylium chloride, 152
6-(3-Methyl-2-butenylamino)-9-β-D-ribofuranosylpurine, 327
Methyl 4-chlorocarbanilate, 237
α-Methylene butyrolactone, 188
Methylmethionine sulfonium, 297
N-Methyphenylurea, 239
N-Methylphenylurea N-demethylase, 228
N-Methylphenylurea herbicides, 226, 237
2-Methylthio-9-β-D-ribofuranosyl, 327
Microsomal enzyme systems, 227
Microsomal protein, *see* Protein
Mimosa, 53
 m. pudica, 52, 82
Minerals, 264
Mirabilis jalapa, 130
Mitochondria, 18, 26, 27, 76, 97, 228, 257, 324
Moisture, loss of 266
Molecular structure and stimulating effectiveness, 210
Molluginaceae, 107
3-Monoglucosides, 269
Monomethyldiuron, 233
Monomethylfenuron, 230
Monomethylfluometuron, 230
Monomethylmonuron, 227, 230
Monomethylphenylurea, 238
Mononucleotide, 321
Monothiols, 242
Monuron, 230, 233, 238
Mougeotia, 52
Mung bean (*Phaseolus aureus*), 84
Mustard oil glucosides, 198, 210
Mycorrhiza of orchids, 188
Myricetin, 269
Myrosulfatase, 170

N

NAD, 20, 319, 321
NADP, 4, 5, 10, 11, 13, 15, 319, 321
NADP-reductase, 13, 15, 17
NADPH, 4, 5, 10, 13, 15
NADPH-cytochrome c reductase, 229
Naphthoquinones, 172
β-Naphthoquinone-4-sulfonate, 297

Naphthols, 171, 175
1-Naphthyl hydroxymethylcarbamate, 231
Naringenin, 136
Necrotic tissue, 185
Nematicidal compounds, 176
Nematodes, 167, 183, 185
Nerol, 294
Neurohumor acetylcholine (ACh), 85
Nicotiana, 318
   N. glauca, 324, 328
   N. langsdorffii, 328
Nicotiana alkaloids, 319, see also Tobacco alkaloids
Nicotinamide, 320, 327
Nicotinamide adenine dinucleotide, 321, 329, 330
Nicotine, 216, 318, 322, 325, 331, 334
Nicotine polynucleotide, 332
Nicotine synthesis, 322, see also Tobacco alkaloids
Nicotinic acid, 328, 329, 319, 331
Nitrate, 322
Nitrogen atmosphere, 55, 295
Nitrogenous compounds, 282
2-Nitro-p-toluidine, 212
3-Nitro-o-toluidine, 212
5-Nitro-o-toluidine, 212
β-1,3-Noncellulosic linkages, 229
Norleucine, 295
Nuclear division, 166
Nuclei, 175, 188
Nucleic acids, 264
Nyctinastic movements, 52

O

Oak leaves, 217
trans-2-Octenal, 294
Oils, 170
Oligomycin, 27
Orchidaceae, 188
Organic acids, 67, 260, 296
   changes during tea seed germination, 261
Organophosphate insecticides, 232
1-Orthophosphoimidazole, 19
Oxalate, 94
Oxalic acid, 261
Oxaloacetate, 324

Oxaloacetic acid, 325
Oxidase, 227
   adventive, 172
   functional, 172
Oxidation
   of tea flavanols, 305
Oxidative bleaching, see Bleaching
Oxidative N-demethylation, see N-Demethylation
Oxidative phosphorylation, 18, 20, 101, 174
   mitochondrial, 9, 19
Oxygen 2, 55
   consumption, 95
   evolution, 5

P

P_{fr}, 42, 47, 48, 52, 65, 73
Palisade parenchyma, 255
Paraoxon, 92, 232
Parasite, 186
Parathion, 232
Parenchyma cells, 174
Parenchymal tissue, 188, 327
Parillin, 170
Parinaric acid, 181
Pathogens, 165, 185
Peach aphid (Myzus persicae), 198, 209
Pectic enzymes, 187
Pectin, 259
Pectinases, 174
Pelargonidin, 135, 143
Pelargonin, 209
Pentamethylneobetanidin, 130
1-Penten-3-ol, 295
cis-2-Pentenol, 295
Pericycle, 175
Peroxidase, 167, 169, 183, 185
9-Phenacyl-5-ketotetrahydroxanthenes, 154
Phenolase, 185
Phenolic compounds, 173, 277, 280, 295
Phenol oxidases, 185
Phenol quinines, polymerization of, 176
Phenols, 169–173, 175, 188, 277, 280, 295
Phenyl compounds, 179
Phenylacetaldehyde, 294, 297
Phenylalanine, 169, 258, 295, 296, 297
Phenylalanine ammonia lyase enzyme, 188

Subject Index

2-Phenylbenzofurans, 148, 151
Phenylethanol, 294, 295
Phenylheptaenetriyne, 178, 179
Phenylmethanesulfonyl fluoride, 242
Phenylurea, 238
    metabolites of, 233
Pheophorbide, 262
Pheophytin a, 262
Pheophytin b, 262
Phlobatannins, 186
Phloem, 174, 184, 185
Phloretin, 170, 173
Phlorizin, 187
Phloroglucinol, 304
Phorate, 232
Phosphatase-phosphorylase, 170
Phosphatases, 170
Phosphoenolpyruvate, 324
Phosphoglyceraldehyde, 324
3-Phosphoglyceric acid, 3
1-Phosphoimidazole, 18, 21, 26
Phosphoimidazolyl radical, 20, 25
Phospholipid (or phosphoprotein), 20
Phosphorus, 63
Phosphorylase, 170
Phosphorylation, 9, 21
    coupled to electron transport, 18
    oxidative, *see* Oxidative phosphorylation
    photosynthetic, 11
Photochemical energy conversion, 11, 13
Photons, 5
Photooxidation, 3, 4
Photoreactions, 8
    system, 11
Photoreceptor, 35
Photoredox reactions, 12
Photosynthesis, 1–27
    bacterial, 10
    early adventures, 1
    energy transfer, 5
    path of carbon, 3
    of electron transport, 8
    photophosphorylation, 4
    reducing power generation, 4
Photosynthetic inhibitors, 226
Photosystem I (PSI), 9
Photosystem II (PSII), 9
Phototransformation, 36, 48
Phymatotrichum omnivorum, 175
Phytoalexins, 188

Phytochrome, 64, 73, 83
    action
        aerobic, 55
        in leaflet closure, 51–77
        mechanism of, 51
        movement of K salts, 61
        temperature effect, 57
    discovery, 36
    membrane permeability, 52
    molecular weight, 36, 40
    nature of purified, 35–48
    purification, 37
    repression and derepression of genes, 52
    turgor changes, 52
Phytolaccaceae, 121
P_i, 18
Pieris brassicae, 198, 205, 209
Pigment, 48, 121, 188, 268, 273
    molecules 5, 12
    photoreversible, 35
    tea leaves, 262
$(-)$-α-Pinene, 199
Plasma membrane, 52, 74
Plasmodesmata, 73
Plastocyanine (PC), 10, 17
Plastiquinone, 9
Polyacetylene, 176, 177, 179
Polyacrylamide, 39
Polyenes, 169, 175, 181
Polyeneynes, 169
Polygonaceae, 107, 132, 172
Polyketide, 253
Polynuclear cells, 183
Polynucleotides, 329
Polypeptides, 42
    chromophoric, 47
Polyphagous army worm (*Prodenia liturata*), 205
Polyphenolase, 167, 169, 186
Polyphenolic compounds, 261, 270, 282, 305
    of tea, 252
Polyphenolic flavylium, 135
Polyphenoloxidase, 175, 185, 186
    Cu-containing, 256
Polyphenols, 151, 169, 187
Polyploidy, 166, 167, 175
Polysaccharides, 259, 282
Polyteny, 175
Poly-N-vinylpyrrolidone (PVP), 16

Polyynes, 169, 175
Porphyrin, 13
Portulaca, 130
Potassium, 59, 61, 71, 76
  K movement and phytochrome, 65
Potassium ferricyanide, 273
Potato leafhoppers, 205
P_r, 39, 42, 47, 48, 52, 76
Proanthrocyanidins, 160, 284
Proendodermis, 166
Proline, 258
Prometryne, 231, 243
Propachlor, 241–243
Propanil, 237
Propionic acid, 295
Protease, 42–44, 47
Protein, 20, 48, 52, 175, 184, 205, 255, 281, 323, 326, 327, 334
  globular, 44
  microsomal, 229
Protocatechuate-3,4-dioxygenase, 130
Proton transfer, 332
Protoplast, 167
Pseudomonas tumefaciens, 175
Psilotum, 180, 189
Purine, 213, 264, 320, 329
Puromycin, 52
Purple bacteria, 7
Purpurogallin, 272, 273
Pyrazole, 23
Pyrethrin, 216
Pyridine, 185, 323
Pyridine-dehydrogenase-flavoprotein-cytochrome, 330
Pyridine nucleotides, 238
Pyridine ring, 318, 328
  biochemical properties, 320
  physical properties, 321
Pyridinium polynucleotide, 332, 335
  of alkaloid biosynthesis, 329
Pyrimidines, 320
Pyrrolidine, 323, 324, 332

Q

Quantasome, 6
Quantum efficiency, 8
Quantum yield of photosynthesis, 8
Quercetin, 269
Quercetin-3-glucose (isoquercitrin), 211

Quercetin-3-rhamnose (quercitrin), 211
Quercetin-3–rutinose (rutin), 211
Quinic acid, 253, 261
Quinol–quinone, 171
Quinone, 3, 20, 169, 174, 175, 185, 188
Quinoncimine, 167
Quinonoidal anhydro bases, 138

R

Raffinose, 259
Red light, 35, 55, 68, 69, 77, 81, 91, 99, 101
Reductive assimilation, 322, 334
Resistance, 166
Respiration, 18, 95
Respiratory electron carriers, 20
Respiratory electron transport, 101
Respiratory oxidative phosphorylation, 11
Rhamnaceae, 172
Rhamnose, 259
Rhodospirillum rubrum, 4, 7
Ribonucleic acids (RNA), 264, 326
Ribose, 259
N-Ribosides, 320
Ribosomal acid, 326
Ribosphosphate, 321
tRNA, 332
Root cultures, 323
Root system, 334
Root tips, 323
Rotenone, 216
Rubiaceae, 172

S

Salicin, 213
Salicylaldehyde, 216
Salicylic acid, 294, 295
Saponin, 213
Scopoletin, 174
Secondary plant substances and insects, 197–218
  chemoperception of plant constituents, 209
  feeding stimulants, 198
  molecular structure and stimulating effectiveness, 210
  nutrients and host plant recognition, 199
  toxicity of, 213

Subject Index

Seeds, 175
  germination, 53
  phytochrome in, 40
Semicrystallinity, 322
Serine, 258
Shikimic acid, 253, 261, 264
Shoot system and alkaloid, 335
Sieve tube(s), 174
Sieve-tube plastids, 131
Silkworm (*Bombyx mori*), 211
  larvae, 198
Sinigrin, 198
Sodium acetate, 64, 69
Sodium azide, 55
Sodium bicarbonate, 273
Sodium borohydride, 155
Sodium dodecyl sulfate, 39
Solanine, 216
Solanum demissum, 184
Solanum tuberosum, 184
Sphenoclea, 121
Sphenocleaceae, 121
Spherosomes, 53
Stachyose, 259
Starch, 170, 259, 260
Stele, 170, 175
Steroids, 205
Stomatal guard cells, 77
Strecker degradation, 296
Subepidermal cell, 174
Succinic acid, 261, 324
Sucrose, 205, 259
Sucrose 6-phosphate, 260
Sugar phosphates, 260
Sugarbeet (*Beta vulgaris*), 130
Sugars, 205, 264, 296
Sulfate residue, 327
Sulfhydryl reagents, 229
Sulfobromophthalein, 243
Sulfoxidation, 231
Sulfur, 2
Sulfur bacteria, 2
Sulfur dioxide, 146
Swep, 237

T

Tamarix, 77
Tanada effect, 94, 99

Tannin, 53, 77, 161, 187, 216
Taste, 266, 304
Tea
  aroma
    composition, 288
    formation, 294
  briskness of, 303
  catechins, 296
  chemical composition of, 251
  classification, 248
  cream, 303
  enzymes, 296
  fermentation, 256
    factors affecting, 303
    mechanism, 302
  flavor, chemical basis for changes, 294
  flush, 249, 251
    amino acids, 258
    caffeine and puvines, 263
    carbohydrates, 259
    chemical composition, 251
    chlorophylls and carotenoids, 262
    enzymes, 255
    minerals, 264
    organic acid, 260
    polyphenolic compounds, 252
  history, 248
  localization of leaf catechol oxidase, 258
  manufacture, 247–305
    chemistry of, 265
      aroma, 288
      black tea, 249, 265
      fermentation, 268
      thermal process, 251
    process, 249
  organoleptic properties, 248
  photosynthetic assimilation by leaves, 260
  production, 248
  seeds, 260
  strength of, 303
Temik, 237
Terpenes, 297
Terpenoid, 199
Terthienyl compounds, 176
Tetraenes, 180
4,2′,4′,6′-Tetrahydroxychalcone-2′-glucoside, 254
Theaceae, 264

Theaflavins, 268, 269, 271, 273, 281, 284, 303
  structure, 273
Theaflavin digallate, 276
Theaflavin gallate, 273
Theanine, 281
Thearubigins, 268, 269, 276, 279–282, 284, 288, 303, 305
Theasaponin, 260
Theligonaceae, 121
Theligonum, 121
Theobromine, 263
Theogallin, 255, 281
Theophylline, 213, 263
Thiobarbituric acid, 186
Threonine, 258, 295
Tmesipteris, 180, 189
Tobacco alkaloid
  biosynthesis of, 317–335
    compartmentalization of synthesis, 322
    hormonal interactions, 325
    properties of pyridine ring, 320
    speculative model building, 328
  methyl group of, 327
Tobacco hornworm, 216
Tobacco mosaic virus, 183, 186
Tobacco pith cells, 326
Tomatine, 213
Tortrix viridana, 216
L-1-Tosylamido-2-phenylethyl chloromethyl ketone, 242
Toxicity of secondary plant substances, 213
Toxins, 184, 185
S-Transferase, 242
Triazine, 243
Tridecadienetetrayne, 178, 179
Trienes, 180
Trigonal-bipyramidal intermediate, 19
Trigonal-bipyramidal phosphoimida-2-dyl radical, 19
5,7,8-Trihydroxyflavylium, 142
7,2′,4′-Trihydroxy-3-methoxyflavylium chloride, 151
4′,5,7-Trimethoxyflavylium perchlorate, 140
Trypsin, 44, 45, 243
Tryptophan, 169, 258, 295

Turgor pressure, 53, 59
Tyrosine, 169, 258, 295

U

Ubiquinones, 172
UDPG, 229
Ultrastructure, 131
Umbelliferae, 176
Uracil, 332
Uridine diphosphoglucose, 229
Uromyces phaseoli, 183

V

Vacuoles, 53, 77, 175, 255, 261
n-Valeraldehyde, 294
Valine, 258, 296
Vascular, 172
Vascular bundle(s), 170, 258
Vascular bundle sheath, 174
Vascular tissue, 166
Vectors, 185
Ventral cells, 67
Ventral pulvinule cells, 71
Ventricular heat of marine clam, 85
Virus-infected tobacco, 174
Vitamin K, 4, 10
Vitamins, 205
Volatile carbonyl compounds, 295

W

Water, 2
Withering, 264, 266, 294
Wound lesions, 166
Wounded tissue, 175
Wounding, 167, 179

X

Xanthines, 263, 264
Xanthylium salts, 160
Xylose, 259

Z

Zn(II)-tetraphenylporphyrin, 14